T0229862

**Chapter 6**    Aluminum-Ion Batteries: New Attractive Emerging
Energy Storage Devices ...............................................138

*Hongsen Li, Huaizhi Wang, Hao Zhang, Zhengqiang Hu and
Yongshuai Liu*

# Preface

In the present scenario, the advancements in energy storage and energy generation technologies are critical to meet the day-to-day requirements. Different types of energies are essential to sustain our day-to-day life. The ever-increasing demands for energy along with the issue of environmental pollution led us to find a way to enhance the performance of the existing energy technologies and/or find newer clean energy systems. Many latest battery technologies such as sodium ion batteries could be a potential clean energy technology in the near future. Thus, expediting the commercialization of such technology is crucial to meet our ever-growing energy needs.

Further, the demand for portable electronic gadgets and electric vehicles is huge. These technologies depend on the performance and availability of energy storage devices. For instance, the mobile phone technology primarily depends on the performance of the lithium-ion batteries (LIBs). For example, weight, cost, and charge storage capacity of the devices depend on the performance of the associated LIBs. Even though the LIB is a world leader in the area of portable devices, the lithium resources are limited, which makes it unsustainable. Further, its electrochemical performance is nearing the theoretical limit, which led the scientists across the world to think of an alternative energy storage systems. Hence, alternate energy storage technologies should be brought into the market to meet the increasing energy demands. Hence, the constant search for an alternate battery system resulted in the emergence of different metal ($Na^+$, $K^+$ and $Mg^+$) ion, metal sulfur and metal air batteries.

The path to commercial success of any battery technology is paved with challenges. Even though the research on different battery chemistries such as sodium-ion battery (SIB) have been investigated since the early 1980s, these technologies are not yet commercialized due to the lack of further follow up in the respective field due to the high popularity and commercialization of LIBs.

However, recently researchers have started focusing more on these alternative battery technologies. The other potential metal ion batteries are also investigated widely to meet the ever-growing energy demands. If we look into the scientific literature, we could see the successful development and application of different anodes, cathodes, electrolytes, binders and additives in various metal ion batteries. Despite all these developments in different active materials and components, there remain numerous challenges, which include full cell design, electrode material balancing, etc., in the practical applications of battery technologies.

In this regard, this book will mainly focus on the current research on materials for advanced battery technologies and propose future directions for different types of batteries to meet the current challenges associated with various metal ion batteries. Furthermore, the book will also provide insights into scientific and practical issues in the development of different batteries. The wide perspective of the book will ensure it is suitable for scientists, professors and students who are working in the energy sector. From an academic and industrial research standpoint, this book intends to bring out a new perspective on the storage technologies, which is beyond LIBs. Hence, the main intention of this book is to introduce such different themes of batteries to the readers and at the same time evaluate the opportunities and challenges of these battery systems from a commercial point of view.

# Editor Bios

**Ranjusha Rajagopalan** received her Ph.D. in Physics with specialization in battery technology from University of Wollongong, Australia, in the year 2017. She received her master's and bachelor's degrees in physics from Calicut University, India. She has also completed another master's (nanosciences) and doctoral (chemical engineering) degrees. During her tenure at College of Chemistry and Chemical Engineering, Central South University, China, she was working on the functionalization of organic/inorganic materials for developing high-performance metal ion batteries, including sodium and potassium ion batteries. Her research interests also include enhancement of power and cycling performances of battery electrodes by designing novel materials and composites. Her current research interest is on the material development for biosensor at Vitaltrace, Australia.

**Haiyan Wang** obtained his Ph.D. in Applied Chemistry at Central South University in 2012. He was also a visiting student to the University of St. Andrews. He worked for two years at The Hong Kong University of Science and Technology. Now he is a full Professor in the College of Chemistry and Chemical Engineering at Central South University. His current research interests are mainly focused on the new energy materials and devices for energy storage, including Li/Na ion batteries and aqueous metal batteries (Al/Zn). He has published over 160 refereed papers in international journals including *Nature Communications*, *Angewandte Chemie International Edition*, *Energy and Environmental Science*, and *Advanced Materials* (over 9,000 citation times, H-index 56). He has received some important awards, for example, the First Science and Technology Progress Award in Chongqing Province and the Second Natural Science Award in Hunan Province. He is also the winner of other awards such as Hong Kong Scholar, Leading Talents of Science and Technology Innovation in Hunan Province, Distinguished Young Scholar of Hunan Natural Science Fund, Young Talents in Hunan Province, and My Favorite Young Teacher in Hunan Province. He is a Fellow of International Association of Advanced Materials (IAAM).

**Yougen Tang** obtained his Ph.D. in Material Science at Central South University. From 2000, he is a full Professor in the College of Chemistry and Chemical Engineering at Central South University. He is also the director of the Institute of Chemical Powder Sources and Materials at the Central South University as well as the head of Hunan Provincial Key Laboratory of Chemical Power Sources. He has published over 200 refereed papers. His current research interests are focused on the new energy materials and advanced batteries.

# Contributors

**Ricardo Alcántara** University of Córdoba, Spain

**Marta Cabello** Centre for Cooperative Research on Alternative Energies (CIC energy GUNE), Basque Research and Technology Alliance (BRTA), Spain

**Sumol V. Gopinadh** Energy Systems Division, Vikram Sarabhai Space Centre, Thiruvananthapuram, Kerala, India

**Dachong Gu** National Engineering Research Center for Magnesium Alloys, College of Materials Science and Engineering, Chongqing University, China

**Zhengqiang Hu** Center for Marine Observation and Communications, Qingdao University, Qingdao, China

**Weibo Hua** School of Chemical Engineering and Technology, Xi'an Jiaotong University, Xi'an, Shaanxi, China

**Bibin John** Energy Systems Division, Vikram Sarabhai Space Centre, Thiruvananthapuram, Kerala, India

**Pedro Lavela** University of Córdoba, Spain

**Dajian Li** National Engineering Research Center for Magnesium Alloys, College of Materials Science and Engineering, Chongqing University, Chongqing, China

**Hongsen Li** College of Physics, Center for Marine Observation and Communications, Qingdao University, China

**Mingtao Li** Department of Chemical Engineering and Technology, Xi'an Jiaotong University, China

**Yongshuai Liu** College of Physics, Center for Marine Observation and Communications, Qingdao University, Qingdao, China

**Yumei Liu** School of Materials Science and Engineering, Peking University, Beijing, China

**Fusheng Pan** National Engineering Research Center for Magnesium Alloys, College of Materials Science and Engineering, Chongqing University, China

**Peddinti V.R.L. Phanendra** Energy Systems Division, Vikram Sarabhai Space Centre, Thiruvananthapuram, Kerala, India

**Ranjusha Rajagopalan** College of Chemistry and Chemical Engineering, Central South University, Changsha, P. R. China

**Wei Tang** School of Chemical Engineering and Technology, Xi'an Jiaotong University, China

**Yougen Tang** College of Chemistry and Chemical Engineering, Central South University, Changsha, P. R. China

**Mercy T.D.** Energy Systems Division, Vikram Sarabhai Space Centre, Thiruvananthapuram, Kerala, India

**Xiaolu Tian** Department of Chemical Engineering and Technology, Xi'an Jiaotong University, China

**José L. Tirado** University of Córdoba, Spain

**Anoopkumar V.** Energy Systems Division, Vikram Sarabhai Space Centre, Thiruvananthapuram, Kerala, India

**Haiyan Wang** College of Chemistry and Chemical Engineering, Central South University, Changsha, P. R. China

**Huaizhi Wang** College of Physics, Center for Marine Observation and Communications, Qingdao University, Qingdao, China

**Jingfeng Wang** National Engineering Research Center for Magnesium Alloys, College of Materials Science and Engineering, Chongqing University, Chongqing, China

**Liang Wu** National Engineering Research Center for Magnesium Alloys, College of Materials Science and Engineering, Chongqing University, China

**Yuan Yuan** National Engineering Research Center for Magnesium Alloys, College of Materials Science and Engineering, Chongqing University, China

**Hao Zhang** College of Physics, Center for Marine Observation and Communications, Qingdao University, Qingdao, China

**Ligang Zhang** School of Materials Science and Engineering, Central South University, P. R. China

**Xingwang Zheng** National Engineering Research Center for Magnesium Alloys, College of Materials Science and Engineering, Chongqing University, China

# 1 Introduction to Metal Ion Batteries

*Mingtao Li and Xiaolu Tian*

## 1.1 WHAT ARE METAL ION BATTERIES?

In order to cope with climate change, it is imperative to develop new energy sources. Among them, solar energy, wind energy and tidal energy are inexhaustible, but they show obvious intermittency, volatility and randomness. Efficient energy conversion and storage methods are urgently needed to enable these energy sources to be utilized constantly. Secondary batteries can achieve high-efficiency storage of energy and show great application potential in the field of large-scale energy storage. Combined with the development of renewable energy, secondary batteries will play an important role in achieving the goal of "zero carbon emission".

Secondary batteries are more economical and environmentally friendly than common primary batteries. Common secondary batteries in the market include lead-acid batteries, nickel–cadmium batteries, nickel–hydrogen batteries, flow batteries and metal ion batteries. Among them, metal-ion batteries (MIBs) have been widely commercialized due to their high energy density, outstanding cycle stability, and environmental friendliness. Especially, lithium-ion batteries (LIBs) have been widely used in small portable electronic devices such as mobile phones and notebook computers, as well as in new energy vehicles since their commercial use. In addition, the research work on a variety of new MIBs has also made positive progress [1, 2].

As the most widely used representative of MIBs, LIB is a type of battery that relies on lithium ions inside the battery and electrons in the external circuit moving between the anode and cathode. The cathode of a LIB is usually made of a lithium-containing compound (such as a lithium transition metal oxide) coated on a metal aluminum foil, while the anode is usually made of graphite coated on a metal copper foil.

Although LIBs are widely used due to their advantages of high energy density and recyclability, the reserves of lithium resources in the earth's crust are limited and unevenly distributed, and the wide production of LIBs has led to a sharp rise in the price of lithium. In the meantime, the various risks of LIBs including capacity decay, internal resistance increase, short circuit, gas production, and thermal runaway restrict their large-scale applications in smart power grids and electric vehicles. Therefore, it is imperative to develop new batteries with lower cost, higher safety and longer cycle life by exploiting more abundant elements of the earth [3, 4].

DOI: 10.1201/9781003208198-1

A possible solution for overcoming the disadvantages of LIBs would be the non-lithium batteries based on alternative metal ions, such as alkali metal ions ($Na^+$ and $K^+$) and multivalent metal ions ($Mg^{2+}$, $Ca^{2+}$, $Al^{3+}$, and $Zn^{2+}$) with abundant reserves and availability at low prices. Among them, sodium and potassium are in the same main group as the lithium element and have similar physical and chemical properties. Sodium-ion batteries are regarded as the most promising energy storage devices to replace LIBs and have received extensive attention from the academic community. However, due to the large sodium-ion radius (1.02 Å) and the high standard electrode potential (–2.71V vs. SHE), the energy density and power density of sodium-ion batteries are still far behind those of LIBs. The same challenge also needs to be overcome for potassium ion batteries before practical use. Rechargeable batteries based on multivalent metal ions have higher volume energy density compared with lithium ions because they carry more electrons during the charge–discharge process. In addition, multivalent MIBs are safer than LIBs for their resistance to dendrites and moisture. However, there are still many problems to be solved due to the late research on multivalent ion batteries. The incomplete system of multivalent ion batteries with high energy density is the main problem that limits its further development, in which the lack of suitable anode and cathode materials and stable electrolytes remain to be a huge obstacle to the construction of multivalent ion batteries.

## 1.2   EVOLUTION OF MIBS

Lithium is the metal with the lowest density and the highest electrochemical potential and energy/weight ratio. The low atomic weight and small size of lithium ions also accelerate the diffusion of lithium, indicating that it is an ideal material for making batteries [5].

In 1912, experiments on lithium batteries began under the leadership of G.N. Lewis, but those batteries were not practical enough to enter the market until the 1970s.

In the 1970s, Exxon's M.S. Whittingham used titanium sulfide as the cathode material and metallic lithium as the anode material to fabricate the first kind of lithium battery.

In 1975, Sanyo developed the $Li/MnO_2$ batteries and then began mass production of lithium secondary batteries.

In 1980, Armand proposed to replace metal lithium with embeddable materials as anode material, in which the lithium ions can be inserted and extracted back and forth. This concept of battery is vividly called "rocking chair battery". The creative design of the "rocking chair battery" avoids the safety problems caused by the formation of lithium dendrites from lithium metal anode.

Also in 1980, American chemist John B. Goodenough discovered the $LiCoO_2$ cathode, and Moroccan scientist Rachid Yazami discovered graphite anode and solid electrolytes.

In 1981, Japanese chemists Tokio Yamabe and Shizukuni Yata discovered a new type of nano-carbonacious-Polyacene and found that it is a very effective anode in traditional liquid electrolytes.

In 1983, M. Thackeray, J. Goodenough, and others discovered that manganese spinel is an excellent cathode material, with low price, high stability, and excellent lithium conductivity. It had high decomposition temperature and much lower oxidation potential than lithium cobalt oxide. Although the short circuit and overcharge risks existed, it could avoid the danger of combustion and explosion.

In 1985, a research team led by Akira Yoshino of Japan's Asahi Chemical Company established the first LIB prototype, which is a rechargeable and more stable version of the lithium battery.

In 1991, Sony corporation commercialized LIBs.

In 1996, Padhi and Goodenough discovered that phosphates with an olivine structure, such as lithium iron phosphate, are more superior than traditional cathode materials, and as a result, $LiFePO_4$ has become the current mainstream cathode material.

As the new generation of secondary batteries with high energy density and long cycle life, LIBs are currently widely used in mobile communications, digital technology, electric vehicles (EVs), energy storage, and other fields. It is difficult to estimate the demand for LIBs and their materials in the future, and huge supporting upstream and downstream industrial chains have been developed. With the technological development of LIBs, various pure EVs, hybrid new EVs, and extended-range new energy vehicles have sprung up. LIBs such as $LiCoO_2$ batteries, $LiMn_2O_4$ batteries, $LiFePO_4$ batteries, and ternary lithium batteries have occupied a major position in their respective fields, and in order to further meet future needs, researches on LIBs continue to achieve new technological progress.

However, during the production, transportation, and practical use of LIBs, many failures can lead to the poor performance, lower reliability, and decreased safety of a single battery or the entire battery pack. These failures include the following:

1.  *Capacity decay*. The capacity decay is closely related to the manufacturing process and using improper environment. Further, the root cause lies in the failure of active materials, such as the structural failure of cathode materials, transitional growth of solid electrolyte interphase (SEI) on the anode surface, decomposition of the electrolyte, and corrosion of the current collector.
2.  *Internal short circuit*. The short circuit may arise from the short circuit between copper/aluminum current collectors and the separator failure due to the puncture because of cathode impurities and lithium dendrites.
3.  *Gas production*. In pouch cells, the excessive consumption of electrolyte and the decomposition of cathode materials can lead to abnormal gas production and cause further safety problems.
4.  *Thermal runaway*. The thermal fun away refers to the rapid rise of local or overall temperature inside the LIB, where huge amount of heat is accumulated leading to further side reactions which could cause explosion.

As a result, the studies on LIBs safety have assumed importance so as to design safer and more reliable LIB-based energy system for EVs. The traditional methods to improve the safety of LIBs mainly involve the application of stable electrolyte/electrode materials and the safe design of battery systems together with external protection devices. In the past decade, the development of non-lithium MIBs with high safety and low cost has also become a hot spot as the next generation energy storage device for EVs.

MIBs such as sodium-ion batteries (SIBs), magnesium ion batteries, and aluminum ion batteries have been identified as strong alternatives to LIBs because of their superior safety, low cost, and environment friendliness. As early as the late 1970s, researches on SIBs were carried out almost at the same time as LIBs. Due to the limitations of research conditions and the enthusiasm for LIBs, the research on SIBs was once slow and stagnant. It was not until 2010 that SIBs gained renewed attention due to the lithium resource bottleneck. At present, SIBs have successfully entered practical stage, but limited to the fields of low-speed EVs and low-energy storage because of the inferior energy density of 160–180 Wh kg$^{-1}$. The research of other MIBs based on $K^+$, $Mg^{2+}$, $Ca^{2+}$, $Zn^{2+}$, and $Al^{3+}$ was conducted in the late 1980s or 1990s, yet they are far from practical use due to many problems including the limited variety of electrode/electrolyte materials, the unclear charge–discharge mechanism, and the poor battery performance. The development of stable cathode materials with high energy density is an especially key challenge. When large metal ions are intercalated in the cathode material, they can form strong interactions and cause structural instability and even collapse, resulting in poor battery capacity and reduced cycle life. Therefore, the development of non-lithium metal ion cathode materials needs to improve in the near future.

In conclusion, although the development speed of LIBs tends to be stable, the research on new materials and preparation processes will bring new breakthroughs in battery performance. In particular, the research on non-lithium MIBs is expected to further improve the cruising range and the safety of new energy vehicles. For specific application scenarios, research and exploration of secondary MIBs also show broad prospects. SIBs, magnesium-ion batteries, and zinc-ion batteries have potential applications in low-speed electric vehicles, large-scale energy storage, and communication base stations. However, the preparation of most MIBs remains only at the laboratory level, and the commercial application is still in its infancy. While in-depth researches on basic mechanisms are being conducted, technological transformation and industrial development should also be promoted in parallel.

## 1.3 THE GENERAL WORKING PRINCIPLE OF THE ROCKING CHAIR BATTERIES

Since 1990s, the researches of rocking chair lithium-ion secondary batteries have been rapidly expanding, in the direction of various synthesis methods of electrode materials, reversible electrode reaction mechanism, electrolyte (especially polymer electrolyte), various electrochemical tests, and structural characterizations.

In a typical rocking chair battery, the active materials of the electrodes are composed of lithium intercalation compounds and the low-potential materials are used as anode, while the high-potential materials are used as cathode [6]. The greater the potential difference between the cathode and anode, the higher the electromotive force of the battery. When this battery is charged, Li ions can be deintercalated from the cathode and intercalated into the anode through the electrolyte. Conversely, when the battery is discharged, Li ions can be deintercalated from the anode and then intercalated into the cathode through the electrolyte. The operating process of the battery is actually the process of back-and-forth Li ion intercalation/deintercalation between the two electrodes, so this kind of battery is called "rocking chair batteries". For example, in a LIB (graphite as the anode and layered metal oxide $LiMO_2$ as the cathode), Li ions are cyclically intercalated and deintercalated between the cathode and anode materials during the charging process.

The reaction formula is as follows:

$$\text{Cathode: } LiMO_2 \rightarrow Li_{1-x}MO_2 + xLi^+ + xe^-$$

$$\text{Anode: } xLi^+ + xC_6 + xe^- \rightarrow xLiC_6$$

During the charging process, the cathode loses electrons, and $Li^+$ is deintercalated from the $LiMO_2$ cathode. Meanwhile, $Li^+$ is intercalated into the graphite anode through the porous separator (Figure 1.1).

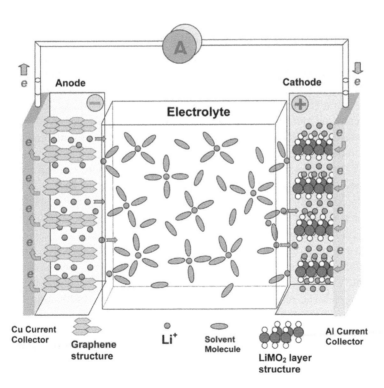

**FIGURE 1.1** Schematic description of a "(Li ion) rocking-chair" cell that employs graphitic carbon as anode and transition metal oxide as cathode [7].

The highly reversible electrochemical reactions of rocking chair LIBs result in long cycle life. This reversibility of the electrochemical reaction is closely related to the special structure of the active materials constituting the electrodes. The cathode and anode active materials of rocking chair batteries are intercalation compounds, such as $Li_xFe_2O_3$, $Li_xWO_2$, $Li_xTiS_2$ and $Li_xC_6$. Cathode materials include $Li_xMO_2$ (M = Co, Ni, Mn, Fe), $Li_xMS_2$ (M = Mo, Ti, V, Fe), $Li_xM_2O_4$ (M = Mn, Ti, V), defective λ-$MnO_2$, chemically modified lithium manganese composite oxide (CDMO), and $H_{4x}Mn_{8-x}$ $O_{16}$·$YH_2O$, and so on. The intercalation compound with low crystal density possesses a layered structure, a three-dimensional network structure, or a tunnel structure, which have special open structures where Li ions "pass in and out" freely. When the intercalation and deintercalation of Li ions occurs, the crystals only expand and contract correspondingly, while the structure type is basically maintained. The electrochemical reactions carried out during the charging and discharging process of the battery are actually a kind of intercalation reaction. When Li ions diffuse between the layers, gaps or tunnels in the crystals, the bond breakage and the structure reconstruction of the electrode materials do not occur. The energy required for diffusion is very small, so the intercalation reaction of Li ions in the two electrodes is easy to proceed.

## 1.4 DIFFERENT COMPONENTS

### 1.4.1 SODIUM-ION BATTERIES (SIBs)

#### 1.4.1.1 Introduction of Sodium-Ion Batteries

SIBs are also a type of rocking chair secondary battery, which is consistent with the principle of LIBs. Sodium and lithium belong to the same main group elements, and both exhibit similar "rocking chair" electrochemical behaviors during battery operation (Figure 1.2). During the charging process of a SIB, sodium ions are deintercalated from the cathode and intercalated into the anode. At the same time, electrons pass through the external circuit. The more sodium ions intercalated into the anode, the higher the charging capacity.

Like LIBs, a SIB is mainly composed of a cathode, an anode, a current collector, an electrolyte, and a separator. Since the radius of sodium ions is relatively large, the regular layered structure is preferred for anode and cathode materials. The design of the layer spacing is a key parameter for the performance of SIBs. The development of SIBs has lasted for more than 50 years. The research on both sodium-ion and lithium-ion batteries started in the 1970s [9]. Due to the increasing demand for energy storage, low-cost energy storage battery technology is becoming more and more urgent.

*Early stage*: Basic research on sodium-ion began in the 1970s and was mainly used for energy storage scenarios. In the late 1970s, researches were carried out on SIBs and LIBs almost simultaneously. However, the early research on SIBs was slow and stagnant due to the limitations of the research conditions at that time and the stronger interest in LIBs. Further, in those time periods, scientists were focusing more on sodium sulfur batteries than SIBs. Sodium sulfur batteries were first proposed in 1966 by Kummer and Weber who worked at Ford in the United States. Early sodium sulfur batteries have been widely studied and applied in large-scale energy storage

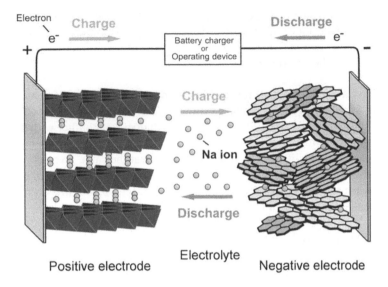

**FIGURE 1.2** Schematic illustration of SIBs [8].

systems, such as electric vehicle, due to their obvious advantages of low cost and high energy density [10].

*Mid-term*: Lithium resource scarcity was becoming prominent, and SIB research began to attract attention. The main reasons were as follows:

1. Lead-acid battery environmental pollution is inevitable: its solid and gaseous pollution may be eliminated, but the pollution of heavy metal ions cannot be avoided.
2. Limited reserves of lithium resources: 70% of the world's lithium resources are currently distributed in South America. It is difficult for LIBs to meet the needs of the two major industries, that is, electric vehicles and grid energy storage.
3. SIB cost advantage: The current price of battery-grade lithium carbonate is rising, while sodium is obviously easier to obtain hence it is more economical.

*Current*: From the laboratory to the practical stage, many companies have deployed SIB technology research. The companies deploying SIBs include:

1. *Faradion in the United Kingdom.* The developed SIB technology comprised of cathode materials based on nickel, manganese, and titanium layered oxides. This company could develop 10 Ah soft pack battery samples with an energy density reaching 140 Wh kg$^{-1}$, the average working voltage of 3.2 V, and the capacity retention rate exceeding 92% after 1,000 cycles.
2. The United States Natron Energy uses Prussian blue materials to develop high-rate aqueous SIBs with a cycle life of 10,000 cycles at 2 C rate.

3. The Battery Research Department of Toyota Corporation of Japan developed a new SIB cathode material in 2015. Currently, due to the low energy density, the application of SIBs is limited to the fields of energy storage power station and two-wheeled vehicles.

SIBs have a complementary relationship with Lithium Nickel Manganese Cobalt Oxide (NCM) batteries and a certain substitution relationship with LFP batteries. The capacity density of SIBs is 70–200 Wh $kg^{-1}$, which does not conflict with the energy density range of 240–350 Wh $kg^{-1}$ of NCM lithium batteries while high-energy sodium batteries and LFP batteries are theoretically at the same level. At this stage, sodium batteries are mainly concentrated in the 130–150 Wh $kg^{-1}$ range. In terms of cycles, the theoretical cycle of sodium batteries can reach 10,000 times, which is currently around 3,000–4,000, a little bit behind LFP lithium batteries.

### 1.4.1.2  SIBs vs. LIBs

The fast charging performance of SIBs is better than that of LIBs because the Stokes diameter of Na ion is smaller, and as a result the sodium salt electrolyte has higher ion conductivity than lithium salt electrolyte of the same concentration [11]. However, sodium ions have a larger radius than Li ions, making them difficult to be intercalated into the electrode crystal structure and therefore leading to a slower movement rate. But this shortcoming can be changed by changing and improving the characteristics of the anode materials.

In terms of safety, SIBs are obviously better than LIBs. Across the world, there have been frequent fire accidents of LIBs in EVs and other energy storage devices. According to incomplete statistics, a total of 32 energy storage power station fires and explosions occurred worldwide between 2011 and 2021, among which ternary LIBs were used in 26 incidents. The electrochemical performances of the SIBs are relatively stable, because they are easy to passivate and deactivate during the thermal runaway process and have shown better performance in safety tests than that of LIBs. It does not emit smoke, fire, or explode during puncture, nor does it catch fire or burn when subjected to short circuit, overcharge–discharge, and squeeze experiments. In comparison, the initial self-heating temperature of LIBs is 165°C, while that of SIBs reaches 260°C. In the ARC test, the maximum self-heating speed of SIBs is significantly lower than that of LIBs. All these indicate that SIBs have better thermal stability.

### 1.4.1.3  Sodium Battery Technology and Materials

Since sodium batteries do not contain lithium ions, and the raw materials of LIBs except for the separator are no longer applicable. Like LIBs, SIBs are also composed of a cathode, an anode, a current collector, a separator, electrolyte, a casing, and a top cover.

The cathode materials mainly include: transition metal oxides, polyanionic compounds, Prussian blue compounds, and amorphous materials [12, 13] Transition metal oxides are currently the most popular cathode materials, such as sodium iron phosphate, sodium iron manganate, sodium titanate manganese, etc. Zhongke Hai sodium, sodium innovation energy, and Faradion are the main companies employing this technology. Prussian blue materials display good electrochemical performance, are low cost, and exhibit good stability, but there are problems in the preparation process such as difficulties in controlling the content of coordination water. Ningde Times, Star Sky Sodium Electric, and Natron Energy are the major users of this technology. Polyanionic materials show good stability, cycle life, and diversity, but the lower intrinsic electronic conductivity limits their practical application.

Anode electrode materials include metal compounds, carbon-based materials, alloy materials, and non-metallic elements [14]. Among metal compounds, metal oxides, sulfides, and selenides are the main representatives. Metal alloy materials can react with sodium at low potential during discharge, and the dealloying reaction occurs in the charge process at high potential. Such materials tend to have high theoretical specific capacity and low output potential (<1 V), but the volume change during the reaction is huge (usually >200%), making these materials easy to crack hence affect the battery performance. The obvious difference from LIBs is that sodium ions have a larger ion radius than lithium ions, and therefore, the volume expansion caused by the formation of metal sodium and negative electrode materials in the alloy is also more obvious. Non-metallic elements of the same family as carbon, phosphorus, and silicon have become the center of focus in recent years. Among them, purple phosphorus is converted to white phosphorus when heated, and white phosphorus has unstable chemical properties. Therefore, both purple phosphorus and white phosphorus cannot be used as electrode materials, while red phosphorus has low conductivity and volume expansion problems which are difficult to solve. Black phosphorus has a wrinkled layered structure and high conductivity, but it is difficult to prepare. The preferable carbon-based material for the anode is hard carbon. Hard carbon is usually used as the anode active material instead of graphite because of its stable structure, while SIBs with graphite anode show poor storage capacity. For soft carbon, low sodium storage capacity and high charging potential limit its use as an ideal anode material in high-energy carbon-based SIBs. Unlike soft carbon, hard carbon is difficult to graphitize even after high temperature treatment. It also exhibits better sodium storage capacity and lower working potential, thus making it more suitable for use as an anode material for SIBs.

In addition to the cathode and anode materials, the current collector plays an important role as a material that carries the activity of cathode and anode and collects electrons. The anode current collector of SIBs can apply aluminum foil instead of copper foil, reducing the cost further. Unlike lithium, sodium does not undergo an electrochemical alloying reaction with aluminum at room temperature, so the copper current collector can be replaced by cheaper aluminum.

## 1.4.2 POTASSIUM ION BATTERIES (PIBs)

### 1.4.2.1 Comparison of Potassium Battery and Lithium Battery
While SIBs have received widespread attention from the academic community, one of the alkali metal counterparts, potassium, has received much less attention. As similar elements in the first group, sodium and potassium have comparable properties and

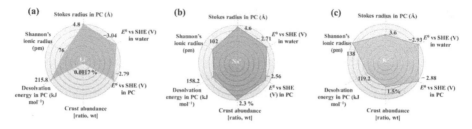

**FIGURE 1.3** Comparison of the physical and chemical properties of (a) Li$^+$; (b) Na$^+$; and (c) K$^+$ [17].

reserves. However, the former seems more promising because of its closer radius with lithium ions. Hence, the application of most of the sodium materials can be directly referred to the research and development results of LIBs. Besides, PIBs are less likely to be industrialized as they have lower energy density than SIBs. However, recent test results for sodium-ion and potassium-ion batteries show that these disadvantages of potassium ion can be compensated by some of its unique advantages [15, 16].

As illustrated in Figure 1.3, PIBs have the following advantages:

1. *Abundant resources:* Potassium resources are abundant on earth, occupying 2.09% of the earth's crust, which is close to sodium resources (2.36%), more than 1,000 times that of lithium resources (0.0017%).
2. In the organic non-aqueous electrolyte solution, the electrode potential of K/K$^+$ is lower than that of Na/Na$^+$ and Li/Li$^+$. For example, the electrode potential of lithium, sodium and potassium in PC solution (for standard hydrogen electrode) is in the order of: K (–2.88V) < Li (–2.79V) < Na (–2.56V). The low electrode potential is beneficial to increase the energy density of the battery.
3. *The diffusion rate of potassium in the carbon anode material is faster.* The research results of the potassium ion-carbon material intercalation reaction show that the diffusion rate of potassium ions in the carbon material is faster than that of sodium, so the rate performance of the battery is higher than that of the SIBs.
4. *The conductivity of potassium ion in electrolyte is higher.* Compared with sodium and lithium, potassium ions in non-aqueous electrolyte have higher ionic conductivity, which is beneficial to improve the power characteristics of PIBs.
5. *Restricted dendrites formation in the alkali metal anode.* The mixture of sodium and potassium becomes liquid at room temperature, but lithium can neither form an alloy with sodium nor potassium. If a sodium–potassium alloy is used as the negative electrode, it is liquid at room temperature, while the liquid anode does not have the problem of dendrite formation, and the metal deposited during the charging process will become liquid. Moreover, the liquid metal anode does not require solid-state electrolytes because it is immiscible with liquid organic electrolyte, so the alloy anode can work

at room temperature. Due to the "relative difference" of sodium–potassium potential (potassium's potential is more negative and relatively easier to lose electrons), the active element in sodium–potassium alloy is metal potassium. Sodium metal does not participate in the battery reaction, but sodium plays a role. By liquefying potassium metal, sodium–potassium alloy as anode has a specific capacity of 579 mAh g$^{-1}$. This gives PIBs another opportunity.

By taking the above advantages into consideration, scientists across the globe have started in-depth investigation on PIBs. By optimizing the electrolyte and finding high-capacity cathode and anode materials, it is possible to construct high-performance, high-safety, and low-cost PIBs.

### 1.4.2.2 The Electrode Materials of PIBs

Similar to SIBs, the cathode materials of PIBs mainly include transition metal oxides, polyanionic compounds, and Prussian compounds [18].

Transition metal oxides $A_xMO_2$ (A=Li, Na, K, etc., M=Co, Ni, Mn, etc.) have the characteristics of low toxicity, low cost, and simple synthesis process, and have been widely used in lithium and sodium-ion batteries. As early as 1975, there were reports on the structure of potassium–cobalt composite oxide ($K_xCoO_2$). However, Co in $K_xCoO_2$ materials is costly and toxic. Compared with Co, Mn is richer in resources and lower in price. Hence, manganese-based compounds as electrodes have received widespread attention.

The polyanionic cathode material can be represented by the general formula $A_xM_y$ $[(XO_m)_n]_z$, where A represents alkaline metals Li, Na, and K, M is transition metal ion, and X represents elements such as P, V, S, Si, etc. The X polyhedron and the M polyhedron are connected by co-edges or common points to form a polyhedral frame, and the A ion is located in the gap. The polyanionic compounds have a stable framework structure, large potassium ion transmission gaps, and long cycle stability, thus achieving improved electronic conductivity of the material and excellent electrochemical performance. By choosing different anionic groups to construct a mixed anion system, new structures and novel material systems can be obtained. Such materials have been extensively studied in LIBs and SIBs. Among them, $K_3V_2(PO_4)_3$ and $KVPO_4$ materials are the most researched cathode materials for PIBs. In addition to the above phosphate-based polyanions, sulfate-based polyanion electrode materials have also been studied and it is found that the electrochemical performance of $KFeSO_4F$ is better than that of the same type of lithium-ion and sodium-ion cathode materials.

Prussian blue (PB) and its analogs (PBAs) have a unique three-dimensional tunnel structure, which allows alkali metal ions to be reversibly deintercalated in their crystal lattice, thus showing good electrochemical performance. Prussian blue compounds can be represented by the general formula $A_xM_a[M_b(CN)_6]\cdot zH_2O$, where A represents an alkali metal element, $M_a$ and $M_b$ represent a transition metal element, and their structures are observed to be face-centered cubic structures. Transition metal ions form six coordination with C and N atoms in cyanide. The alkali metal ions are present in the three-dimensional channel structure and coordination pores. Prussian blue compounds have the following advantages as cathode materials for PIBs:

1. Low cost. Their raw materials are mainly Fe and Mn, with abundant resources and low prices.
2. Good electrochemical performance such as good cycle stability and rate capability. Therefore, Prussian blue compounds are considered as ideal cathode materials for PIBs.

The anode materials include carbon-based materials, titanium-based materials and alloy materials [19].

Carbon-based materials have been widely studied as anode materials for the advantages of low cost, environmentally benign performance, and renewability. As a member of carbon-based materials, graphite has a theoretical capacity of 372 mAh $g^{-1}$ in LIBs and has been commercialized. Potassium is another alkali element that can directly intercalate with graphite to form $KC_8$, with a theoretical capacity of 319 mA $g^{-1}$. Potassium ions can also be embedded in graphite by electrochemical methods, and the reversible capacity can reach 273 mAh/g. Similar to the intercalation reaction of lithium in graphite, potassium ions undergo a step change ($KC_{24}$->$KC_{16}$->$KC_8$), corresponding to the transition from the third order to the second order and then the first order. As an anode material for PIBs, graphite has high reversible capacity and good rate. However, the graphite structure will deform due to the excessive volume of potassium ions, which has the disadvantage of rapid capacity decay. To conclude, graphite as an anode material presents two major problems: (1) The volume change caused by the potassium ion insertion/de-intercalation process is very large, resulting in rapid capacity attenuation and low initial coulombic efficiency; (2) The potassium ion diffusion rate is low due to the large potassium ion radius, which has a negative impact on its rate capability.

Titanium-based materials have the advantages of low cost, good cycle stability, and stable structure. They can be used as anode materials for PIBs because potassium ions can be intercalated in the electrode with a low potential. The widely studied titanium-based materials for PIBs include titanium oxide $TiO_2$, titanate (such as $K_2Ti_8O_{17}$), titanium phosphate ($KTi_2(PO_4)_3$) and titanium carbide compounds ($Ti_3C_2$), etc. But these materials have poor conductivity and need to be modified by means such as doping and carbon coating.

Alloy materials have attracted increasing attention for their advantages of good electrical conductivity and high theoretical capacity. The potassium storage mechanism of alloy anode materials is different from that of carbon materials and titanium-based materials. The major problem associated with alloying reaction of potassium is the huge volume expansion, which will cause the capacity deterioration. At present, there are only a few researches on alloy anode materials for PIBs, which include Sb-based, Sn-based, and P-based alloy anode materials.

### 1.4.2.3 The Electrolyte of PIBs

The potassium salt in the organic electrolyte of the PIBs basically use KFSI, $KPF_6$, and $KClO_4$, while the solvents are usually esters (propylene carbonate (PC), ethylene carbonate (EC), diethyl carbonate (DEC), etc.) and ethers dimethoxyethane (DME), and the functional additives are fluoroethylene carbonate (FEC) [20]. At present, PIBs are still in the early stages of development, and there is little research on their electrolytes. In EC/DEC electrolyte, KFSI has higher solubility than $KPF_6$ and $KClO_4$. But KFSI-based electrolyte can cause Al corrosion, leading to irreversible changes under high voltage conditions. The coulombic efficiency and cycle stability of PIBs can be improved when the electrolyte solute is replaced by $KPF_6$ and FEC is added. Different electrolyte solvents also have a certain impact on the electrochemical performance of PIBs. Compared with EC/DEC and EC/DMC solvents, PIBs with EC/PC have higher coulombic efficiency and stability, because DEC and DMC may decompose under low voltage conditions. Compared with organic electrolytes, solid electrolytes have better mechanical strength and thermal stability, but solid electrolytes for PIBs have rarely been studied.

## 1.4.3 MAGNESIUM ION BATTERIES (MGIBS)

### 1.4.3.1 Introduction to MgIBs

The magnesium in the earth's crust is more abundant than lithium, which greatly reduces the cost of MgIBs [21]. Besides, magnesium ion can provide two electrons in the redox reaction, which leads to a high theoretical volume capacity density of 3,833 $mAh \cdot cm^{-3}$, almost twice as that of lithium (Figure 1.4). Also, metallic magnesium is relatively stable in the atmosphere, and it is not easy to form dendritic deposits. Therefore, the use of magnesium metal as an anode in batteries is much safer than lithium metal anodes.

In 1990, the first MgIBs system was constructed based on $Mg(BR_4)$ solution (where R can be various organic groups) as the electrolyte and $Mg_xCoO_y$ as the cathode. However, due to the insufficient stability of the cathode to the electrolyte, these $Mg$-$Mg_xCoO_y$ batteries were not indeed practical. The study on MgIBs made substantial progress in 2000, when new MgIBs are assembled by keeping dissolved $Mg(AlCl_2BuEt)_2$ in tetrahydrofuran as the electrolyte, and $Mg_xMo_3O_4$ as cathode. The battery can work at $0.1$–$1$ mA $cm^{-2}$ in a wide temperature range ($-20$~$80$ °C) for more than 2,000 cycles with a capacity retention rate exceeding 85%.

Although MgIBs are considered as a next-generation battery system, which is alternative to LIBs, their development is still hindered by the following issues [23, 24]:

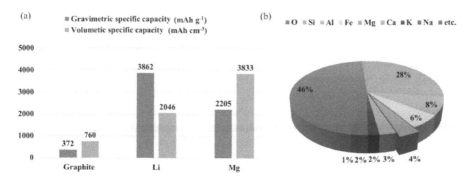

**FIGURE 1.4**    (a) The theoretical capacity of graphite, Li, and Mg anodes. (b) The element abundance in the crust [22].

1. Limited choice of cathode materials

   Cathode materials with a high degree of reversibility are very limited. Although transition metal compounds and organic materials have been explored to be suitable for MgIB cathode materials, the slow kinetics of magnesium ions in these cathode materials result in poor battery performance. This could be mainly attributed to the divalent nature of magnesium ions which weakens the reversibility of magnesium ions, leading to slow diffusion of magnesium ions and poor electrochemical performance. In addition, there are strong interactions between the electrode host materials and magnesium ions, which leads to a decrease in reversible capacity.

2. Development of suitable electrolyte

   It is very difficult to develop electrolytes with high compatibility with electrodes. Electrolytes in LIBs are not suitable for MgIBs because metal magnesium reacts with $ClO_4^-$, $BF_4^-$, and other anions to form a passivation layer on the surface of magnesium metal, and magnesium ions cannot pass through this passivation layer, resulting in unstable cycle performance. Besides, the stability of electrolytes to metal magnesium anode is limited [25].

### 1.4.3.2 The Electrode Materials of MgIBs

At present, the research directions of MgIBs are mainly concentrated in cathode materials, such as transition metal sulfides and transition metal oxides, while the most explored anode materials are magnesium alloy materials [26].

Titanium-based materials ($TiS_2$ and $TiO_2$-B) are investigated as an electrode in MgIBs, because of their advantages such as high volumetric energy density, environmentally benign nature, and low cost. $TiS_2$ nanotubes can reversibly intercalate a large amount of magnesium ions. MgIBs using $TiS_2$ cathode at a current density of 10 mA g$^{-1}$ and a voltage range of 0.5–2.0 V can reach a discharge capacity of 236 mAh g$^{-1}$.

Among vanadium-based materials, vanadium pentoxide ($V_2O_5$) and hydrated oxide ($V_2O_5.nH_2O$) have been studied as cathode materials for MgIBs. The magnesium ion pre-intercalated hydrated vanadium oxide has high electronic conductivity and excellent structural stability. Further, in this material a rapid magnesium ion migration can be realized due to the charge shielding effect of the lattice water.

Manganese-based oxides have the advantages of low cost, high stability, and environmental friendliness. Due to these advantages Mn-based materials such as $MnO_2$, $Mn_3O_4$, and manganese-based three-dimensional materials are explored as cathode materials in MgIBs. Molybdenum-based materials such as $MoO_3$, $Mo_6S_8$, and $MoS_2$ are also used in MgIBs.

From the perspective of the development history of MgIBs, the current research on cathode materials is mainly based on vanadium-based compounds. At the same time, metal sulfides are also investigated as anodes due to their higher theoretical specific capacity. However, MgIBs generally show low voltage and poor cycle stability. In future, the effective improvements on the magnesium storage performance of active materials mainly focus on the nano-engineering of electrode materials and composite formation with carbon. In addition, the modification of the metal magnesium anode and the design of a safe and stable electrolyte system are the keys to improve the performance of MgIBs.

## 1.4.4 Calcium Ion Batteries (CIBs)

Among multivalent ion batteries, CIBs have the following characteristics: the standard reduction potential of calcium (2.87 V, vs. SHE) is the closest to that of lithium metal (3.04 V, vs. SHE). The volumetric specific capacity of the calcium anode is 2072 mAh cm$^{-3}$ and the mass specific capacity is 1,337 mAh g$^{-1}$. Calcium is the fifth most abundant element in the earth's crust, 2,500 times more than lithium, and is widely distributed worldwide. Since the charge density and polarization intensity of Ca$^{2+}$ are less than that of other multivalent metal ions (Al$^{3+}$, Mg$^{2+}$, and Zn$^{2+}$), Ca$^{2+}$ exhibits better diffusion kinetics and higher power density. At present, the research on CIBs is still in its infancy. It is difficult to look for stable calcium storage electrode materials because of the quick formation of passivation layer on the surface of calcium metal anode while using traditional organic electrolytes, which makes it difficult to realize the reversible deposition of calcium ions. Therefore, the development of high-performance calcium electrode materials is of great significance.

In 1991, calcium ions were found to be difficult to penetrate the passivation film on the surface of the calcium metal anode in traditional organic electrolytes, which hindered the reversible oxidation–reduction reaction of calcium ions [27]. In 2016, a new type of calcium ion liquid battery was developed using a mixture of molten CaCl$_2$ and LiCl as the electrolyte, and molten Ca-Mg alloy and metallic bismuth as the anode and cathode, respectively [28]. Although its working voltage was not high (<1 V), the battery exhibited an excellent cycling performance (228 Wh L$^{-1}$) at high temperatures (550–700°C). Coulomb efficiency remained above 99% even after 1,400 cycles at a current density of 200 mA·cm$^{-2}$. Later it was observed that the electrolytes containing Ca(ClO$_4$)$_2$ led to less reversible redox process, while electrolytes containing Ca(BF$_4$)$_2$ could show typical reversible metal plating/stripping in cyclic voltammograms [29]. At 75–100°C, the reversible deposition reaction of calcium ions could occur in the Ca(BF$_4$)$_2$-EC/PC electrolyte, and the battery circulated for more than 30 cycles at 100°C. This finding provided proof to the concept of calcium rechargeable batteries. In addition, theoretical calculations were used to reveal the potential of CaMn$_2$O$_4$ as a cathode material for CIBs, which provided a guide for future research on cathode materials.

In terms of the anode materials of CIBs, Harvard University's Alán Aspuru–Guzik team and Northwestern University's Chris Wolverton's team used density functional theory calculations to explore the relationship between the driving force of the reaction and calcium content. These studies predicted that many metals (Si, Sb, Ge), transition metals (Al, Pb, Cu, Cd), and precious metals (Ag, Au, Pt, Pd) can be considered as promising and low-cost anode candidates for CIBs [30].

Most of the currently reported anode materials for CIBs are calcium metal and calcium alloys; the cathode materials include Prussian blue compound A$_x$MFe(CN)$_6$·$y$H$_2$O(A=Li, Na, Mg, Ca, etc., M=Ba, Ti, Mn, Fe, Co, Ni), vanadium oxide $V_2O_5$, spinel $AM_2O_4$ (M=Ti, V, Cr, Mn, Fe, Co, Ni), perovskite, and layered transition metal sulfides. Electrolytes for reversible dissolution/deposition of calcium ions include $Ca(PF_6)_2$/EC+PC, $Ca(BF_4)_2$/EC+PC, $Ca(BH_4)_2$/THF, $Ca(ClO_4)_2$/ AN(acetonitrile), $Ca(ClO_4)_2$+$Ca(BF_4)_2$+$Ca(TFSI)_2$/EC+PC, etc. Although the research on CIBs is still in its infancy, the current research work has laid an important foundation and opened up new ways for the development of positive and negative electrode materials for rechargeable CIBs.

In summary, the calcium-based ion batteries have attracted widespread attention due to their high theoretical capacity and better diffusion kinetics. Recently, more and more researches have been focusing on the development of electrode materials and electrolytes for CIB applications. Based on the performance comparison and research progress of different CIB systems, it is found that although CIBs have made encouraging progress, they are far from satisfying the practical application in large-scale energy storage. Therefore, the possible ways to improve electrochemical performance of CIBs in the future include (1) prediction of electrode materials through theoretical calculations; (2) surface modification to build a stable SEI film; and (3) heteroatom doping and nanostructure design [31]. In addition, it is possible to develop calcium-based dual-ion batteries, which has been proven to be a type of room-temperature CIBs with good reversibility.

### 1.4.5 ZINC-ION BATTERIES (ZIBS)

#### 1.4.5.1 Introduction to ZIBs

ZIBs are an emerging energy storage technology and have shown good application prospects because of their high specific capacity, rich resources, and safety of metallic zinc [32]. The charge storage mechanism of ZIBs depends on the migration of zinc ions ($Zn^{2+}$) between the Zn anode and the cathode material that can reversibly intercalate $Zn^{2+}$. Different from the widely used zinc-manganese primary battery (Zn-MnO$_2$ battery), ZIBs use weakly acidic or neutral zinc salts such as $ZnSO_4$ and $Zn(CF_3SO_3)_2$ as the electrolyte [33]. Therefore, in ZIBs, the production of ZnO on the Zn anode and the irreversible formation of $Mn(OH)_2$ and MnOOH on the cathode can be significantly suppressed. ZIBs belong to a multivalent metal battery system. Like Mg and Ca, the oxidation–reduction reaction of metal Zn can transfer two electrons when it is converted to $Zn^{2+}$ and therefore, ZIBs have a high theoretical volume specific capacity of 5851 mAh cm$^{-3}$, and their mass specific capacity can reach 820 mAh g$^{-1}$. Although multivalent systems can theoretically provide higher specific capacities, most metals (Mg, Al) as the anode will form inert products in the aqueous electrolyte system, leading to the passivation of the metal surface and poor charge and discharge performance. Therefore, in water-based multivalent metal battery systems, carbon-based materials are usually used as the anode. Due to the relatively low oxidation–reduction potential (–0.76 V vs. SHE), metal Zn has strong corrosion resistance. As a result, ZIB is one of the few metal battery systems that can use both aqueous and non-aqueous electrolytes. In addition, ZIBs can be assembled in air, so the manufacturing cost can be greatly reduced. The above advantages of ZIBs make them more appealing for power grids, household batteries, and other distributed power sources compared to other energy storage technologies.

For ZIB electrolytes, the water-based electrolytes are safe, environmentally friendly, easy to prepare, and lower in cost compared to organic electrolytes. Although organic electrolytes have higher oxygen evolution stability and can provide a wider electrochemical window than aqueous electrolytes, the redox potential of Zn is much smaller than that of metal Li (–3.045 V vs. SHE), and Na (–2.714 V vs. SHE). Therefore, the output voltage of organic ZIBs after matching the existing high-potential cathode materials can only reach 2–3 V. Hence, the energy density of which is not significantly improved compared to aqueous ZIBs. In addition, the ionic conductivity of the aqueous electrolyte (1 S/cm) is two orders of magnitude higher than that of the organic electrolyte (1–10 mS/cm). Further, the transfer resistance at the interface between aqueous electrolyte and the electrodes is much lower than that of organic electrolyte.

The discovery of aqueous ZIBs can be traced back to 1980s, when Shoji used a weakly acidic 2 mol/L $ZnSO_4$ electrolyte to obtain a rechargeable Zn‖$MnO_2$ battery. In 2012, Kang Feiyu team first proposed the concept of ZIBs and revealed the working mechanism of their cathodes and anodes. Subsequently, aqueous ZIBs have received extensive attention from researchers in recent years, and the number of research reports on aqueous ZIBs has increased significantly. At present, research on aqueous ZIBs mainly focuses on the development of cathode materials, modification of Zn anodes, and the development/optimization of new electrolytes. However, aqueous ZIB technology is still in the initial research stage, and it is facing the following challenges before its commercialization and wide application: (1) the dissolution of the cathode materials into the electrolyte; (2) the side reaction of the co-intercalation of $H^+$ and $Zn^{2+}$ in cathode materials; (3) the slow $Zn^{2+}$ intercalation kinetics in the cathode due to high electrostatic repulsion; (4) the desolvation of the electrode–electrolyte interface; (5) the dendrite growth on Zn anode; (6) the low coulombic efficiency caused by the hydrogen evolution reaction on the anode and the generation of by-products; (7) the narrow electrochemical window of the aqueous electrolyte limiting the energy density; (8) the unclear effects of the concentration of zinc salt and additives on the electrochemical performance.

### 1.4.5.2   The Compositions of ZIBs

In order to meet the application requirements of large-scale energy storage, the cathode materials of aqueous ZIBs are supposed to have high reversible charge capacity and good cycle stability. At the same time, resource-rich and environmentally friendly characteristics are also required. At present, the cathode materials mainly include manganese-based oxides, vanadium-based oxides, polyanionic compounds, Prussian blue compounds, organic materials, and transition metal disulfides.

The transition metal manganese (Mn) element has different valence states ($Mn^{2+}$, $Mn^{3+}$, $Mn^{4+}$, $Mn^{7+}$), which result in excellent ion storage performance. Among them, manganese dioxide ($MnO_2$) with +4 valence is suitable for large-scale production because it is easy to prepare, rich in resources, low in price, and environmentally friendly. In commercial alkaline primary batteries, the $MnO_2$ cathode reaction will generate $Mn(OH)_2$ and MnOOH products irreversibly. However, in the acidic aqueous electrolyte of ZIBs, $MnO_2$ is the first-used cathode material because it undergoes reversible electrochemical reactions (Figure 1.5). $MnO_2$ has rich structures including

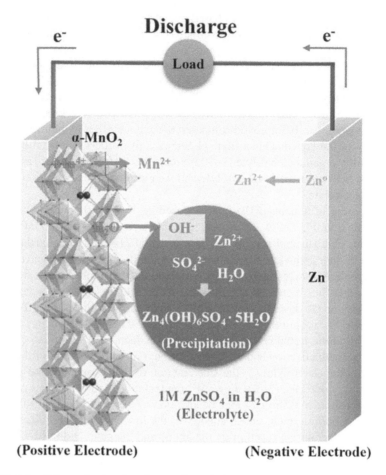

**FIGURE 1.5** Schematic showing the reactions during the discharge process of α-MnO$_2$/Zn cell, employing aqueous ZnSO$_4$ electrolyte [34].

tunnel structures (α, β, γ and rhombohedral structures), layered structure (δ type), and spinel structure (λ type), which are connected by regular octahedrons (MnO$_6$) through common edges or vertex angles. But during the electrochemical charge and discharge process, the MnO$_2$ structures of different crystal configurations are easily transformed into each other, which will cause an increase in stress and the destruction of the crystal structure, leading to the capacity decline during the prolonged cycling. In addition, other problems which limit its practical application include (1) the Mn$^{2+}$ formation during the process of MnO$_2$ dissolving in water; (2) the poor conductivity of MnO$_2$ negatively affecting the ion diffusion; and (3) the overall electrochemical performance.

As a transition metal element, vanadium (V) can exist as V$^{2+}$, V$^{3+}$, V$^{4+}$, and V$^{5+}$ during oxidation reaction, and the valence state can be transformed during the reaction. V-based oxides have been studied as aqueous ZIB cathode because the oxidation reaction of intermediate V can realize the reversible intercalation of $Zn^{2+}$. The current research on V-based cathode materials is mainly focusing on layered $XV_2O_5$ (X=Zn, Mg, Ca, Na), $V_2O_5$, and other V oxides. In general, V-based oxide cathodes have higher discharge capacity, excellent rate discharge performance, and good cycle life under high current density. However, it also faces dissolution problems under low current density conditions. In addition, although V-based oxides have different morphologies and structures (layered structure or tunnel structure), the materials based on the electrochemical reaction of $V^{5+}/V^{4+}$ redox couple show low discharge potential. It needs to be further investigated in depth on how to adjust the electronic structure and optimize the electrolyte structure to increase the oxidation potential.

Initially, metal Zn was used as an anode material for ZIBs in 1799. The Italian physicist Alessandro Volt used metal Zn and Ag circular plates sandwiching a wet rag containing salt water to fabricate the earliest voltaic battery. In addition to ZIBs and $Zn–MnO_2$ batteries, metal Zn is employed widely in other alkaline electrolyte batteries such as zinc silver batteries ($Zn–Ag_2O$), zinc nickel batteries ($Zn–NiOOH$), and zinc air batteries ($Zn–O_2$). But in the alkaline aqueous solutions, metal Zn is prone to participate in side chemical reactions, and there are also problems such as dendrite growth and water consumption, causing irreversible charge–discharge and poor coulomb efficiency. In acidic or neutral aqueous electrolytes, metal Zn can stably undergo oxidation–reduction reactions for its high hydrogen evolution potential. Moreover, the growth of Zn dendrites and the formation of passivation products can be suppressed to a certain extent, and thus the ZIBs show better reversibility and stability. A recent study proposed that the electrochemical performance of Zn anode can be improved by the following ways: (1) improve the structure of current collector; (2) construct a functional interface layer; and (3) optimize the electrolyte composition and concentration. These measurements can help to regulate the dissolution/deposition process of $Zn^{2+}$, inhibit the generation of Zn dendrites, and improve the coulombic efficiency [35].

The type and concentration of Zn salts and additives in the aqueous ZIB electrolytes also greatly affect the charge/discharge performance. At present, the reported Zn salts that can be used in the aqueous ZIBs include $ZnSO_4$, $Zn(CF_3SO_3)_2$, $ZnCl_2$, $Zn(NO_3)_2$, $Zn(ClO_4)_2$, $Zn(CH_3COO)_2$, and $ZnF_2$, of which $ZnSO_4$ and $Zn(CF_3SO_3)_2$ are the two most commonly used Zn salts [36].

## 1.4.6 ALUMINUM ION BATTERIES (AIBs)

The working principle of AIBs includes the plating/stripping process of aluminum on the aluminum anode, and the simultaneous intercalation/deintercalation of chloroaluminate ions or $Al^{3+}$ on the cathode [37]. The optimization of electrolyte and the exploration of cathode materials are currently the key directions of AIB research.

At present, the energy density of AIBs is mainly limited by the cathode materials, because the specific capacity of the cathodes in AIBs is much lower than that of the metal aluminum anode [38]. The reported cathode materials mainly include carbon-based materials, metal oxides, metal sulfides, and metal selenides. Among them, carbon-based materials are widely used in AIBs due to their high electrical conductivity, low cost, and abundant reserve. Especially, the layered and tubular structures of some carbon materials are more conducive to the intercalation of ions. However, due to the fact that carbon-based cathodes conduct the single-electron $[Al_xCl_y]^-$ instead of the three-electron $Al^{3+}$, the specific capacity of the carbon-based materials has been at low level. So, methods to realize the intercalation of trivalent $Al^{3+}$ in carbon-based cathodes remain to be further explored. Since the electronegativity of oxygen atoms is greater than that of sulfur atoms, the electrostatic interaction between metal sulfides and $Al^{3+}$ is smaller than that of transition metal oxides. Therefore, transition metal sulfides as cathode materials show much better electrochemical performance than metal oxides in AIBs.

At present, the commonly used electrolyte for AIBs is room temperature ionic liquids with wide electrochemical window, excellent oxidation resistance and good electrical conductivity. It is generally anhydrous $AlCl_3$ and BMImCl (1-butyl-3-methylimidazolium chloride), where the BMImCl ionic liquid mixes in a specific ratio. In addition, the researches of using ionic liquid mixed with anhydrous $AlCl_3$ and urea as the electrolyte have also been studied.

In summary, AIBs have received more and more attention due to their unique multi-electron transfer reactions and high theoretical specific capacity. The exploration of high-performance cathode materials is the most important development area for AIBs. Among them, low-cost carbon materials have become the most concerned cathode materials for AIBs [39]. In the future, their cycle stability and specific capacity can be further improved by doping heteroatoms, increasing specific surface area and modifying defect active sites. The commonly used electrolytes are room temperature molten salt ionic liquids, which possess high cost and sensitivity to water/air. In the future, it is still necessary to design and prepare relatively safe and stable electrolytes to realize the reversible deposition and dissolution of aluminum in electrode materials. In addition, the new electrolytes are expected to dissolve the dense aluminum oxide film on the metal aluminum surface without corroding current collectors and battery shells.

## 1.4.7 DUAL METAL-ION BATTERIES (DMIBs)

Various metal ions can be simultaneously involved in storing energy in DMIBs, which provide a new perspective for advanced energy storage. In should be noted that the performance of a mixed-ion battery is not just a superposition of the performance of a single-ion battery [40, 41]. By connecting the respective advantages of different MIBs, DMIBs have recently attracted widespread attention for their novel properties. Although research on mixed-ion batteries is still in its early stage, this strategy will be an excellent option to overcome the inherent shortcomings of single-ion batteries in the near future.

Unlike conventional single-ion batteries where only one active ion shuttles back and forth between the two electrodes, DMIBs involve the transfer process of different metal ions. In order to achieve simultaneous energy storage, the battery must meet the following requirements: (1) presence of dual ions in double-salt electrolyte or combined electrodes; (2) maintained chemical potential of the double metal ions during charging–discharging process.

As a representative, magnesium-based DMIBs have higher capacity and energy density [42, 43] Magnesium-based DMIBs (mainly Mg/Li and Mg/Na hybrid

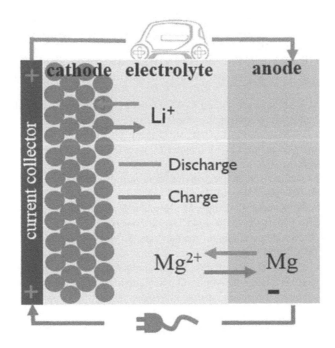

**FIGURE 1.6**   The operating mechanism of an Mg–Li hybrid battery [43].

batteries) usually consist of Li/Na as the cathode, metallic Mg as the anode, and electrolytes containing both $Mg^{2+}$ and $Li^+/Na^+$. Since the mobility of $Li^+/Na^+$ is better than $Mg^{2+}$, the intercalation reaction of $Li^+/Na^+$ usually takes place on the cathode side, while the deposition and dissolution of Mg take place on the anode due to its higher redox potential (Figure 1.6). Since the cathode materials preferentially accommodate monovalent ions because of their superior kinetics. The Mg–Li DMIBs can avoid the strong interactions between $Mg^{2+}$ and the cathode host and exhibit enhanced rate performance.

In general, compared with previous works on single-ion batteries, DMIBs provide a new approach to improve the electrochemical performance of batteries through the synergistic effect between various active ions. DMIBs simultaneously exhibit the advantages of safety, excellent kinetics, and low cost, and play an important role in the design of energy storage devices. To understand the synergies and mechanisms involved, more future researches remains to be conducted, thus promoting the development of a new generation MIBs with high energy density.

## REFERENCES

[1]   Goodenough, J. B.; Kim, Y., Challenges for rechargeable Li batteries. *Chem Mater* 2010, *22* (3), 587–603.

[2]   Tarascon, J. M.; Armand, M., Issues and challenges facing rechargeable lithium batteries. *Nature* 2001, *414* (6861), 359–367.

[3] Kim, J. G.; Son, B.; Mukherjee, S.; Schuppert, N.; Bates, A.; Kwon, O.; Choi, M. J.; Chung, H. Y.; Park, S., A review of lithium and non-lithium based solid state batteries. *J Power Sources* 2015, *282*, 299–322.

[4] Wang, Y. R.; Chen, R. P.; Chen, T.; Lv, H. L.; Zhu, G. Y.; Ma, L. B.; Wang, C. X.; Jin, Z.; Liu, J., Emerging non-lithium ion batteries. *Energy Storage Mater* 2016, *4*, 103–129.

[5] Reddy, M. V.; Mauger, A.; Julien, C. M.; Paolella, A.; Zaghib, K., Brief history of early lithium-battery development. *Materials* 2020, *13* (8), 1884.

[6] Etacheri, V.; Marom, R.; Elazari, R.; Salitra, G.; Aurbach, D., Challenges in the development of advanced Li-ion batteries: a review. *Energy Environ Sci* 2011, *4* (9), 3243–3262.

[7] Xu, K., Nonaqueous liquid electrolytes for lithium-based rechargeable batteries. *Chem Rev* 2004, *104* (10), 4303–4417.

[8] Yabuuchi, N.; Kubota, K.; Dahbi, M.; Komaba, S., Research development on sodium-ion batteries. *Chem Rev* 2014, *114* (23), 11636–11682.

[9] Delmas, C., Sodium and sodium-ion batteries: 50 years of research. *Adv Energy Mater* 2018, *8* (17), 1703137.

[10] Wang, C.; Zhou, D.; Palomares, V.; Shanmukaraj, D.; Sun, B.; Tang, X.; Wang, C.; Armand, M.; Rojo, T.; Wang, G., Revitalising sodium-sulfur batteries for non-high-temperature operation: a crucial review. *Energy Environ Sci* 2012, *13* (11), 3848–3879.

[11] Nayak, P. K.; Yang, L. T.; Brehm, W.; Adelhelm, P., From lithium-ion to sodium-ion batteries: advantages, challenges, and surprises. *Angew Chem Int Ed* 2018, *57* (1), 102–120.

[12] Bai, Q.; Yang, L. F.; Chen, H. L.; Mo, Y. F., Computational studies of electrode materials in sodium-ion batteries. *Adv Energy Mater* 2018, *8* (17), 1702998.

[13] Qian, J. F.; Wu, C.; Cao, Y. L.; Ma, Z. F.; Huang, Y. H.; Ai, X. P.; Yang, H. X., Prussian blue cathode materials for sodium-ion batteries and other ion batteries. *Adv Energy Mater* 2018, *8* (17), 1702619.

[14] Kim, S. W.; Seo, D. H.; Ma, X. H.; Ceder, G.; Kang, K., Electrode materials for rechargeable sodium-ion batteries: potential alternatives to current lithium-ion batteries. *Adv Energy Mater* 2012, *2* (7), 710–721.

[15] Eftekhari, A.; Jian, Z. L.; Ji, X. L., Potassium secondary batteries. *ACS Appl Mater Interfaces* 2017, *9* (5), 4404–4419.

[16] Hwang, J. Y.; Myung, S. T.; Sun, Y. K., Recent progress in rechargeable potassium batteries. *Adv Functional Mater* 2018, *28* (43), 1802938.

[17] Liu, S.; Kang, L.; Henzie, J.; Zhang, J.; Ha, J.; Amin, M. A.; Hossain, M. S. A.; Jun, S. C.; Yamauchi, Y., Recent advances and perspectives of battery-type anode materials for potassium ion storage. *ACS Nano* 2021, *15* (12), 18931–18973.

[18] Sha, M.; Liu, L.; Zhao, H. P.; Lei, Y., Review on recent advances of cathode materials for potassium-ion batteries. *Energy Environ Mater* 2020, *3* (1), 56–66.

[19] Zhang, C. L.; Zhao, H. P.; Lei, Y., Recent research progress of anode materials for potassium-ion batteries. *Energy Environ Mater* 2020, *3* (2), 105–120.

[20] Liu, Y. W.; Gao, C.; Dai, L.; Deng, Q. B.; Wang, L.; Luo, J. Y.; Liu, S.; Hu, N., The features and progress of electrolyte for potassium ion batteries. *Small* 2020, *16* (44), 2004096.

[21] Shah, R.; Mittal, V.; Matsil, E.; Rosenkranz, A., Magnesium-ion batteries for electric vehicles: Current trends and future perspectives. *Adv Mech Eng* 2021, *13* (3), 16878140211003398.

[22] Wu, D.; Ren, W.; NuLi, Y. N.; Yang, J.; Wang, J. L., Recent progress on selenium-based cathode materials for rechargeable magnesium batteries: A mini review. *J Mater Sci Technol* 2021, *91*, 168–177.

[23] Huie, M. M.; Bock, D. C.; Takeuchi, E. S.; Marschilok, A. C.; Takeuchi, K. J., Cathode materials for magnesium and magnesium-ion based batteries. *Coordin Chem Rev* 2015, *287*, 15–27.

[24] Deivanayagam, R.; Ingram, B. J.; Shahbazian-Yassar, R., Progress in development of electrolytes for magnesium batteries. *Energy Storage Mater* 2019, *21*, 136–153.

[25] Shuai, H. L.; Xu, J.; Huang, K. J., Progress in retrospect of electrolytes for secondary magnesium batteries. *Coordin Chem Rev* 2020, *422*, 213478.

[26] Kuang, C. W.; Zeng, W.; Li, Y. Q., A review of electrode for rechargeable magnesium ion batteries. *J Nanosci Nanotechno* 2019, *19* (1), 12–25.

[27] Aurbach, D.; Skaletsky, R.; Gofer, Y., The electrochemical-behavior of calcium electrodes in a few organic electrolytes. *J Electrochem Soc* 1991, *138* (12), 3536–3545.

[28] Ouchi, T.; Kim, H.; Spatocco, B. L.; Sadoway, D. R., Calcium-based multi-element chemistry for grid-scale electrochemical energy storage. *Nat Comm* 2016, *7*, 10999.

[29] Ponrouch, A.; Frontera, C.; Barde, F.; Palacin, M., Towards a calcium based recharge-able battery. *Nat Mater* 2016, *15*(2), 169–172.

[30] Yao, Z. P.; Hegde, V. I.; Aspuru-Guzik, A.; Wolverton, C., Discovery of calcium-metal alloy anodes for reversible Ca-ion batteries. *Adv Energy Mater* 2019, *9* (9), 1802994.

[31] Ji, B. F.; He, H. Y.; Yao, W. J.; Tang, Y. B., Recent advances and perspectives on calcium-ion storage: Key materials and devices. *Adv Mater* 2021, *33* (2).

[32] Xu, W. W.; Wang, Y., Recent progress on zinc-ion rechargeable batteries. *Nano-Micro Lett* 2019, *11* (1), 90.

[33] Ming, J.; Guo, J.; Xia, C.; Wang, W. X.; Alshareef, H. N., Zinc-ion batteries: Materials, mechanisms, and applications. *Mat Sci Eng R* 2019, *135*, 58–84.

[34] Lee, B.; Seo, H. R.; Lee, H. R.; Yoon, C. S.; Kim, J. H.; Chung, K. Y.; Cho, B. W.; Oh, S. H., Critical role of pH evolution of electrolyte in the reaction mechanism for rechargeable zinc batteries. *ChemSusChem* 2016, *9* (20), 2948–2956.

[35] Wang, F. H.; Liu, H. B., Research progress of zinc anode materials for aqueous zinc ion recharge battery. *Chinese J Inorg Chem* 2019, *35* (11), 1999–2012.

[36] Huang, S.; Zhu, J. C.; Tian, J. L.; Niu, Z. Q., Recent progress in the electrolytes of aqueous zinc-ion batteries. *Chem-Eur J* 2019, *25* (64), 14480–14494.

[37] Leisegang, T.; Meutzner, F.; Zschornak, M.; Munchgesang, W.; Schmid, R.; Nestler, T.; Eremin, R. A.; Kabanov, A. A.; Blatov, V. A.; Meyer, D. C., The aluminum-ion battery: A sustainable and seminal concept? *Front Chem* 2019, *7*, 268.

[38] Zhang, Y.; Liu, S. Q.; Ji, Y. J.; Ma, J. M.; Yu, H. J., Emerging nonaqueous aluminum-ion batteries: challenges, status, and perspectives. *Adv Mater* 2018, *30* (38), 1706310.

[39] Pan, W. D.; Liu, C.; Wang, M. Y.; Zhang, Z. J.; Yan, X. Y.; Yang, S. C.; Liu, X. H.; Wang, Y. F.; Leung, D. Y. C., Non-aqueous Al-ion batteries: cathode materials and corresponding underlying ion storage mechanisms. *Rare Metals* 2022, *41* (3), 762–774.

[40] Yao, H. R.; You, Y.; Yin, Y. X.; Wan, L. J.; Guo, Y. G., Rechargeable dual-metal-ion batteries for advanced energy storage. *Phys Chem Chem Phys* 2016, *18* (14), 9326–9333.

[41] Cheng, Y. W.; Choi, D. W.; Han, K. S.; Mueller, K. T.; Zhang, J. G.; Sprenkle, V. L.; Liu, J.; Li, G. S., Toward the design of high voltage magnesium-lithium hybrid batteries using dual-salt electrolytes. *Chem Commun* 2016, *52* (31), 5379–5382.

[42] Cheng, Y. W.; Shao, Y. Y.; Zhang, J. G.; Sprenkle, V. L.; Liu, J.; Li, G. S., High per-
     formance batteries based on hybrid magnesium and lithium chemistry. *Chem Commun*
     2014, *50* (68), 9644–9646.

[43] Gao, T.; Han, F. D.; Zhu, Y. J.; Suo, L. M.; Luo, C.; Xu, K.; Wang, C. S., Hybrid $Mg^{2+}$
     /Li+ battery with long cycle life and high rate capability. *Adv Energy Mater* 2015, *5*
     (5), 1401507.

# 2 Sodium-Ion Batteries

## A Potential Replacement to Lithium-Ion Batteries

*Yumei Liu and Weibo Hua*

## 2.1 INTRODUCTION

It is important to pursue efficient and green new energy alternatives because of the gradual exhaustion of fossil energy and the increasing energy demand. With the rapid uptake of renewable energy and the emerging market for grid-scale battery applications, the demand for electrochemical energy storage technology is increasing rapidly [1, 2] The pumped-hydro installations, which account for 95% of the total rated power globally, are declining due to the geographic and geological requirements [3, 4]. Besides, the new hydropower installations cannot meet the ever-increasing demand for green energy. It is therefore greatly important for the development of flexible, low-cost, energy-saving, and environmentally benign electrical energy storage (EES) technology [3]. Batteries are an ideal carrier for energy storage. The commercial success of lithium-ion batteries (LIBs) began as early as 1991 by Sony. Also, it has become the main power source storage system in the past 30 years [5], especially in electronic products and electric vehicles [6,7]. However, the high cost and the limitation of lithium sources hinder the penetration of LIBs into large-scale energy storage systems [8]. There is no denying that energy security (such as fossil fuels) plays a great role in global politics and the economy [9]. The problem now is whether the resource availability of LIBs can satisfy the demand for large-scale energy storage, which remains uncertain and worrisome.

Owing to the limited lithium resources and the ever-increasing energy demand, scientific researchers are committed to developing alternative and scalable battery technologies. Sodium, as one of the most abundant elements, contains up to 2.36% in the earth's crust, where sodium reserves are much higher than that of lithium reserves (only 0.006% in the crust). In addition, sodium salinity in seawater that can be used as sodium reserve is also as high as 2.7% [10, 11]. Therefore, in order to solve the plight of the unstable supply of raw material and meet the rapidly growing energy demand, priority should be given to the material abundance and environmentally benign nature while designing new power sources, for example, sodium ion batteries (SIBs). Based on the resource reserve strategy and the impendency of large-scale EES, it is crucial to replace the scarce lithium with cheap and abundant sodium. In addition, the working principle of SIBs is similar to that of LIBs. Hence, it is considered as one of the new battery systems for the large-scale EES. With the continuous advancement of

DOI: 10.1201/9781003208198-2

battery technologies, SIBs are treated as promising candidates due to their advantages such as low cost, abundant reserves, safety, and environmental friendliness [6].

In this chapter, the recent key research on the potential electrodes and electrolytes in SIBs systems is summarized. And the bottlenecks in their commercialization as well as the challenges are highlighted. Their advantages over LIBs and the future directions are also discussed. This chapter aims to shed light on different potential electrode materials for SIBs and the possible solutions to promote their performance in an economical way. It is expected that this chapter will be of interest to a broader readership in the battery community.

## 2.2 POTENTIAL ELECTRODES AND ELECTROLYTES

### 2.2.1 CATHODE MATERIALS FOR SIBs

At present, progress in the research of cathode materials is remarkable as compared to other components of SIBs. A wide variety of compounds have been explored as cathode materials for SIBs. The promising candidates that can enable the stability of sodium-ion diffusion are mainly classified into four types: layered oxides, polyanionic systems, Prussian blue analogs (PBA), and organic compounds. Generally, polyanionic compounds and PBA show a lower volumetric energy density because of their heavy molecular weights. Therefore, they are more suitable for large-scale energy storage systems where the requirement of volumetric energy density is moderate. The layered oxides possess a relatively high volumetric energy density and are more suitable for smart storage applications [2]. The newly developed research on organics as cathode materials has attracted much attention recently. For comparison, the biggest advantage of organic material as electrodes is its natural abundance, contributing to its high sustainability [12].

#### 2.2.1.1 Transition Metal Oxides

The characteristics of transition metal oxide materials are the simple process of their synthesis and adjustable composition as well as high energy density. Furthermore, these cathode materials are usually divided into layered and tunnel structures according to their crystal structure properties. Generally, the tunnel-structured cathode materials show lower capacity because of their poor initial sodium ion contents. In contrast, the layered oxide materials exhibit higher theoretical capacity, faster kinetics properties, and smaller electrode polarization. Intriguingly, the structure of layered oxides could be tunable. Through component modulation and process handling, there is more chance to synthesize the target products with structural modulation. It is reported that small differences in the composition of transition metal or Na content could cause the structural transformation from P2- to O3-type structures, and vice versa [13].

The typical layered framework of $Na_xTMO_2$ is constructed by the $TMO_6$ octahedra along the $c$ axis (Figure 2.1). The Na ions are located between the stacked layers, where the $TMO_6$ octahedra are connected to each other through an edge-sharing configuration. Generally, $Na_xTMO_2$ layered oxides are classified as P2-type or O3-type cathodes, where P2-type and O3-type are defined according to their sodium stacking order [14]. The local arrangement of sodium ions is further divided into the prismatic

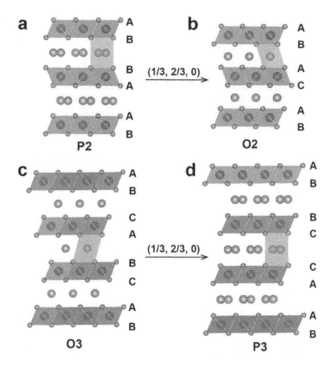

**FIGURE 2.1** The crystallographic structure of the layered $NaTMO_2$ phases for (a) P2-type; (b) O2-type; (c) O3-type; and (d) P3-type stackings. The blue balls are transition metal ions, and the yellow balls represent $Na^+$. Reproduced with permission [14]. Copyright 2018, Wiley-VCH.

coordination (denoted as symbol P) and the octahedral coordination (denoted as symbol O). Numbers "2" and "3" represent the numbers of transition metal layers or stacks in a repeat unit of the $Na_xTMO_2$ crystal structure. If the repetition is not possible due to the distortion of an in-plane crystal lattice, the $Na_xTMO_2$ are described as O'3e-type and P'2-type for comparison. The scheme of the lattice configurations in P2, O2, P3, and O3 crystal structures is demonstrated in Figure 2.1. It classifies the crystal structures by the two or three nonequivalent $TMO_2$ sheets (AB BA for P2, AB AC for O2, AB CA BC for O3, AB BC CA for P3) [14].

NaCoO$_2$ exhibits attractive electrochemical properties due to its ability to generate various polymorphs [15]. However, its poor cycling stability, sloping voltage profiles, and propensity to react with $NaPF_6$-containing electrolytes are the problems for practical application [2]. The expensive cobalt resources are considered as a critical problem that greatly restricts its large-scale applicability. Regarding environmental friendliness and low cost, NaMeOs using manganese and iron as the redox couples are highly favored for SIBs. The research on $NaFeO_2$, $NaNiO_2$, and $NaMnO_2$ cathode materials, therefore, has attracted attention. In order to stabilize the framework, transition metal elements (e.g., Ti, Zr, Li, Mg, Zn, Cu, etc.) are introduced into the crystal structure during the synthesis of NaMeOs [16]. Herein, according to the numbers of transition metals in the P2- and O3-type layered structures, they are usually defined as single-metal-based $Na_xMeO_2$, binary-metal-based $Na_xMeO_2$, ternary-metal-based $Na_xMeO_2$, and multi-metal-based $Na_xMeO_2$.

Y. Guo et al. successfully obtained a series of inactive $Ti^{4+}$ doped O3-type layered $NaNi_{0.5}Mn_{0.5-x}Ti_xO_2$ ($0 \leq x \leq 0.5$) cathodes in 2017 [17], as shown in Figure 2.2. This study aimed at suppressing the complex phase evolutions and improving the slow Na ion diffusion kinetics of O3-type $NaNi_{0.5}Mn_{0.5}O_2$ materials. At 0.05 C between 2.0 and 4.0 V, a relatively high energy density of 432 Wh $kg^{-1}$ with a reversible capacity of 135 mAh $g^{-1}$ and an average voltage of 3.2 V (*vs.* $Na^+/Na$) in $NaNi_{0.5}Mn_{0.2}Ti_{0.3}O_2$ were achieved [17]. O3-type $NaFe_{0.55}Mn_{0.45-x}Nb_xO_2$ ($x = 0$, 0.01, 0.02, 0.03) were synthesized by introducing an appropriate amount of Nb by T. Yuan et al. [18] with an energy density as high as 365 Wh $kg^{-1}$ in 2020.

Ceder's group explored quaternary layered $Na(Mn_{0.25}Fe_{0.25}Co_{0.25}Ni_{0.25})O_2$ (O3-type) material for SIB application. This material could demonstrate a high reversible capacity of 180 mAh $g^{-1}$ and an excellent specific energy density of 578 Wh $kg^{-1}$ [19]. It is proved that the cycling durability of $Na(Mn_{0.25}Fe_{0.25}Co_{0.25}Ni_{0.25})O_2$ was better than that of binary oxides at a high cut-off voltage. Sun et al. prepared O3-type layered $Na[Li_{0.05}(Ni_{0.25}Fe_{0.25}Mn_{0.5})_{0.95}]O_2$ cathodes via a co-precipitation route [20]. $3d$ or $4d$ transition metal dopants such as Zr are also introduced during the material preparation to enhance the energy density [21]. It is reported that an element with a larger ionic radius could efficiently inhibit the migration pathways of $Fe^{3+}$. It is expected that more advanced work should be conducted to enhance the electrochemical properties of these multiple iron-based/containing oxides, so as to achieve their practical application in different energy storage facilities.

### 2.2.1.2 Polyanionic Systems

Taking a cue from the great progress of olivine $LiFePO_4$ cathode material and the successful intercalation of sodium in $NaTi_2(PO_4)_3$, there is fast progress in polyanionic compounds as cathodes for SIBs. In general, the most widely investigated systems are those containing phosphate $(PO_4)^{3-}$, sulfate $(SO_4)^{2-}$, and pyrophosphate $(P_2O_7)^{4-}$ [2]. Among them, phosphate has been taken into prime consideration in recent years. Generally, most polyanionic-type materials exhibit superior structural stability and adjustable working potential because of the inductive effect [22–24]. In detail, the framework of polyanionic compounds is constructed by corner- or edge-sharing $MO_6$ (M = transition metal) octahedra and $XO_4$ (X = P, S, Si, B, C, etc.) tetrahedra. The robust bond of X–O offers sufficient stability and thus increases the ionicity of $MO_6$ octahedra. Finally, the "induced effect" leads to higher operational potentials for the 3D framework via adopting a more electronegative $XO_4$ group [25, 26]. Besides, another advantage of these stable 3D frameworks is that only small volume changes occur during cycling. For the polyanionic compounds, the bare and micrometer-sized particles rarely show encouraging electrode and thermal properties. Fortunately, advancement has been achieved by the modification, such as carbon wrapping [27, 28], nanostructure fabrication, and structural modulation via ion doping [29]. Therefore, the improvement of electronic conductivity and structural stability finally contributes to superior electrode performance of polyanionic compounds.

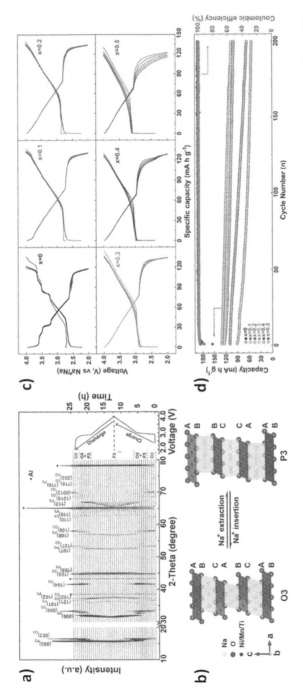

**FIGURE 2.2** (a) In situ XRD patterns of $Na/NaNi_{0.5}Mn_{0.2}Ti_{0.3}O_2$ during the first charge/discharge at 0.05 C between 2.0 and 4.0 V. (b) The scheme of O3-P3 phase transition processes during Na insertion/extraction. (c) Voltage profiles of $O3-NaNi_{0.5}Mn_{0.5-x}Ti_xO_2$ ($x = 0, 0.1, 0.2, 0.3, 0.4,$ and $0.5$) at 0.05 C. (d) Cycling stability of $O3-NaNi_{0.5}Mn_{0.5-x}Ti_xO_2$ at 1 C (1 C = 240 mA g$^{-1}$). Reproduced with permission [17]. Copyright 2017, Wiley-VCH.

Among the crystal structure of polyanion materials, the Na superionic conductor (NASICON) phase $Na_3V_2(PO_4)_3$ is particularly important and popular [30–32]. In 2002, electrode properties of $Na_3V_2(PO_4)_3$ (NVP) were first investigated by Uebou et al. [2] and are renowned for its impressively flat voltage plateau, as shown in Figure 2.3a [33]. The theoretical capacity of NVP is 118 mAh $g^{-1}$ and each molecular formula unit can transfer 2 moles of Na at a higher potential. In addition, the various redox potentials of $V^{n+}/V^{m+}$ enable NVP to be considered as both cathode and anode in a symmetrical cell [34]. Both excellent rate capability as well as stable cycling durability in NASICON-type $Na_3FeV(PO_4)_3$ were obtained via Fe-doping-engineering strategies reported by Goodenough [35]. It is proposed that element V is responsible for lifting the working potential, while Fe contributes to a long cycling life by stabilizing the structure. Another classical cathode material in SIBs is $NaFePO_4$ with three polymorphs (olivine, maricite, and amorphous $NaFePO_4$) [36–39]. Among them, the most promising electrode with attractive electrochemical properties is amorphous $NaFePO_4$. From the perspective of practical applications, iron-based materials are highly promising as scalable and cost-effective cathode materials.

Following the success of lithium iron(II) pyrophosphate ($Li_2FeP_2O_7$) as a cathode in LIBs [40, 41], the focus of cost-saving electrodes for SIBs gradually shifted from the phosphate to pyrophosphate [42]. $Na_2FeP_2O_7$ shows favorable kinetics properties for sodium diffusion due to its open 3D crystal structure. Pyrophosphate-based materials using other transition-metals as the redox couples have also been studied [43]. For example, $Na_2MnP_2O_7$ is isostructural with $Na_2FeP_2O_7$ and exhibits good electrochemical activity in SIBs [44]. The mixed polyanion compound with phosphate and pyrophosphate as the ligands were also investigated as the electrodes. For instance, $Na_4Fe_3(PO_4)_2(P_2O_7)$ is impressive for high voltage plateau of 3.2 V (vs. $Na^+/Na$) in Figure 2.3b [45]. Moreover, $Na_4Fe_3(PO_4)_2(P_2O_7)$ can achieve an improved and stable reversible discharge capacity of 122 mAh $g^{-1}$, much higher than that in $Na_2FeP_2O_7$ (theoretical capacity is 97 mAh $g^{-1}$) [2]. Combining operando synchrotron-based X-ray diffraction (SXRD) and in situ X-ray absorption near-edge spectroscopy (XANES), the electrochemical process of Na-deficient $Na_{3.32}Fe_{2.11}Ca_{0.23}(P_2O_7)_2$ was revealed to be a highly reversible single-phase solid-solution reaction. More importantly, a strong correlation between the voltage profile and lattice parameters was observed (Figure 2.3c). Especially, the lattice parameter $b$ could indicate the Na content/specific capacity of the electrode during cycling [46]. A new material, alluaudite, was explored as an electrode and investigated in detail by Yamada [26]. This pioneering contributions of a new alluaudite-type framework, $Na_2Fe_2(SO_4)_3$, in Yamada's group should be mentioned. As shown in Figure 2.3d–g, $Na_2Fe_2(SO_4)_3$ with space group $P2_{1/c}$ and surprisingly high redox potential up to 3.8 V were discovered and studied. The reason why the $Fe^{2+}/Fe^{3+}$ redox couple shows such a high voltage is that the $SO_4^{2-}$ group has stronger electronegativity than other polyanionic systems. Regarding fluorophosphate, vanadium- and Fe-based materials such as $NaVPO_4F$ [47] and $Na_2FePO_4F$ [48] exhibit promising electrochemical performance.

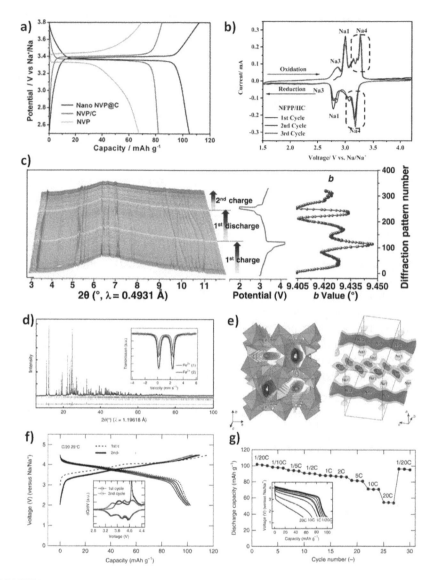

**FIGURE 2.3** (a) Charge–discharge curves of Nano NVP@C, NVP/C, and NVP. Reproduced with permission [33]. Copyright 2014, Royal Society of Chemistry. (b) Cyclic voltammetry plots of NFPP/HC. Reproduced with permission [45]. Copyright 2021, American Chemical Society. (c) In situ high-resolution SXRD patterns of $Na_{3.32}Fe_{2.11}Ca_{0.23}(P_2O_7)_2$ with voltage–composition profiles and the change in lattice parameter b (wavelength $\lambda = 0.4931$ Å). Reproduced with permission [46]. Copyright 2021, Elsevier. (d) Rietveld refinement pattern against powder XRD data for $Na_2Fe_2(SO_4)_3$, the inset is Mössbauer spectrum of pristine $Na_2Fe_2(SO_4)_3$ at room temperature, showing two different $Fe^{II}$ sites in 1:1 ratio. (e) Na-ion diffusion in the crystal structure of $Na_2Fe_2(SO_4)_3$ (green and yellow polyhedra are $FeO_6$ and $SO_4$, respectively). (f) Voltage profiles of $Na_{2-x}Fe_2(SO_4)_3$ cathode at 0.05 C between 2.0 and 4.5 V. (g) Rate capabilities of $Na_{2-x}Fe_2(SO_4)_3$. Reproduced with permission [26]. Copyright 2021, Nature.

### 2.2.1.3 Prussian Blue and Its Analogues

Although layered oxides and polyanionic composites have been recognized as dominant SIB cathode materials, the cheap Prussian blue (PB) and its analogs (PBA) are also identified as promising options for next-generation SIB. Figure 2.4a demonstrates the cubic morphology of PB and PBA, which contain a large family of transition metal cyanides. Their corresponding chemical formulas are $Fe[Fe(CN)_6] \cdot nH_2O$ (PB) and $Na_xM[M'(CN)_6]_{(1-y)} \cdot \square_y \cdot nH_2O$ (PBAs) ($0 \leq x \leq 2$, $0 \leq y < 1$), respectively, where M and M' are transition metals coordinating with nitrogen and carbon, and $\square$ represents the $[M'(CN)_6]$ vacancies that are occupied by coordinating water [49]. The evolution direction of these cathodes may lie in synthesizing high-quality PB and PBA crystals. Here, the high-quality means that the Na-rich PB and/or PBAs should possess low lattice vacancies, low water content, and high crystallinity. Their three-dimensional open framework allows PB and PBA to provide a large number of sodium diffusion channels. Hence, they can intercalate large $Na^+$ and $K^+$ with high reversibility. Different transition metals, including Ni, Cu, Co, and Zn, are capable of occupying the M' sites in PBAs, while Fe is still the most popular element used in the electrode materials for SIBs [49]. Guo's group made substantive attempts to obtain a high-quality PB [50]. They obtained the $Na_{0.61}Fe[Fe(CN)_6]_{0.94}$ crystal containing few $[Fe(CN)_6]$ vacancies and low $H_2O$ molecules by a slow co-precipitation route. Low-quality PB was also prepared for comparison. The illustration of the redox mechanism of high-quality PB is shown in Figure 2.4b. It is confirmed that the slow growth process of a crystal offers sufficient pathways for the migration of Na ion and the electron transfer, hence ensuring structural integrity during cycling. Under

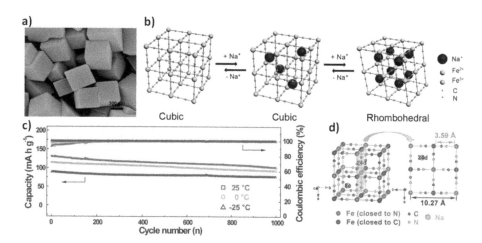

**FIGURE 2.4** (a) Typical morphology of PB and/or PBAs crystals. (b) Schematic of the sodium ion storage mechanism of high-quality PB. Reproduced with permission [50]. Copyright 2014, Royal Society of Chemistry. (c) Cycling stability of PB/CNT at 2.4 C under three different temperatures. (d) Possible $Na^+$ occupied sites including face-center (24d) and body-center site (8c) in the PB crystal obtained by DFT calculations. Reproduced with permission [51]. Copyright 2016, Wiley-VCH.

low temperature, remarkably stable cycling performance was demonstrated by nucleating PB nanoparticles on CNTs (PB/CNT) [51]. Benefiting from the fast sodium-diffusion and charge-transfer kinetics, high capacity retention in PB/CNT composite was obtained at 2.4 C, reaching 81% and 86% after 1,000 cycles at 0 and −25°C (Figure 2.4c), respectively. Its electrode architecture, including sufficient electrolyte penetration, strain accommodation, and intimate interface contact, is also conducive to the overall performance of SIB. Na-rich PBAs are gaining increasing attention [52, 53]. Na-rich $Na_{1.95}Fe[Fe(CN)_6]_{0.93}·\square_{0.07}$ was successfully prepared via sodium citrate addition and temperature control [52]. According to the results of theoretical calculations (Figure 2.4d), it is concluded that the most favorable $Na^+$ interstitial site in $Fe[Fe(CN)_6]$ should be the face-centered 24$d$ site instead of the body-centered 8$c$ site [54]. Based on the in situ growing PB@C, the influence of vacancies on Na-storage capability and reaction kinetics was clarified [55]. By utilizing the synergistic effect of different transition metal elements, a new type of PBAs such as $Na_2Ni_xMn_yFe(CN)_6$ compounds were prepared to improve their electrochemical performance [56, 57]. For instance, Shibata et al. investigated the synergistic effect of Fe, Ni, and Co substitution on the electrochemical performance of manganese hexacyanoferrate [58].

Due to the synergistic effect between different metal elements, PBA containing multiple metals may show different potential plateaus. Although Mn-based PBAs show higher working voltages, they are more likely to exhibit short cycling life stemming from the Jahn–Teller distortion of $Mn^{3+}$. To enhance their cycling performance, future studies should persistently decrease the coordinated water and content of vacancies in the PBAs framework.

### 2.2.1.4 The Organics

Organic compounds used as SIBs electrodes are an ideal candidate for the sustainable storage of renewable energy because of their outstanding characteristics such as economy, sustainability, eco-friendliness, and abundance. The possibility of adopting organics compounds for Na-storage started in the 1980s. For many years, such materials are always ignored as potential electrode candidates. The reason lies in the successful commercialization of inorganic electrode materials in mature LIBs technology attracting much attention. Recently, progress has been made in the research on both cathodes and anodes as far as organic electrodes are concerned.

Many organic electrodes suffer from poor cycling life, poor rate capability, and inferior specific capacity [59]. One way to solve this issue is to optimize these compounds based on the category of free radicals. These radicals are unstable intermediates generated during cycling. Additionally, the intermediate radicals can immediately interact with the electrolyte molecules or radicals to form inactive compounds. These adverse reactions significantly decrease the reversibility of the redox process. The reversibility of carbonyl-based organic materials, such as dichloroisocyanuric acid (DCCA), is unsatisfactory due to the formation of inactive compounds [12]. The intermediate free radicals usually stem from the reduction reactions of the unsaturated bonds (C=C, C=N, N=N, and C=O), accompanied by the unpaired electron (located on C) and the negative charge (situated on N and O). Employing pairs of unsaturated groups can stabilize free radicals and thus modify conjugated compounds. The key is to introduce reversible cleavage/reformation of

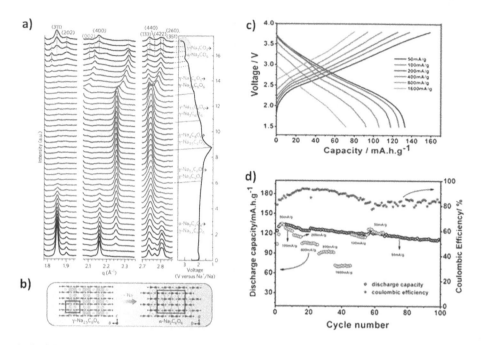

**FIGURE 2.5** (a) In situ synchrotron XRD patterns of $Na_2C_6O_6$ electrodes and its corresponding voltage profile collected during cycling in a pouch cell. (b) Phase transformation from $\gamma$-$Na_{2.5}C_6O_6$ to $\alpha$-$Na_2C_6O_6$. Reproduced with permission [64]. Copyright 2017, Nature. (c) Voltage curves of the PPy/FC cathode at different rates. (d) Rate capabilities and cycling performance at 50 mA $g^{-1}$ of the PPy/FC cathode. Reproduced with permission [66]. Copyright 2012, Royal Society of Chemistry.

unsaturated bonds from the internal-consumption process of unpaired electrons [60, 61]. Another method is to reduce the solubility of intermediate radicals. This can be achieved by adopting the crosslinks to increase the formula weight of the compound [62]. However, each category of organic compound behaves differently during the sodium storage process because of its versatile nature [63]. Based on the redox activity and functionality, organic materials are commonly classified as carbonyl, radical, Schiff base, azo materials, conjugated polymers, and biologically derived compounds. In the disodium rhodizonate ($Na_2C_6O_6$) compound, a reversible 4-Na storage reaction can be achieved (Figure 2.5a,b) [64]. It is revealed that the effective four-electron redox reaction requires a reversible phase evolution even if it is kinetically controlled upon charging. Diphenylamine-4-sulfonate (DS)-doped PPy (PPy/DS) as a cathode in SIBs could provide a capacity of 115 mAh $g^{-1}$, excellent rate, and long cycling performances [65]. Furthermore, the cathode using PPy doping with $Fe(CN)_6^{4-}$ (denoted as PPy/FC) demonstrated a capacity of 135 mAh $g^{-1}$ with 85% capacity retention after 100 cycles between 1.5 and 3.8 V (Figure 2.5c,d). Upon discharge, the initial $PPy_{17.1}^{4+}Fe(CN)_6^{4-}$ material experiences the following reduction reaction [66]:

$$PPy_{17.1}{}^{4+}Fe(CN)_6{}^{4-} + 4Na^+ + 4e \rightarrow PPy_{17.1}{}^{4+}Fe(CN)_6{}^{4-}.4Na^+ \qquad (1)$$

To date, there are some severe challenges to be solved in organic electrodes. For instance, the low electronic conductivity of most carbonyls results in poor rate capability. Further, the high solubility of small carbonyls in organic electrolytes leads to limited cycling life. The low doping amount causes an inferior specific capacity in the materials with doping mechanism. The poor theoretical capacity should be enhanced in the compounds associated with the C=N bond reaction. The high solubility in organic electrolytes significantly restricts the performance of pure N≡N azo materials [12]. Therefore, in order to further enhance the electrode properties, a more fundamental understanding of the reaction mechanism in each organic compound is required.

## 2.2.2 ANODES

### 2.2.2.1 Organic Molecules, Soft and Hard Carbon

Organic compounds are also considered as the promising anode candidates in SIBs because their redox-active functional groups can be easily tunable [67]. The current mechanisms of their sodium storage only involve the structural evolution during the electrochemical cycling process. Further, the reasons for the formation and stability of Na-C compounds are still unknown [67, 68]. Two major problems remain to be solved: (a) which carbon species is the most attractive for Na storage and (b) what is their theoretical capacity limitation. Sodium vapor reacting with $CH_3CH_2Cl$ and $CHCl_2CHCl_2$ has been confirmed to generate $NaC≡CNa$ and $Na_2C=CNa_2$, respectively, but neither of them was sufficiently characterized and investigated in detail [69]. Therefore, the ultimate formula of $Na_xC_y$ that is formed under ambient conditions in SIBs is still unknown.

The multi-ring aromatics with the conductive additives appear to have a lower activity to store sodium chemically [70]. For instance, with the addition of acetylene black, pyrene shows a lower capacity of $< 50$ mAh g$^{-1}$ (Figure 2.6a). However, when the carbonyl group is bonded with the carbon ring in the molecule, the transformation from −C=O into −C−O−Na could enhance the capacity. As in the case of perylene 3,4,9,10-tetracarboxylic dianhydride (PTCDA), a high capacity of up to 400 mAh g$^{-1}$ can be obtained (Figure 2.6b) [70]. Even if oxygen species are straightly connected to carbon rings, the carbonyl groups also have sodium storage activity. For example, the disodium rhodizonate ($Na_2C_6O_6$) can accommodate four Na$^+$ and form $Na_6C_6O_6$ [64].

Oxygen-rich carbonaceous compounds are usually pyrolyzed under an inert atmosphere to remove oxygen-containing groups and improve electronic conductivity. Notably, soft carbons and hard carbons should not be identified by their mechanical hardness, but by the atomic ratio of the residual hydrogen to carbon [71]. Generally, only soft carbon can be graphitized not hard carbon. Researchers often empirically estimate the types of carbon according to their parent materials. The soft carbon materials are usually considered to be derived from fossil fuels (pitches, cokes, etc.). Typical hard carbon materials mainly originate from those biomass compounds having carbohydrates and lignin [71]. Among the investigated carbon materials, graphite as

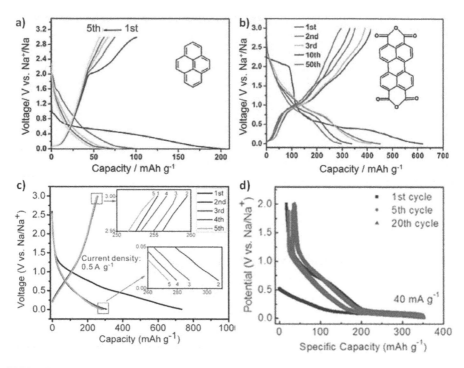

**FIGURE 2.6** Discharge and charge profiles of various carbon-containing systems: (a) pyrene [67]. (b) PTCDA, Reproduced with permission [67]. Copyright 2020, Wiley-VCH. (c) 3D porous carbon frameworks (3D PCFs), Reproduced with permission [73]. Copyright 2018, Wiley-VCH. (d) hard carbon. Reproduced with permission [71]. Copyright 2018, Wiley-VCH.

anode for SIBs might be the least possible. Because it is incapable of accommodating $Na^+$ using conventional electrolytes, and it also shows poor chemical activity under the co-intercalation of certain solvents and $Na^+$ [72]. In addition, a low capacity of less than 200 mAh $g^{-1}$ with ICE below 50% is a common problem in soft carbon [71, 73]. It is worth noting that the absence of voltage plateau cannot always be accounted for some undelivered capacity in soft carbon. For example, a high capacity of above 300 mAh $g^{-1}$ at 0.1 A $g^{-1}$ was achieved in the soft carbon material, which is derived from acetone with typical sloping voltage curves [74]. It is proposed that the considerably large surface area (467 $m^2$ $g^{-1}$) is responsible for its high capacity. However, it is still not a promising commercial material for SIBs due to the low ICE around 35% (Figure 2.6c). In contrast, the surface area of another soft carbon material prepared by anthracite was low (< 10 $m^2$ $g^{-1}$). However, it surprisingly provided a voltage plateau and delivered a capacity of 222 mAh $g^{-1}$ as well as an ICE of 81% at 60 mA $g^{-1}$. Generally, due to the larger surface area, there are more active sites and faster kinetics properties, thus facilitating the improvement of the capacity and rate capability. Nevertheless, it often comes at a cost of lower ICE, which is discouraging in the commercialization aspect. In addition to the surface area, heteroatom doping also encounters this problem. Doping nitrogen, oxygen, sulfur, and phosphor into carbon materials is an effective way to further improve the electrode performance. However, these carbon materials are not qualified for commercialization because of their low ICE. Therefore, the attempts on endowing carbon materials with porosity and defects via heteroatom doping are also discouraged if ICE is too low. Thus, the most promising candidate in practical application may be hard carbon, which shows a capacity of over 300 mAh g$^{-1}$ and an ICE above 80% in Figure 2.6d [71, 75]. Besides, hard carbon can be prepared from the carbonization of carbohydrates, lignin, cellulose, and polymers.

### 2.2.2.2 Conversion-Type Materials

During the specific sodiation process, the conversion-type materials will be converted into recognized products. On this base, their theoretical capacity can be obtained by calculation. In this section, the sodium storage characteristics of materials based on oxygen, phosphorus, sulfur, and selenium will be discussed.

The combination of sodium with oxygen (Na-O$_2$), sulfur (Na-S), and selenium (Na-Se) is regarded as a possible solution to achieve high-energy-density batteries. Taking a cue from the established technologies in lithium–sulfur (Li-S) batteries, much advancement has been made in Na-S and Na-Se battery systems. Unlike Na-S and Na-Se batteries, Na-O$_2$ batteries require an air electrode that enables reversible oxygen reduction and evolution. Pure oxygen is commonly used as their cathode material. Therefore, whether pure oxygen can be replaced by air is still not clear. Additionally, almost every component requires tremendous effort to get an in-depth understanding of the Na-O$_2$ battery at its early stage [76]. Specifically: (a) design new cathode materials to obtain high energy density, reversibility, and columbic efficiency; (b) elucidate the NaO$_2$ growth mechanisms and identify the discharge product Na$_2$O vs. Na$_2$O$_2$; and (c) investigate new electrolytes such as solid-electrolyte to avoid the side reactions. The key is to disclose how oxides are generated/decomposed during discharge/charge, so as to design and optimize the catalytic air electrode and electrolyte as well as sodium anode. Oxygen is undoubtedly a cheap cathode material, but they are far away from the practical application in Na-O$_2$ batteries.

TMX (X = O, S, and Se) emerged as anodes with high capacity for SIBs. However, they commonly suffer from "voltage penalty", that is, high sodiation voltage. Their high operational voltage will cause a decrease in energy density when fabricating a full battery. According to the literature, the delivered capacities of metal oxides based on Fe, Co, Ni, Mn, and Mo are far below their theoretical values, even with carbon composites modification [10]. Transition metal sulfides (TMS) and transition metal selenides (TMSe) do not suffer such poor sodiation activity as TMO in SIBs [77]. High theoretical capacities of TMS and TMSe were demonstrated in SIBs applications, [78] such as FeS$_2$ ($\approx$871 mAh g$^{-1}$) and NiSe$_2$ ($\approx$500 mAh g$^{-1}$), and the high sodium storage capacity makes them promising as electrodes. However, their voltage regions are undesirable because they are high as anodes and low as cathodes. It is acknowledged that they are more likely to be regarded as anodes to couple with high-voltage cathodes [78].

Among the various TMS, including MnS [79], FeS [80], CoS [81], CuS [82], VS$_2$ [83], FeS$_2$ [84], CoS$_2$ [85], NiS$_2$ [86], and MoS$_2$ (Figure 2.7a,b) [87]. MoS$_2$ has been

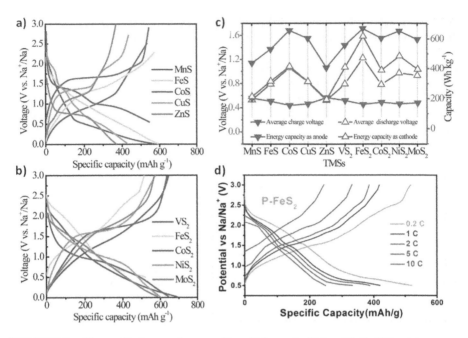

**FIGURE 2.7**   Charge-discharge voltage profiles of (a) metal sulfides; (b) disulfides; (c) average charge and discharge voltages of TMS obtained on the base of (a) and (b). Reproduced with permission [67]. Copyright 2020, Wiley-VCH. (d) Voltage profiles of Na/P-FeS$_2$ cell with various rates. Reproduced with permission [84]. Copyright 2018, American Chemical Society.

intensively studied in the battery community. However, it is not the most promising anode because of the relatively low energy density (Figure 2.7c). A promising candidate for sodium storage might be FeS$_2$ owing to the low cost and high specific capacity (895 mAh g$^{-1}$). FeS$_2$ has been long considered as the anode material. Some work transferred its application from the anode to the cathode because of its relatively high average charge voltage of 1.7 V (vs. Na$^+$/Na) [84, 88] (Figure 2.7d). Compared with these sulfides, the sodiation voltage of TMSe is similar while their theoretical capacity is slightly lower. For example, the pure FeSe has a theoretical energy density of only 356 Wh kg$^{-1}$ when employed as an anode for SIBs. However, their remarkable rate capability and cycle stability have become attractive. More impressively, TMSe-based materials showed excellent cycling stability, with only a small percentage capacity fading over thousands of cycles [89, 90]. Conversion metal chalcogenides, including oxides, sulfides, and selenides, usually work based on a combination of alloying and conversion reactions. Consequently, their products are alloy compounds and sodium chalcogenide.

### 2.2.2.3   Alloy-Type Compounds, Phosphorus, and Phosphides

The typical alloy-type materials include tin, silicon, germanium, antimony, and bismuth (Sn, Si, Ge, Sb, and Bi). Their sodium storage process is conducted through alloying reactions. Generally, silicon shows negligible activity for sodium-ion storage in the SIB systems [91, 92]. The sodiation process is more energetically favorable in amorphous silicon, and reduction of the particle size below 10 nm may benefit the utilization of silicon [93]. Germanium appears to be less appealing because of its low theoretical capacity (369 mAh g$^{-1}$), high voltage polarization ($V_{discharge} - V_{charge} = 0.47$ V), and quite limited resource availability (<1% of Li abundance). Although tin, antimony, and bismuth are also not earth-abundant, and even toxic to some extent, many efforts have been made on these materials due to their impressive sodium storage capabilities [94, 95]. However, the corresponding large volume expansion during sodiation is quite challenging, resulting in a very limited cycle life. To ensure stable cycling and high utilization in sodiation reaction, an effective way is to design a nanoparticle. This is particularly important for tin-based anodes, mainly because their large volume expansion after full sodiation can be up to 420%, larger than those in antimony (390%) and bismuth (250%) [96]. In addition, their detailed oxidation–reduction reaction mechanisms are still controversial. Compared with the theoretical value (711 mAh g$^{-1}$) of $SnO_2$, its experimental capacity is much lower. Even with tiny particles (1 nm), it only exhibited a poor capacity of about 400 mAh g$^{-1}$ at 20 mA g$^{-1}$ [97, 98]. Similarly, $Sb_2O_3$ also shows a poor capability to accommodate sodium ions [99]. The rapid capacity decay is also difficult to be alleviated in $Bi_2O_3$, even if it reached a high capacity of 600 mAh g$^{-1}$ close to the theoretical value (690 mAh g$^{-1}$) [100, 101]. Possible solutions to accommodate the volume expansion and enhance the cycle reversibility depend largely on carbon modification and electrolytes optimization. Tin phosphide ($Sn_4P_3$) is quite compelling with its low sodiation voltage and high reversible capacity [102]. Most capacity is delivered below 1.0 V (vs. Na$^+$/Na) in $Sn_4P_3$ (Figure 2.8a). However, coupling with two elements (Sn and P) inevitably causes a larger volume expansion during de-sodiation reactions. Phase segregation in $Sn_4P_3$ might occur and it transforms to Sn and P during cycles (Figure 2.8b,c). Technically, $Sn_4P_3$ can be regarded as a mixture of elements Sn and P [102, 103]. As a consequence, it is quite challenging to stabilize $Sn_4P_3$ against cycling.

Phosphorus is attractive as an anode material due to its low theoretical sodiation voltage of around 0.46 V vs. Na$^+$/Na and high specific capacity of around 2594 mAh g$^{-1}$. It also can be considered as an alloy-type material [96, 105]. It should be noted that the reaction product $Na_3P$ is more like an ionic salt than an alloy compound since phosphorous is not a metal. Therefore, phosphorus as a conversion-type material is also under consideration from this perspective. Here, we summarized the phosphorous as an alloy-type material [106]. White phosphorus (WP) with tetrahedral $P_4$ configuration is highly pyrophoric and toxic (Figure 2.9a), and this composite is basically out of the question for battery applications. Generally, the most popular ones are red and black phosphorus (RP and BP) [107, 108] (Figure 2.9b,c). The average sodiation voltages for the conversion of BP to NaP (0.80 V vs. Na$^+$/Na) and $Na_3P$ (0.47 V vs. Na$^+$/Na) are slightly higher than those of RP. Currently, the priority of research is given to solve its large volume change. For example, the volume expansion of BP converted into $Na_3P$ can reach 449%, the corresponding sodiation mechanism is shown in Figure 2.9d–f. Also, BP is too costly for SIB applications and needs more effort to reduce its price [107, 108]. RP should be more appealing from economic considerations. However, it is also an arduous task to achieve long-term

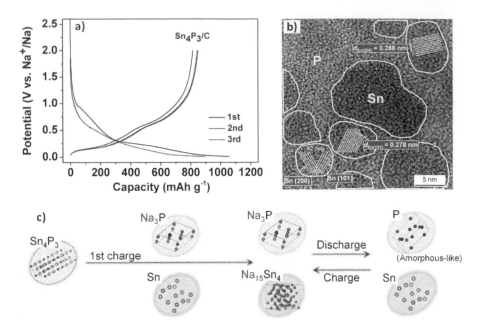

**FIGURE 2.8** (a) Initial galvanostatic charge and discharge profiles of $Sn_4P_3$ at 50 mA g$^{-1}$ [102]. Reproduced with permission. Copyright 2014, American Chemical Society. (b) HR-TEM image of the $Sn_4P_3$ particles after the first desodiation. (c) Schematics of phase transformation of $Sn_4P_3$ during cycling. Reproduced with permission [104]. Copyright 2017, American Chemical Society.

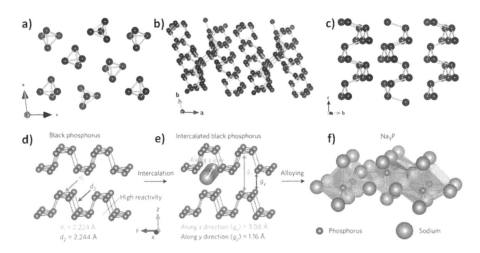

**FIGURE 2.9** (a–c) Different structures of P allotropes: (a) white P; (b) red P; and (c) black P. Reproduced with permission [108]. Copyright 2018, American Chemical Society. (d–f) Sodiation/desodiation mechanism in black P: (d) pristine state; (e) initial Na$^+$ intercalation; and (f) alloy reaction in Na$_3$P. Reproduced with permission [107]. Copyright 2018, Wiley-VCH.

stability in RP anode because of its insulated nature and large volume expansion (358%) upon discharge.

### 2.2.2.4   Intercalation-Type Materials

For some types of sodiation reactions, the definition of an intercalation process is vague. Therefore, there is no clear boundary to identify the intercalation and conversion process. Typical intercalation-type materials, such as Ti-based materials, provide interlayer space or tunnels for $Na^+$ uptake and release, where the original chemical bonds are affected but not destroyed. Accordingly, their crystal structure changes are reversible during the cycling. This is different from conversion-type materials whose crystallinity damage is permanent [109, 110]. On one hand, most researchers believed that the $Ti^{4+}/Ti^{3+}$ redox reaction is responsible for delivering capacity in $TiO_2$ with a theoretical capacity of 335 mAh g$^{-1}$. On the other hand, $TiO_2$ could also be considered as a conversion-type material with theoretical capacity of 1342 mAh g$^{-1}$. In fact, the experimental capacity of $TiO_2$ is below 300 mAh g$^{-1}$. Hence, we still consider $TiO_2$ as an intercalation-type material. The rationale behind this is that $TiO_2$ accommodates sodium associated with $Ti^{4+}$ changing to $Ti^{3+}$ without Ti–O bonds broken.

The study of crystalline $TiO_2$, including anatase, rutile, and bronze (bronze is referred as $TiO_2$(B)), as SIB anodes may be compared with amorphous $TiO_2$ [111]. Amorphous $TiO_2$ nanotubes exhibited a capacity of about 150 mAh g$^{-1}$ between 0.9 and 2.5 V (vs. $Na^+$/Na) [112]. However, the behavior of $Na^+$ in amorphous $TiO_2$ is still unclear because of structural uncertainty. The theoretical calculation is in conflict with the experimental observation, as in the case of rutile $TiO_2$. In detail, sodium accommodation in rutile $TiO_2$ is thermodynamically unfavored, but practical sodiation activity has been confirmed [113, 114]. Both in theory and practice, anatase and bronze $TiO_2$ ($TiO_2$(B)) are active [111, 115]. Unfortunately, the preparation of pure $TiO_2$(B) is challenging because $TiO_2$(B) is metastable and will turn into the anatase phase (at > 550°C) [116]. It is demonstrated that three types of vacancies exist in the $TiO_2$(B) structure and no matter which site is occupied firstly, the final compound will become $NaTiO_2$ [115]. Figure 2.10a shows that the anatase $TiO_2$ nanoparticles were modified through the combination of CNT and PPY coating. Similar anodic peak voltages between anatase and bronze $TiO_2$ are found in Figure 2.10b,c, while slightly lower capacity is exhibited in $TiO_2$ (B) (around 225 mAh g$^{-1}$ at 0.2 C) compared to anatase $TiO_2$ (around 240 mAh g$^{-1}$) [111, 117], as can be seen in Figure 2.10d,e. Therefore, the transition from anatase $TiO_2$ to $TiO_2$ (B) requires further consideration. Regarding the crystalline $TiO_2$ (rutile, anatase, and bronze), anatase titanium dioxide should be the most widely studied phase. It may also be the most promising candidate for SIB as a sodium anode. In terms of the amorphous and crystalline phase, it is difficult to draw general conclusions about which type of $TiO_2$ anode is better. The reason for this is that the atomic arrangements in the amorphous phase are theoretically unlimited, and therefore it is very hard to compare. Replacing carbonate electrolyte with diglyme electrolyte boosted both the capacity and ICE in $TiO_2$ composite (capacity increases from 121 to 252 mAh g$^{-1}$, and ICE improves from 30% to 70%) [118]. To improve the poor capacity and low ICE of $TiO_2$ anode, carbon modification, structural construction, and crystallographic modulation are also effective [111]. Interestingly, $TiS_2$ can provide different theoretical capacities based on different

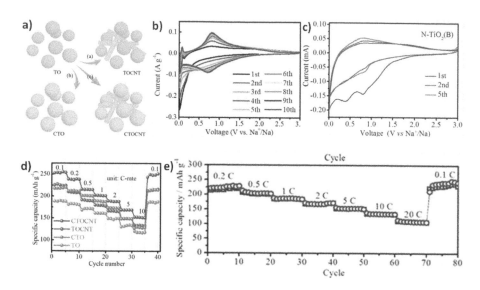

**FIGURE 2.10** (a) Modification of anatase TiO$_2$ nanoparticles by combining CNT and PPY coating (denoted as CTOCNT composite). Reproduced with permission [118]. Copyright 2019, Wiley-VCH. (b–c) CV plots of CTOCNT (b) [118] and N-TiO$_2$(B) (c) [117] electrodes at 0.1 mV s$^{-1}$. Reproduced with permission [117]. Copyright 2017, Royal Society of Chemistry. (d–e) Rate capabilities of CTOCNT (d) [118] and N-TiO$_2$(B) (e) [117] electrodes.

numbers of electron intercalation. Capacities of around 239, 479, and 957 mAh g$^{-1}$ respectively correspond to one electron, two electrons, and four electrons participating in the reactions [110, 119, 120]. It is worth mentioning that the highest theoretical value (957 mAh g$^{-1}$) involving the four-electron reaction originates from the conversion reaction. Undoubtedly, TiS$_2$ anode will become particularly competitive if the highest capacity is accessible.

Another typical category of intercalation material is sodium titanate (Na$_2$Ti$_3$O$_7$ and Na$_2$Ti$_6$O$_{13}$) and sodium titanium phosphate (NaTi$_2$(PO$_4$)$_3$. As sodium accommodates within these compounds, the valence state of Ti will change from Ti$^{4+}$ to Ti$^{3+}$ [111]. Two sodium ions can be inserted into Na$_2$Ti$_3$O$_7$, exhibiting a theoretical capacity of 178 mAh g$^{-1}$ and a low average voltage of around 0.3 V (vs. Na$^+$/Na) [121, 122]. Impressively, Na$_2$Ti$_3$O$_7$/C composites show a reversible capacity above 178 mAh g$^{-1}$. This capacity deviation could be attributable to the contribution of carbon materials as part of the delivered capacity is below 0.3 V [123, 124]. In general, the low sodiation voltage of Na$_2$Ti$_3$O$_7$ is advantageous, but more efforts are required to promote their low ICE and poor electronic conductivity. Na$_2$Ti$_6$O$_{13}$ usually provides a low theoretical capacity of 49.5 mAh g$^{-1}$ by intercalating one sodium ion per unit at 0.8 V (vs. Na$^+$/Na) [125, 126]. Lowering the cut-off voltage to 0 V will insert more than three sodium ions and achieve a higher capacity of around 198 mAh g$^{-1}$. However, the capacity fades fast and leads to a poor cycle life [126]. Two sodium ions can be accommodated in NaTi$_2$(PO$_4$)$_3$ with a theoretical capacity of 133 mAh g$^{-1}$, corresponding to the voltage at 2.1 V. Therefore, the irreversible capacity caused by electrolyte decomposition could be reduced by rising cut-off voltage (more than 1.5 V vs. $Na^+/Na$). This also permits it to work in an aqueous electrolyte [127, 128]. However, as an anode, the high operating voltage of $NaTi_2(PO_4)_3$ is not suitable and favorable. Similarly, $V_2O_5$ material also has this problem [129, 130] When it comes to the redox couple of $V^{m+}/V^{n+}$, the presence of $V^{3+}/V^{2+}$ and $V^{4+}/V^{3+}$ allows the insertion and extraction of $Na^+$ in the respective voltage ranges, as in the case of $Na_3V_2(PO_4)_3$ compound [131, 132]. It can work as both cathode and anode to construct a symmetric SIB. Nonetheless, it is a query whether the dual functions of an electrode material can endow the battery with the advantage of high energy density and low cost.

## 2.2.3 ELECTROLYTES

The majority of current electrolytes are liquids containing salts dissolved in a solvent at room temperature. Generally, electrolytes are considered as a passive component, which is an integral part of any battery system. Besides, there will be some general and basic requirements for electrolytes and solvents for any practical device. In theory, an ideal electrolyte for advanced batteries should satisfy the criteria below [2]:

a. Electrochemical stability that keeps the electrochemical system stable without the adverse reactions occurring at the interface between electrolyte and electrodes in the voltage window;
b. Wide working temperature which maintains the electrolytes liquid within the required temperature range;
c. High dielectric permittivity allows solvent to dissolve the required solutes;
d. Low viscosity which endows ionic conductivity;
e. And environmental friendliness as well as low cost should be required.

Obviously, there is no electrolyte with a single solvent that can meet all the above requirements. In addition to the general properties, the interphase in situ growth between the electrodes and electrolyte usually plays a vital role in the overall performance of SIBs. Therefore, an in-depth understanding of the chemistry of electrolytes and the properties of the electrolyte–electrode interphase is essential for the development of SIBs.

The pursuit of high energy density lies in the utilization of cathodes and anodes with high and low voltage, respectively. Therefore, the electrochemical and chemical stability of electrolytes is often challenged by the strong redox reactions triggered by the electrode materials. It is found that the thick SEI at the anode and unstable CEI at the cathode are usually formed as most used electrolytes are metastable. Consequently, the overall performance is degraded. Besides, chemical stability requires no side reactions occurrence between electrolyte and other components (including separators, active materials, current collectors, and battery packaging set). Current electrolytes used in SIBs are organic solvents in a wide range of liquids. These liquid electrolytes take a risk of high vapor pressures and high flammability, thus the problem of thermal runaway should be carefully considered to ensure battery safety. When solvated ions transfer in the electrolyte, the dynamic viscosity of electrolytes determines their drag forces imposed by the nearby solvent molecules.

According to the Stokes–Einstein equation, the ionic mobility is generally inversely proportional to the solvent viscosity. Therefore, low viscosity is conducive to high ion mobility in most electrolytes [133]. Generally, the ionic conductivity in a battery is required to be higher than 1 mS cm$^{-1}$ [133]. This in turn ensures low Ohmic resistance in the battery. Low viscosity also brings another advantage, wettability of the solvent. It is critical to provide good contacted interphases between the solid components and the liquid phases. This would finally promote the smooth diffusion of sodium ions between different phases.

The formation of close ion pairings will impede the dissolution of salt. At a given temperature and salt concentration, the high dielectric permittivity in a solvent can reduce the occurrence of ion pairs. So far, there is no practical solvent that has both low dynamic viscosity and high dielectric permittivity [134]. However, the optimized electrolytes have succeeded with binary, ternary, and quaternary mixtures of various solvents. The mixture of ethylene carbonate (EC, high dielectric permittivity) and dimethyl carbonate (DMC, low viscosity) is a successful example [135]. Currently, electrolytes in various forms have been studied for SIBs, including liquid, gel, and solid-state forms. The properties of common electrolytes for SIB applications are reviewed below. Organic liquid–based liquids have become the main solvents in the past decade. Propylene carbonate (PC) is a promising candidate for battery operation because of its low cost and high dielectric permittivity as well as large electrochemical stability window (ESW). It is concluded that about 60% of the electrolytes used in SIBs is PC solvent, and mixing PC with other solvents to optimize electrolyte is particularly popular [133]. Another successful alkyl carbonate electrolyte for SIBs is ethylene carbonate (EC). It exhibits remarkable properties, such as low viscosity and high dielectric permittivity (around 90 at 40°C). In addition, at ambient temperature, EC is a solid with a melting point as high as 36.4°C, while the melting point of PC is much lower (−48.8°C). So, the combination of EC:PC seems to be a good choice regardless of which salt (NaClO$_4$ or NaPF$_6$) is used [136]. This is confirmed in previous literature [137].

Ether-based electrolytes are also attractive due to their low viscosity and high ionic conductivity. The ether-based solvents commonly used for SIBs application are tetra-hydrofuran (THF), dimethoxyethane (DME), polymethoxy ether, dimethoxy propane, diethyl ether, and diethoxyethane [138] Ethers usually show remarkable stability when it is employed as the anode electrolytes, but it is rarely used in cathodes. This is because ether is prone to decompose at a relatively low potential in cathodes (such as above 3.2 V) compared to the alkyl carbonate. Similarly, the limited practical ESW greatly restricts the use of THF on the cathode and even full cells [139]. Linear carbonates have the advantage of low viscosity. However, as electrolytes, their disadvantages (including low dielectric permittivity and boiling point as well as flash point temperature) lead to poor safety, which should be further considered. Taking DMC as an example, the temperatures of its boiling point and flash point are only 91°C and 18°C, respectively [135]. Furthermore, to compare with the PC electrolyte (1 M NaClO$_4$ as salt), galvanostatical charge/discharge test was conducted by the Na/hard carbon half-cell assembled with different formulation of EC:DEC, EC:EMC, and EC:DMC with a ratio of 1:1 [136]. The result demonstrated that only the EC:DEC mixture can be cycled over 100 times reversibly, similar to the pure PC electrolyte, while other formulas are unfortunately incompatible with metallic sodium. In Figure 2.11, the

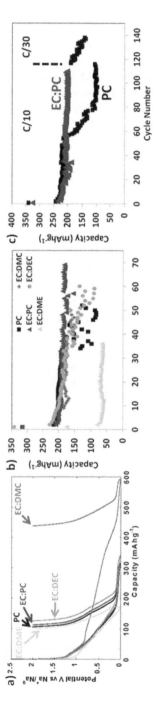

**FIGURE 2.11** The cycling performance of hard carbon using 1 M NaClO$_4$ in various solvent: (a) Voltage versus capacity profiles at 0.05 C-rate; (b) Capacity retentions; (c) Capacity retentions for tape-cast hard carbon electrodes cycled in PC and in EC:PC at 0.1 C-rate up to 110 cycles and further at 0.033 C-rate. Reproduced with permission [136]. Copyright 2012, Royal Society of Chemistry.

performance of the above electrolytes in SIB applications with hard carbon as anode is illustrated and compared. Compared to the poor coulombic efficiencies in the case of EC:DMC and EC:DME, higher values were obtained in PC-, EC:PC-, and EC:DEC-containing electrolytes. In addition, it appears that the best cycling stability is obtained using EC:PC as the solvent [136].

Room temperature ionic liquids (IL) could be used as salts in the liquid state. As excellent electrolytes, IL usually shows favorable chemical, electrochemical, and thermal stability. The most attractive capability of IL is that its vapor pressure is negligible or undetectable, thus it is known as a safe and non-flammable electrolyte [140, 141]. The preparation of IL has been improved, but the high price of manufacturing IL products and difficulty in purification make it difficult to be used in commercial application. This is the most serious setback for IL. Attempts have been made to optimize the formulation, and a breakthrough might be expected in the large-scale application of IL as a fireproof SIB electrolyte for the future. Recently, the focus has been gradually shifted to solid-state electrolytes (SSEs), especially the solid polymer electrolytes (SPE). SPE commonly consists of a polymer matrix in which the salt is soluble resulting from the coordination between salt ions and polymer chains. Currently, much attention has been paid to develop SSEs. This is mainly due to their capability to prevent dendritic growth and their inherent safety owing to non-flammable nature. The suppression to dendrite growth in turn allows the use of metal as anodes. Therefore, the use of SPE promises innovations in battery design and fabrication, such as the manufactory of thin-film batteries and high-energy-density solid-state batteries.

The currently reported SSE for SIBs could be categorized into inorganic and organic compounds. The composition of SSE-type SIBs and the premise for SSE are presented in Figure 2.12. The most important requirements for the advanced SSEs are high ionic conductivity, high chemical stability, high transference number, and good mechanical properties as well as compatible interphase [142,143]. $b$-alumina should be the first reported inorganic electrolyte

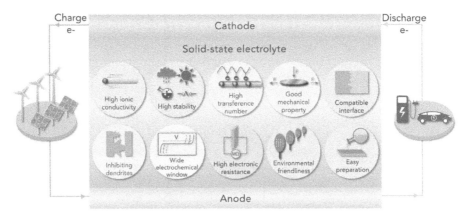

**FIGURE 2.12** The components of Solid-State SIBs and the prerequisites of SSE. Reproduced with permission [144]. Copyright 2018, Cell Press.

for sodium-sulfur batteries in the 1960s [144]. So far, Na superionic conductors (NASICON), sulfides and complex hydrides have been widely used as inorganic SSEs for SIB [145]. Benefiting from the high ionic conductivity (above $10^{-4}$ S $cm^{-1}$), large ionic transference number (close to unity) and robust mechanical properties as well as good thermal stability, inorganic electrolytes show great potential to tackle the safety issues [146]. However, their drawback is the poor contact with electrode materials caused by intrinsic roughness and rigidity. To achieve advanced solid-state SIBs, in addition to developing SSE with good ionic conductivity, the requirements of effective interphase contact and good chemical stability of SSE/electrode should be also satisfied. The insufficient contact area between SSE and electrode commonly results in a large interphase resistance. Besides, the poor chemical compatibility between electrolyte and electrodes also needs to be improved. For instance, the insufficient electrochemical properties in many sulfide SSE-type batteries are mainly due to the instability of sulfide SSE and Na metal [146, 147]. In contrast, the majority of research on NASICON-type SSE is based on $Na_3Zr_2Si_2PO_{12}$, where the number of mobile $Na^+$ ions is an obvious bottleneck for high-performance SSE [148, 149].

Undoubtedly, solid-state SIBs hold great promise for practical applications due to the advantages of satisfactory stability, absence of leakage, and easy-to-direct stacking [150, 151]. However, it is challenging to construct high-performance solid-state SIBs due to the lack of advanced SSE and compatible interphase. Regarding the large-scale application, how to prepare the high-performance SSE through a simple, cost-saving, and scalable method is the precondition.

## 2.3  ADVANTAGES OVER LITHIUM-ION BATTERIES

In terms of large-scale applications, cost-effectiveness and sustainability must be prioritized in batteries. SIBs may be slightly inferior in performance compared to LIBs but more competitive in price as sodium is cheaper than lithium. Unlike LIBs, SIBs only hold a small share of the electric vehicles and electric tools market. As energy storage systems, they are mainly aimed at large-scale applications such as the smart grid. This is the intrinsic incentive of current research of SIBs. Although there are demonstrations in grid- and vehicle-level applications [152, 153], there is still a long way to go for commercialization of SIBs. However, regarding the global lithium reserves, the price of lithium is likely to go up with its gradual depletion. Besides, due to political or technical reasons, the mining cost of lithium sources might also go up, posing a risk to its stable supply. Technically, sodium reserves are inexhaustible and price stability of sodium can be guaranteed [11, 154]. Furthermore, in order to reduce the manufacturing and operating costs, using cost-saving raw materials will highly contribute to the cheaper battery systems. Keeping this in mind, the merit of replacing expensive lithium with cheaper sodium is obvious. From the point of view of battery manufacturers, the cost of large-scale energy storage devices is also a paramount factor. Therefore, selection of electrode materials based on earth-abundant elements should be a basic criterion (for example, Na, C, O, S, P, Si, Al, Fe, Mn, Ti).

## 2.4  THE OBSTACLES TO COMMERCIALIZATION

Though great progress on SIBs has been made, there is a long way to go for their commercialization [155]. On one hand, the first commercial anode material may be hard carbon, as they have been employed in many full SIBs prototypes, where cost analysis has also been performed [11, 156]. At present, the focus is on developing an efficient, cost-saving, and sustainable synthesis route for hard carbon. To overcome the obstacles to the commercialization of hard carbon, several efforts are required: (i) pursuing natural-abundant and low-cost precursors; (ii) achieving high yields through developing sustainable and industrial processes; and (iii) getting an in-depth understanding between microstructural properties of hard carbon and processing parameters to boost sodium storage capability. Among the above requirements, the low output in current technological process not only results in high price, but also poses a challenge to the overall sustainability of hard carbon as anode materials. On the other hand, the prospects for cathode materials are more encouraging, in which a wide variety of promising options have been approved. Layered oxides and polyanionic materials have been considered as the most potential materials for SIBs cathodes, and PBAs as a low-cost option are getting increasing attention. However, each category of Na-storage materials has its shortcomings. Poor cycling stability has been demonstrated in the layered oxides, while the low capacities greatly limit the energy density of polyanionic compounds, and the inherent lattice vacancies lead to inferior electrode properties in PBAs. For many years, the challenges for commercialization are not only how to overcome these shortcomings, but how to surmount them in the most economical and eco-friendly way. Similar to anodes, commercial-scale processing technology is still also a challenge to commercialize the cathode materials. Pursuing high energy density of SIBs is one of the most important and intrinsic motivation. From the point of view of the working mechanism of SIBs, the core of SIB technologies lies in electrode materials, while the key to improving their energy density rests with the cathode materials. Therefore, the research and development of cathode materials is crucial for the commercialization of SIBs.

## 2.5  CHALLENGES AND ADVANTAGES

To commercialize the SIBs, the issues that need to be addressed are clear: satisfactory cycle life of electrodes and low-cost manufacturing of energy storage devices. Besides, there are still some environment, safety, and air stability challenges for cathode materials. In the face of these requirements, it is critical to carefully select electrodes with suitable characteristics and low-cost elements. Each type of cathode materials has its own advantages and disadvantages. Therefore, when designing electrode materials, the difficulties lie in how to achieve a balance between cost, structure, property, and practicability. In general, the layered oxides exhibit high storage capacity, but they are limited by their low cycling performance. While polyanionic compounds demonstrate a stable sodium storage capability, but unfortunately, their low theoretical capacities become a tough obstacle. Regarding PB and PBAs, effort needs to be continuously focused on how to reduce the lattice vacancies and $H_2O$ content in the crystallographic structure. Moreover, increasing the Na content and crystallinity can enhance their electrode performance. Organic electrode materials offer a new chance to design suitable cathodes and anodes for SIBs, but an in-depth understanding between the structure and properties should be first elucidated.

In the foreseeable future, hard carbon seems promising as SIB anode material. The advantages of high delivered capacity (300 mAh $g^{-1}$), low potential plateau (0.0–0.1 V), and abundant precursors make hard carbon a strong competitor for SIBs. However, a critical problem with hard carbon is the low ICE caused by the large irreversible capacity of 20–30%. This is detrimental to the full battery when using HC as the anode [157]. Another critical problem is the Na nucleation on HC when it charged at −0.015 V (vs. Na$^+$/Na) [158, 159]. This behavior will induce safety issues related to sodium plating when charging the HC electrode in a low-voltage range. Therefore, this is still the maximum motivation to replace HC anodes with other suitable anodes that have higher operation potentials. One option is the alloy-based anodes. They are capable of exceeding the energy density of HC electrodes with their high theoretical capacity. From the perspective of capacity and average voltage, alloy-based materials, especially Sn and Sb, might be promising alternatives to replace hard carbon, but what remains to be ensured is their long cycle life [105]. Additionally, according to the cost analysis, the current price of SIBs is close to that of LIBs [11, 160]. Since the definition of price is the capital per capacity ($ kWh$^{-1}$), the cost advantage of SIBs over LIBs could be manifested via increasing capacity. This is also a challenge for a given material but the development of new cathode materials holds the key.

## 2.6   FUTURE DIRECTIONS

The direction for future study lies in how to overcome the above-mentioned problems in the most economical way. Iron and manganese are naturally abundant elements on the earth's crust. They have been widely explored and applied in LIBs and SIBs to maximize their cost advantage. The most common redox couples, for example, Fe$^{2+}$/Fe$^{3+}$, Mn$^{2+}$/Mn$^{3+}$, and Mn$^{3+}$/Mn$^{4+}$, exhibit high electrochemical activity in various structures, and they behave differently in different compounds. In addition to the single redox couples, Fe-based and Mn-based redox couples also hold promise in mixed binary, ternary, and quaternary compounds. Nevertheless, substantial efforts need to be made to clarify their actual concerns, such as low operating potential, unstable structure, and hygroscopicity. As far as the Coulomb efficiency of the full cell is concerned, advanced anode materials should be a prerequisite. The reason is that the excess cathode mass should be kept to a minimum in practical application so as to achieve high energy density. Meanwhile, the output voltage and capacity of anodes should be taken into account. In addition, it is also necessary to consider whether the optimized electrolyte is suitable for both the anode and cathode. According to the literature, the preferred electrolyte formulation for HC anodes may be a mixed electrolyte of EC:PC. Hence, the electrolyte of EC:PC has been employed in most full cell prototypes. Regarding various processing conditions, scalability is a paramount factor because it is highly related to economic efficiency and environmental impact. In conclusion, whether SIB electrode materials hold commercial prospects depends on their electrode performance, raw material cost, manufacturing complexity, safety factor, and scalability.

## REFERENCES

[1] B. Dunn, H. Kamath, J.M. Tarascon. Electrical energy storage for the grid: A battery of choices. *Science* 2011, *334* (6058), 928.

[2] K. Chayambuka, G. Mulder, D.L. Danilov, P.H.L. Notten. Sodium-ion battery materials and electrochemical properties reviewed. *Adv. Energy Mater.* 2018, *8* (16), 1800079 .

[3] Z. Yang, J. Zhang, M.C.W. Kintner-Meyer, X. Lu, D. Choi, J.P. Lemmon, J. Liu. Electrochemical energy storage for green grid. *Chem. Rev.* 2011, *111* (5), 3577.

[4] C.-J. Yang, R.B. Jackson. Opportunities and barriers to pumped-hydro energy storage in the United States. *Renewable Sustainable Energy Rev.* 2011, *15* (1), 839.

[5] G. Chen, Q. Huang, T. Wu, L. Lu. Polyanion sodium vanadium phosphate for next generation of sodium-ion batteries—A review. *Adv. Funct. Mater.* 2020, *30* (34), 2001289.

[6] J.M. Tarascon, M. Armand. Issues and challenges facing rechargeable lithium batteries. *Nature* 2001, *414* (6861), 359.

[7] R. Schmuch, R. Wagner, G. Hörpel, T. Placke, M. Winter. Performance and cost of materials for lithium-based rechargeable automotive batteries. *Nat. Energy* 2018, *3* (4), 267.

[8] H.S. Hirsh, Y. Li, D.H.S. Tan, M. Zhang, E. Zhao, Y.S. Meng. Sodium-ion batteries paving the way for grid energy storage. *Adv. Energy Mater.* 2020, *10* (32), 2001274.

[9] J.-M. Tarascon. Is lithium the new gold? *Nat. Chem.* 2010, *2* (6), 510.

[10] J.-Y. Hwang, S.-T. Myung, Y.-K. Sun. Sodium-ion batteries: Present and future. *Chem. Soc. Rev.* 2017, *46* (12), 3529–3614.

[11] C. Vaalma, D. Buchholz, M. Weil, S. Passerini. A cost and resource analysis of sodium-ion batteries. *Nat. Rev. Mater.* 2018, *3* (4), 18013.

[12] R. Rajagopalan, Y. Tang, C. Jia, X. Ji, H. Wang. Understanding the sodium storage mechanisms of organic electrodes in sodium ion batteries: issues and solutions. *Energy Environ. Sci.* 2020, *13* (6), 1568.

[13] C. Zhao, Q. Wang, Z. Yao, J. Wang, B. Sánchez-Lengeling, F. Ding, X. Qi, Y. Lu, X. Bai, B. Li, H. Li, A. Aspuru-Guzik, X. Huang, C. Delmas, M. Wagemaker, L. Chen, Y.-S. Hu. Rational design of layered oxide materials for sodium-ion batteries. *Science* 2020, *370* (6517), 708.

[14] P.-F. Wang, Y. You, Y.-X. Yin, Y.-G. Guo. Layered oxide cathodes for sodium-ion batteries: phase transition, air stability, and performance. *Adv. Energy Mater.* 2018, *8* (8), 1701912.

[15] H. Yoshida, N. Yabuuchi, S. Komaba. $NaFe_{0.5}Co_{0.5}O_2$ as high energy and power positive electrode for Na-ion batteries. *Electrochem. Commun.* 2013, *34*, 60.

[16] F. Wei, Q. Zhang, P. Zhang, W. Tian, K. Dai, L. Zhang, J. Mao, G. Shao. Review–Research progress on layered transition metal oxide cathode materials for sodium ion batteries. *J. Electrochem. Soc.* 2021, *168* (5), 050524.

[17] P.-F. Wang, H.-R. Yao, X.-Y. Liu, J.-N. Zhang, L. Gu, X.-Q. Yu, Y.-X. Yin, Y.-G. Guo. Ti-substituted $NaNi_{0.5}Mn_{0.5-x}Ti_xO_2$ cathodes with reversible O3–P3 phase transition for high-performance sodium-ion batteries. *Adv. Mater.* 2017, *29* (19), 1700210.

[18] L. Zhang; T. Yuan; L. Soule, H. Sun, Y. Pang, J. Yang, S. Zheng. Enhanced ionic transport and structural stability of Nb-Doped O3-$NaFe_{0.55}Mn_{0.45-x}Nb_xO_2$ cathode material for long-lasting sodium-ion batteries. *ACS Appl. Energy Mater.* 2020, *3* (4), 3770.

[19] X. Li, D. Wu, Y.-N. Zhou, L. Liu, X.-Q. Yang, G. Ceder. O3-type $Na(Mn_{0.25}Fe_{0.25}Co_{0.25}Ni_{0.25})O_2$: A quaternary layered cathode compound for rechargeable Na ion batteries. *Electrochem. Commun.* 2014, *49*, 51.

[20] S.-M. Oh, S.-T. Myung, J.-Y. Hwang, B. Scrosati, K. Amine, Y.-K. Sun. High capacity O3-type Na[Li$_{0.05}$(Ni$_{0.25}$Fe$_{0.25}$Mn$_{0.5}$)$_{0.95}$]O$_2$ cathode for sodium ion batteries. *Chem. Mater.* 2014, *26* (21), 6165.

[21] Z.-Y. Li, R. Gao, L. Sun, Z. Hu, X. Liu. Zr-doped P2-Na$_{0.75}$Mn$_{0.55}$Ni$_{0.25}$Co$_{0.05}$Fe$_{0.10}$Zr$_{0.05}$ O$_2$ as high-rate performance cathode material for sodium ion batteries. *Electrochim. Acta* 2017, *223*, 92.

[22] A. Manthiram, J.B. Goodenough. Lithium insertion into Fe$_2$(MO$_4$)$_3$ frameworks: Comparison of M = W with M = Mo. *J. Solid State Chem.* 1987, *71* (2), 349.

[23] A. Manthiram, J.B. Goodenough. Lithium insertion into Fe$_2$(SO$_4$)$_3$ frameworks. *J. Power Sources* 1989, *26* (3–4), 403.

[24] C. Masquelier, L. Croguennec. Polyanionic (phosphates, silicates, sulfates) frameworks as electrode materials for rechargeable Li (or Na) batteries. *Chem. Rev.* 2013, *113* (8), 6552.

[25] P. Barpanda. Sulfate chemistry for high-voltage insertion materials: Synthetic, structural and electrochemical insights. *Isr. J. Chem.* 2015, *55* (5), 537.

[26] P. Barpanda, G. Oyama, S.-i. Nishimura, S.-C. Chung, A. Yamada. A 3.8-V earth-abundant sodium battery electrode. *Nat. Commun.* 2014, *5* (1), 4358.

[27] J. Kang, S. Baek, V. Mathew, J. Gim, J. Song, H. Park, E. Chae, A.K. Rai, J. Kim. High rate performance of a Na$_3$V$_2$(PO$_4$)$_3$/C cathode prepared by pyro-synthesis for sodium-ion batteries. *J. Mater. Chem.* 2012, *22* (39), 20857.

[28] Q. Zhou, L. Wang, W. Li, S. Zeng, K. Zhao, Y. Yang, Q. Wu, M. Liu, Q.-a. Huang, J. Zhang, X. Sun. Carbon-decorated Na$_3$V$_2$(PO$_4$)$_3$ as ultralong lifespan cathodes for high-energy-density symmetric sodium-ion batteries. *ACS Appl. Mater. Interfaces* 2021, *13* (21), 25036.

[29] C. Guo, J. Yang, Z. Cui, S. Qi, Q. Peng, W. Sun, L.-P. Lv, Y. Xu, Y. Wang, S. Chen. In-situ structural evolution analysis of Zr-doped Na$_3$V$_2$(PO$_4$)$_2$F$_3$ coated by N-doped carbon layer as high-performance cathode for sodium-ion batteries. *J. Energy Chem.* 2022, *65*, 514.

[30] J. Lee, S. Park, Y. Park, J. Song, B. Sambandam, V. Mathew, J.-Y. Hwang, J. Kim. Chromium doping into NASICON-structured Na$_3$V$_2$(PO$_4$)$_3$ cathode for high-power Na-ion batteries. *Chem. Eng. J.* 2021, *422*, 130052.

[31] H. Xiong, G. Sun, Z. Liu, L. Zhang, L. Li, W. Zhang, F. Du, Z.A. Qiao. Polymer stabilized droplet templating towards tunable hierarchical porosity in single crystalline Na$_3$V$_2$(PO$_4$)$_3$ for enhanced sodium-ion storage. *Angew. Chem, Int. Ed.* 2021, *60* (18), 10334.

[32] T.-H. Yu, C.-Y. Huang, M.-C. Wu, Y.-J. Chen, T. Lan, C.-L. Tsai, J.-K. Chang; R.-A. Eichel, W.-W. Wu. Atomic-scale investigation of Na$_3$V$_2$(PO$_4$)$_3$ formation process in chemical infiltration via in situ transmission electron microscope for solid-state sodium batteries. *Nano Energy* 2021, *87*, 106144.

[33] W. Duan, Z. Zhu, H. Li, Z. Hu, K. Zhang, F. Cheng, J. Chen. Na$_3$V$_2$(PO$_4$)$_3$@C core–shell nanocomposites for rechargeable sodium-ion batteries. *J. Mater. Chem. A* 2014, *2* (23), 8668.

[34] L.S. Plashnitsa, E. Kobayashi, Y. Noguchi, S. Okada, J.-i. Yamaki. Performance of NASICON symmetric cell with ionic liquid electrolyte. *J. Electrochem. Soc.* 2010, *157* (4), A536.

[35] W. Zhou, L. Xue, X. Lü, H. Gao, Y. Li, S. Xin, G. Fu, Z. Cui, Y. Zhu, J.B. Goodenough. Na$_x$MV(PO$_4$)$_3$ (M = Mn, Fe, Ni) structure and properties for sodium extraction. *Nano Lett.* 2016, *16* (12), 7836.

[36] M. Casas-Cabanas, V.V. Roddatis, D. Saurel, P. Kubiak, J. Carretero-González, V. Palomares, P. Serras, T. Rojo. Crystal chemistry of Na insertion/deinsertion in $FePO_4$–$NaFePO_4$. *J. Mater. Chem.* 2012, 22 (34), 17421.

[37] J. Kim, D.-H. Seo, H. Kim, I. Park, J.-K. Yoo, S.-K. Jung, Y.-U. Park, W.A. Goddard Iii, K. Kang. Unexpected discovery of low-cost maricite $NaFePO_4$ as a high-performance electrode for Na-ion batteries. *Energy Environ. Sci.* 2015, 8 (2), 540.

[38] Y. Liu, N. Zhang, F. Wang, X. Liu, L. Jiao, L.-Z. Fan. Approaching the downsizing limit of maricite $NaFePO_4$ toward high-performance cathode for sodium-ion batteries.*Adv. Funct. Mater.* 2018, 28 (30), 201801917.

[39] F. Xiong, Q. An, L. Xia, Y. Zhao, L. Mai, H. Tao, Y. Yue. Revealing the atomistic origin of the disorder-enhanced Na-storage performance in $NaFePO_4$ battery cathode. *Nano Energy* 2019, 57, 608.

[40] S.-i. Nishimura, M. Nakamura, R. Natsui, A. Yamada. New lithium iron pyrophosphate as 3.5 V class cathode material for lithium ion battery. *J. Am. Chem. Soc.* 2010, 132 (39), 13596.

[41] H. Zhou, S. Upreti, N.A. Chernova, G. Hautier, G. Ceder, M.S. Whittingham. Iron and manganese pyrophosphates as cathodes for lithium-ion batteries. *Chem. Mater.* 2010, 23 (2), 293.

[42] N. Yabuuchi, K. Kubota, M. Dahbi, S. Komaba. Research development on sodium-ion batteries. *Chem. Rev.* 2014, 114 (23), 11636.

[43] P. Barpanda, L. Lander, S.-i. Nishimura, A. Yamada. Polyanionic insertion materials for sodium-ion batteries. *Adv. Energy Mater* 2018, 8 (17), 1703055.

[44] P. Barpanda, T. Ye, M. Avdeev, S.-C. Chung, A. Yamada. A new polymorph of $Na_2MnP_2O_7$ as a 3.6 V cathode material for sodium-ion batteries. *J. Mater. Chem. A* 2013, 1 (13), 4194.

[45] L.-m. Zhang, X.-d. He, S. Wang, N.-q. Ren, J.-r. Wang, J.-m. Dong, F. Chen, Y.-x. Li, Z.-y. Wen, C.-h. Chen. Hollow-sphere-structured $Na_4Fe_3(PO_4)_2(P_2O_7)/C$ as a cathode material for sodium-ion batteries. *ACS Appl. Mater. Interfaces* 2021, 13 (22), 25972.

[46] Y. Liu, Z. Wu, S. Indris, W. Hua, N.P.M. Casati, A. Tayal, M.S.D. Darma, G. Wang, Y. Liu, C. Wu, Y. Xiao, B. Zhong, X. Guo. The structural origin of enhanced stability of $Na3.32Fe_{2.11}Ca_{0.23}(P_2O_7)_2$ cathode for Na-ion batteries. *Nano Energy* 2021, 79, 105417.

[47] M. Ling, Q. Jiang, T. Li, C. Wang, Z. Lv, H. Zhang, Q. Zheng, X. Li. The mystery from tetragonal $NaVPO_4F$ to monoclinic $NaVPO_4F$: Crystal presentation, phase conversion, and Na-storage kinetics. *Adv. Energy Mater.* 2021, 11 (21), 2100627.

[48] B.L. Ellis, W.R.M. Makahnouk, Y. Makimura, K. Toghill, L.F. Nazar. A multifunctional 3.5 V iron-based phosphate cathode for rechargeable batteries. *Nat. Mater.* 2007, 6 (10), 749.

[49] M. Chen, Q. Liu, S.W. Wang, E. Wang, X. Guo, S.L. Chou. High-abundance and low-cost metal-based cathode materials for sodium-ion batteries: Problems, progress, and key technologies. *Adv. Energy Mater.* 2019, 9 (14), 1803609.

[50] Y. You, X.-L. Wu, Y.-X. Yin, Y.-G. Guo. High-quality Prussian blue crystals as superior cathode materials for room-temperature sodium-ion batteries. *Energy Environ. Sci.* 2014, 7 (5), 1643.

[51] Y. You, H.-R. Yao, S. Xin, Y.-X. Yin, T.-T. Zuo, C.-P. Yang, Y.-G. Guo, Y. Cui, L.-J. Wan, J.B. Goodenough. Subzero-temperature cathode for a sodium-ion battery. *Adv. Mater.* 2016, 28 (33), 7243.

[52] Y. Huang, M. Xie, J. Zhang, Z. Wang, Y. Jiang, G. Xiao, S. Li, L. Li, F. Wu, R. Chen. A novel border-rich Prussian blue synthetized by inhibitor control as cathode for sodium ion batteries. *Nano Energy* 2017, 39, 273.

[53] W.-J. Li, S.-L. Chou, J.-Z. Wang, Y.-M. Kang, J.-L. Wang, Y. Liu, Q.-F. Gu, H.-K. Liu, S.-X. Dou. Facile method to synthesize Na-enriched $Na_{1+x}FeFe(CN)_6$ frameworks as cathode with superior electrochemical performance for sodium-ion batteries. *Chem. Mater.* 2015, *27* (6), 1997.

[54] C. Ling, J. Chen, F. Mizuno. First-principles study of alkali and alkaline earth ion intercalation in iron hexacyanoferrate: The important role of ionic radius. *J. Phys. Chem. C* 2013, *117* (41), 21158.

[55] Y. Jiang, S. Yu, B. Wang, Y. Li, W. Sun, Y. Lu, M. Yan, B. Song, S. Dou. Prussian blue@C composite as an ultrahigh-rate and long-life sodium-ion battery cathode. *Adv. Funct. Mater.* 2016, *26* (29), 5315.

[56] R. Chen, Y. Huang, M. Xie, Z. Wang, Y. Ye, L. Li, F. Wu. Chemical inhibition method to synthesize highly crystalline Prussian blue analogs for sodium-ion battery cathodes. *ACS Appl. Mater. Interfaces* 2016, *8* (46), 31669.

[57] D. Yang, J. Xu, X.-Z. Liao, Y.-S. He, H. Liu, Z.-F. Ma. Structure optimization of Prussian blue analogue cathode materials for advanced sodium ion batteries. *Chem. Commun.* 2014, *50* (87), 13377.

[58] Y. Moritomo, S. Urase, T. Shibata. Enhanced battery performance in manganese hexacyanoferrate by partial substitution. *Electrochim. Acta* 2016, *210*, 963.

[59] F. Wu, C. Zhao, S. Chen, Y. Lu, Y. Hou, Y.-S. Hu, J. Maier, Y. Yu. Multi-electron reaction materials for sodium-based batteries. *Mater. Today* 2018, *21* (9), 960.

[60] S. Wu, W. Wang, M. Li, L. Cao, F. Lyu, M. Yang, Z. Wang, Y. Shi, B. Nan, S. Yu, Z. Sun, Y. Liu, Z. Lu. Highly durable organic electrode for sodium-ion batteries *via* a stabilized $\alpha$-C radical intermediate. *Nat. Commun.* 2016, *7* (1), 13318.

[61] S. Wang, L. Wang, Z. Zhu, Z. Hu, Q. Zhao, J. Chen. All organic sodium-ion batteries with $Na_4C_8H_2O_6$. *Angew. Chem, Int. Ed.* 2014, *53* (23), 5892.

[62] Z. Song, H. Zhan, Y. Zhou. Polyimides: Promising energy-storage materials. *Angew. Chem, Int. Ed.* 2010, *49* (45), 8444.

[63] T.B. Schon, B.T. McAllister, P.-F. Li, D.S. Seferos. The rise of organic electrode materials for energy storage. *Chem. Soc. Rev.* 2016, *45* (22), 6345.

[64] M. Lee, J. Hong, J. Lopez, Y. Sun, D. Feng, K. Lim, W.C. Chueh, M.F. Toney, Y. Cui, Z. Bao. High-performance sodium–organic battery by realizing four-sodium storage in disodium rhodizonate. *Nat. Energy* 2017, *2* (11), 861.

[65] M. Zhou, Y. Xiong, Y. Cao, X. Ai, H. Yang. Electroactive organic anion-doped polypyrrole as a low cost and renewable cathode for sodium-ion batteries. *J. Polym. Sci. Part B: Polym. Phys.* 2013, *51* (2), 114.

[66] M. Zhou, L. Zhu, Y. Cao, R. Zhao, J. Qian, X. Ai, H. Yang. $Fe(CN)_6^{4-}$-doped polypyrrole: a high-capacity and high-rate cathode material for sodium-ion batteries. *RSC Adv.* 2012, *2* (13), 5495.

[67] X. Yang, A.L. Rogach. Anodes and sodium-free cathodes in sodium ion batteries. *Adv. Energy Mater.* 2020, *10* (22), 2000288.

[68] M. Wahid, D. Puthusseri, Y. Gawli, N. Sharma, S. Ogale. Hard carbons for sodium-ion battery anodes: Synthetic strategies, material properties, and storage mechanisms. *ChemSusChem* 2018, *11* (3), 506.

[69] J. Sangster. C-Na (carbon-sodium) system. *J. Phase Equilib. Diffus.* 2007, *28* (6), 571.

[70] H.-g. Wang, S. Yuan; Z. Si, X.-b. Zhang. Multi-ring aromatic carbonyl compounds enabling high capacity and stable performance of sodium-organic batteries. *Energy Environ. Sci.* 2015, *8* (11), 3160.

[71] D. Saurel, B. Orayech, B. Xiao, D. Carriazo, X. Li, T. Rojo. From charge storage mechanism to performance: A roadmap toward high specific energy sodium-ion batteries through carbon anode optimization. *Adv. Energy Mater.* 2018, *8* (17), 1703268.

[72] B. Jache, P. Adelhelm. Use of graphite as a highly reversible electrode with superior cycle life for sodium-ion batteries by making use of Co-intercalation phenomena. *Angew. Chem, Int. Ed.* 2014, *53* (38), 10169.

[73] X. Yao, Y. Ke, W. Ren, X. Wang, F. Xiong, W. Yang, M. Qin, Q. Li, L. Mai. Defect-rich soft carbon porous nanosheets for fast and high-capacity sodium-ion storage. *Adv. Energy Mater.* 2018, *9* (6), 1803260.

[74] H. Hou, C.E. Banks, M. Jing, Y. Zhang, X. Ji. Carbon quantum dots and their derivative 3D porous carbon frameworks for sodium-ion batteries with ultralong cycle life. *Adv. Mater.* 2015, *27* (47), 7861.

[75] X. Dou, I. Hasa, D. Saurel, C. Vaalma, L. Wu, D. Buchholz, D. Bresser, S. Komaba, S. Passerini. Hard carbons for sodium-ion batteries: Structure, analysis, sustainability, and electrochemistry. *Mater. Today* 2019, *23*, 87.

[76] I. Landa-Medrano, C. Li, N. Ortiz-Vitoriano, I. Ruiz de Larramendi, J. Carrasco, T. Rojo. Sodium–oxygen battery: Steps toward reality. *J. Phys. Chem. Lett.* 2016, *7* (7), 1161.

[77] P.K. Nayak, L. Yang, W. Brehm, P. Adelhelm. From lithium-ion to sodium-ion batteries: Advantages, challenges, and surprises. *Angew. Chem, Int. Ed.* 2018, *57* (1), 102.

[78] J. Kim, H. Kim, K. Kang. Conversion-based cathode materials for rechargeable sodium batteries. *Adv. Energy Mater.* 2018, *8* (17), 1702646.

[79] D.-H. Liu, W.-H. Li, Y.-P. Zheng, Z. Cui, X. Yan, D.-S. Liu, J. Wang,Y. Zhang, H.-Y. Lü, F.-Y. Bai, J.-Z. Guo, X.-L. Wu. In situ encapsulating $\alpha$-MnS into N,S-codoped nanotube-like carbon as advanced anode material: $\alpha \rightarrow \beta$ phase transition promoted cycling stability and superior Li/Na-storage performance in half/full cells. *Adv. Mater.* 2018, *30* (21), 1706317.

[80] Y.-X. Wang, J. Yang, S.-L. Chou, H.K. Liu, W.-x. Zhang, D. Zhao, S.X. Dou. Uniform yolk-shell iron sulfide–carbon nanospheres for superior sodium–iron sulfide batteries. *Nat. Commun.* 2015, *6* (1), 8689.

[81] S. Peng, X. Han, L. Li, Z. Zhu, F. Cheng, M. Srinivansan, S. Adams, S. Ramakrishna. Unique cobalt sulfide/reduced graphene oxide composite as an anode for sodium-ion batteries with superior rate capability and long cycling stability. *Small* 2016, *12* (10), 1359.

[82] J. Li, D. Yan, T. Lu, W. Qin, Y. Yao, L. Pan. Significantly improved sodium-ion storage performance of CuS nanosheets anchored into reduced graphene oxide with ether-based electrolyte. *ACS Appl. Mater. Interfaces* 2017, *9* (3), 2309.

[83] J. Zhou, L. Wang, M. Yang, J. Wu, F. Chen, W. Huang, N. Han, H. Ye, F. Zhao, Y. Li, Y. Li. Hierarchical VS$_2$ nanosheet assemblies: A universal host material for the reversible storage of alkali metal ions. *Adv. Mater.* 2017, *29* (35), 1702061.

[84] K. Chen, Y. Zhang, C. Li. High-rate nanostructured pyrite cathodes enabled by fluorinated surface and compact grain stacking *via* sulfuration of ionic liquid coated fluorides. *ACS Nano* 2018, *12* (12), 12444.

[85] Q. Guo, Y. Ma, T. Chen, Q. Xia, M. Yang, H. Xia, Y. Yu. Cobalt sulfide quantum dot embedded N/S-doped carbon nanosheets with superior reversibility and rate capability for sodium-ion batteries. *ACS Nano* 2017, *11* (12), 12658.

[86] D. Zhang, W. Sun, Y. Zhang, Y. Dou, Y. Jiang, S.X. Dou. Engineering hierarchical hollow nickel sulfide spheres for high-performance sodium storage. *Adv. Funct. Mater.* 2016, *26* (41), 7479.

[87] Q. Pan, Q. Zhang, F. Zheng, Y. Liu, Y. Li, X. Ou, X. Xiong, C. Yang, M. Liu. Construction of MoS$_2$/C hierarchical tubular heterostructures for high-performance sodium ion batteries. *ACS Nano* 2018, *12* (12), 12578.

[88] K. Zhang, M. Park, L. Zhou, G.-H. Lee, J. Shin, Z. Hu, S.-L. Chou, J. Chen, Y.-M. Kang. Cobalt-doped $FeS_2$ nanospheres with complete solid solubility as a high-performance anode material for sodium-ion batteries. *Angew. Chem, Int. Ed.* 2016, *55* (41), 12822.

[89] K. Zhang, Z. Hu, X. Liu, Z. Tao, J. Chen. $FeSe_2$ microspheres as a high-performance anode material for Na-ion batteries. *Adv. Mater.* 2015, *27* (21), 3305.

[90] H. Yin, H.-Q. Qu, Z. Liu, R.-Z. Jiang, C. Li, M.-Q. Zhu. Long cycle life and high rate capability of three dimensional $CoSe_2$ grain-attached carbon nanofibers for flexible sodium-ion batteries. *Nano Energy* 2019, *58*, 715.

[91] L. Zhang, X. Hu; C. Chen, H. Guo, X. Liu, G. Xu, H. Zhong, S. Cheng, P. Wu, J. Meng, Y. Huang, S. Dou, H. Liu. In Operando mechanism analysis on nanocrystalline silicon anode material for reversible and ultrafast sodium storage. *Adv. Mater.* 2017, *29* (5), 1604708.

[92] Y. Xu, E. Swaans, S. Basak, H.W. Zandbergen, D.M. Borsa, F.M. Mulder. Reversible Na-ion uptake in Si nanoparticles. *Adv. Energy Mater.* 2016, *6* (2), 1501436.

[93] S.C. Jung, D.S. Jung, J.W. Choi, Y.-K. Han. Atom-level understanding of the sodiation process in silicon anode material. *J. Phys. Chem. Lett.* 2014, *5* (7), 1283.

[94] J.M. Stratford, M. Mayo, P.K. Allan, O. Pecher, O.J. Borkiewicz, K.M. Wiaderek, K.W. Chapman, C.J. Pickard, A.J. Morris, C.P. Grey. Investigating sodium storage mechanisms in tin anodes: A combined pair distribution function analysis, density functional theory, and solid-state NMR approach. *J. Am. Chem. Soc.* 2017, *139* (21), 7273.

[95] Z. Li, J. Ding, D. Mitlin. Tin and tin compounds for sodium ion battery anodes: Phase transformations and performance. *Acc. Chem. Res.* 2015, *48* (6), 1657.

[96] H. Tan, D. Chen, X. Rui, Y. Yu. Peering into alloy anodes for sodium-ion batteries: Current trends, challenges, and opportunities. *Adv. Funct. Mater.* 2019, *29* (14), 1808745.

[97] X. Yang, F. Xiao, S. Wang, J. Liu, M.K.H. Leung, D.Y.W. Yu, A.L. Rogach. Confined annealing-induced transformation of tin oxide into sulfide for sodium storage applications. *J. Mater. Chem. A* 2019, *7* (19), 11877.

[98] J. Patra; H.-C. Chen, C.-H. Yang, C.-T. Hsieh,; C.-Y. Su, J.-K. Chang. High dispersion of 1-nm $SnO_2$ particles between graphene nanosheets constructed using supercritical $CO_2$ fluid for sodium-ion battery anodes. *Nano Energy* 2016, *28*, 124.

[99] M. Deng, S. Li, W. Hong, Y. Jiang, W. Xu, H. Shuai, G. Zou, Y. Hu, H. Hou, W. Wang, X. Ji. Octahedral $Sb_2O_3$ as high-performance anode for lithium and sodium storage. *Mater. Chem. Phys.* 2019, *223*, 46.

[100] W. Zuo, W. Zhu, D. Zhao, Y. Sun, Y. Li, J. Liu, X.W. Lou. Bismuth oxide: A versatile high-capacity electrode material for rechargeable aqueous metal-ion batteries. *Energy Environ. Sci.* 2016, *9* (9), 2881.

[101] W. Fang, L. Fan, Y. Zhang, Q. Zhang, Y. Yin, N. Zhang, K. Sun. Synthesis of carbon coated $Bi_2O_3$ nanocomposite anode for sodium-ion batteries. *Ceram. Int.* 2017, *43* (12), 8819.

[102] J. Qian, Y. Xiong, Y. Cao, X. Ai, H. Yang. Synergistic Na-storage reactions in $Sn_4P_3$ as a high-capacity, cycle-stable anode of Na-ion batteries. *Nano Lett.* 2014, *14* (4), 1865.

[103] J. Choi, W.-S. Kim, K.-H. Kim, S.-H. Hong. $Sn_4P_3$–C nanospheres as high capacitive and ultra-stable anodes for sodium ion and lithium ion batteries. *J. Mater. Chem. A* 2018, *6* (36), 17437.

[104] H. Usui, Y. Domi, K. Fujiwara, M. Shimizu, T. Yamamoto, T. Nohira, R. Hagiwara, H. Sakaguchi. Charge–discharge properties of a $Sn_4P_3$ negative electrode in ionic liquid electrolyte for Na-ion batteries. *ACS Energy Lett.* 2017, *2* (5), 1139.

[105] M. Lao, Y. Zhang, W. Luo, Q. Yan, W. Sun, S.X. Dou. Alloy-based anode materials toward advanced sodium-ion batteries. *Adv. Mater.* 2017, *29* (48), 1700622.

[106] G.-L. Xu, Z. Chen, G.-M. Zhong, Y. Liu, Y. Yang, T. Ma, Y. Ren, X. Zuo, X.-H. Wu, X. Zhang, K. Amine. Nanostructured black phosphorus/ketjenblack–multiwalled carbon nanotubes composite as high performance anode material for sodium-ion batteries. *Nano Lett.* 2016, *16* (6), 3955.

[107] J. Pang, A. Bachmatiuk, Y. Yin, B. Trzebicka, L. Zhao, L. Fu, R.G. Mendes, T. Gemming, Z. Liu, M.H. Rummeli. Applications of phosphorene and black phosphorus in energy conversion and storage devices. *Adv. Energy Mater.* 2018, *8* (8), 1702093.

[108] J. Ni, L. Li, J. Lu. Phosphorus: An anode of choice for sodium-ion batteries. *ACS Energy Lett.* 2018, *3* (5), 1137.

[109] M.G. Boebinger, D. Yeh, M. Xu, B.C. Miles, B. Wang, M. Papakyriakou, J.A. Lewis, N.P. Kondekar, F.J.Q. Cortes, S. Hwang, X. Sang, D. Su, R.R. Unocic, S. Xia, T. Zhu, M.T. McDowell. Avoiding fracture in a conversion battery material through reaction with larger ions. *Joule* 2018, *2* (9), 1783.

[110] Z. Hu, Z. Tai, Q. Liu, S.W. Wang, H. Jin, S. Wang, W. Lai, M. Chen, L. Li, L. Chen, Z. Tao, S.L. Chou. Ultrathin 2D TiS$_2$ nanosheets for high capacity and long-life sodium ion batteries. *Adv. Energy Mater.* 2019, *9* (8), 1803210.

[111] N. Wang, C. Chu, X. Xu, Y. Du, J. Yang, Z. Bai, S. Dou. Comprehensive new insights and perspectives into Ti-based anodes for next-generation alkaline metal (Na$^+$, K$^+$) ion batteries. *Adv. Energy Mater.* 2018, *8* (27), 1801888.

[112] H. Xiong, M.D. Slater, M. Balasubramanian, C.S. Johnson, T. Rajh. Amorphous TiO$_2$ Nanotube anode for rechargeable sodium ion batteries. *J. Phys. Chem. Lett.* 2011, *2* (20), 2560–2565.

[113] F. Legrain, O. Malyi, S. Manzhos. Insertion energetics of lithium, sodium, and magnesium in crystalline and amorphous titanium dioxide: A comparative first-principles study. *J. Power Sources* 2015, *278*, 197.

[114] H. He, D. Huang, W. Pang; D. Sun, Q. Wang, Y. Tang; X. Ji, Z. Guo, H. Wang. Plasma-Induced amorphous shell and deep cation-site S doping endow TiO$_2$ with extraordinary sodium storage performance. *Adv. Mater.* 2018, *30* (26), 1801013.

[115] J.A. Dawson, J. Robertson. Improved calculation of Li and Na intercalation properties in anatase, rutile, and TiO$_2$(B). *J. Phys. Chem. C* 2016, *120* (40), 22910.

[116] M. Fehse, E. Ventosa. Is TiO$_2$(B) the future of titanium-based battery materials? *ChemPlusChem* 2015, *80* (5), 785–795.

[117] Y. Yang, S. Liao, W. Shi, Y. Wu, R. Zhang, S. Leng. Nitrogen-doped TiO$_2$(B) nanorods as high-performance anode materials for rechargeable sodium-ion batteries. *RSC Adv.* 2017, *7* (18), 10885.

[118] X. Yang, S. Wang, X. Zhuang, O. Tomanec, R. Zboril, D.Y.W. Yu, A.L. Rogach. Polypyrrole and carbon nanotube Co-composited titania anodes with enhanced sodium storage performance in ether-based electrolyte. *Adv. Sustainable Syst.* 2019, *3* (4), 1800154.

[119] A. Chaturvedi, E. Edison, N. Arun, P. Hu, C. Kloc, V. Aravindan, S. Madhavi. Two dimensional TiS$_2$ as a promising insertion anode for Na-ion battery. *ChemistrySelect* 2018, *3* (2), 524.

[120] H. Tao, M. Zhou, R. Wang, K. Wang, S. Cheng, K. Jiang. TiS$_2$ as an advanced conversion electrode for sodium-ion batteries with ultra-high capacity and long-cycle life. *Adv.Sci.* 2018, *5* (11), 1801021.

[121] P. Senguttuvan, G. Rousse, V. Seznec, J.-M. Tarascon, M.R. Palacín. $Na_2Ti_3O_7$: Lowest voltage ever reported oxide insertion electrode for sodium ion batteries. *Chem. Mater.* 2011, *23* (18), 4109.

[122] J. Nava-Avendaño, A. Morales-García, A. Ponrouch, G. Rousse, C. Frontera, P. Senguttuvan, J.M. Tarascon, M.E.A.-d. Dompablo, M.R. Palacín. Taking steps forward in understanding the electrochemical behavior of $Na_2Ti_3O_7$. *J. Mater. Chem. A* 2015, *3* (44), 22280.

[123] J. Ni, S. Fu, C. Wu, Y. Zhao, J. Maier, Y. Yu, L. Li. Superior sodium storage in $Na_2Ti_3O_7$ nanotube arrays through surface engineering. *Adv. Energy Mater.* 2016, *6* (11), 1502568.

[124] S. Dong, L. Shen, H. Li, G. Pang, H. Dou, X. Zhang. Flexible sodium-ion pseudocapacitors based on 3D $Na_2Ti_3O_7$ nanosheet arrays/carbon textiles anodes. *Adv. Funct. Mater.* 2016, *26* (21), 3703.

[125] A. Rudola, K. Saravanan, S. Devaraj, H. Gong, P. Balaya. $Na_2Ti_6O_{13}$: A potential anode for grid-storage sodium-ion batteries. *Chem. Commun.* 2013, *49* (67), 7451.

[126] K. Shen, M. Wagemaker. $Na_{2+x}Ti_6O_{13}$ as potential negative electrode material for Na-ion batteries. *Inorg. Chem.* 2014, *53* (16), 8250.

[127] Y. Fang, L. Xiao, J. Qian, Y. Cao, X. Ai, Y. Huang, H. Yang. 3D graphene decorated $NaTi_2(PO_4)_3$ microspheres as a superior high-rate and ultracycle-stable anode material for sodium ion batteries. *Adv. Energy Mater.* 2016, *6* (19), 1502197.

[128] S.I. Park, I. Gocheva, S. Okada, J.-i. Yamaki. Electrochemical properties of $NaTi_2(PO_4)_3$ anode for rechargeable aqueous sodium-ion batteries. *J. Electrochem. Soc.* 2011, *158* (10), A1067.

[129] D. Su, G. Wang. Single-crystalline bilayered $V_2O_5$ nanobelts for high-capacity sodium-ion batteries. *ACS Nano* 2013, *7* (12), 11218.

[130] X. Liu, B. Qin, H. Zhang, A. Moretti, S. Passerini. Glyme-based electrolyte for Na/bilayered-$V_2O_5$ batteries. *ACS Appl. Energy Mater.* 2019, *2* (4), 2786.

[131] S. Li, Y. Dong, L. Xu, X. Xu, L. He, L. Mai. Effect of carbon matrix dimensions on the electrochemical properties of $Na_3V_2(PO_4)_3$ nanograins for high-performance symmetric sodium-ion batteries. *Adv. Mater.* 2014, *26* (21), 3545.

[132] L. Zhang, S.X. Dou, H.K. Liu, Y. Huang, X. Hu. Symmetric electrodes for electrochemical energy-storage devices. *Adv.Sci.* 2016, *3* (12), 1600115.

[133] A. Ponrouch, D. Monti, A. Boschin, B. Steen, P. Johansson, M.R. Palacín. Non-aqueous electrolytes for sodium-ion batteries. *J. Mater. Chem. A* 2015, *3* (1), 22.

[134] K. Xu. Electrolytes and interphases in Li-ion batteries and beyond. *Chem. Rev.* 2014, *114* (23), 11503.

[135] K. Xu. Nonaqueous liquid electrolytes for lithium-based rechargeable batteries. *Chem. Rev.* 2004, *104* (10), 4303.

[136] A. Ponrouch, E. Marchante, M. Courty, J.-M. Tarascon, M.R. Palacín. In search of an optimized electrolyte for Na-ion batteries. *Energy Environ. Sci.* 2012, *5* (9), 8572.

[137] J.Y. Jang, H. Kim, Y. Lee, K.T. Lee, K. Kang, N.-S. Choi. Cyclic carbonate based-electrolytes enhancing the electrochemical performance of $Na_4Fe_3(PO_4)_2(P_2O_7)$ cathodes for sodium-ion batteries. *Electrochem. Commun.* 2014, *44*, 74.

[138] Y. Huang, L. Zhao, L. Li, M. Xie, F. Wu, R. Chen. Electrolytes and electrolyte/electrode interfaces in sodium-ion batteries: From scientific research to practical application. *Adv. Mater.* 2019, *31* (21), 1808393.

[139] R. Alcántara, P. Lavela, G.F. Ortiz, J.L. Tirado. Carbon microspheres obtained from resorcinol-formaldehyde as high-capacity electrodes for sodium-ion batteries. *Electrochem. Solid-State Lett.* 2005, *8* (4), A222.

[140] S. Brutti, M.A. Navarra, G. Maresca, S. Panero, J. Manzi, E. Simonetti, G.B. Appetecchi. Ionic liquid electrolytes for room temperature sodium battery systems. *Electrochim. Acta* 2019, *306*, 317.

[141] X. Lin, C. Chu, Z. Li, T. Zhang, J. Chen, R. Liu, P. Li, Y. Li, J. Zhao, Z. Huang, X. Feng, Y. Xie, Y. Ma. A high-performance, solution-processable polymer/ceramic/ ionic liquid electrolyte for room temperature solid-state Li metal batteries. *Nano Energy* 2021, *89*, 106351.

[142] A. Manthiram, X. Yu, S. Wang. Lithium battery chemistries enabled by solid-state electrolytes. *Nat. Rev. Mater.* 2017, *2* (4), 16103.

[143] X. Yu, A. Manthiram. Electrochemical energy storage with mediator-ion solid electrolytes. *Joule* 2017, *1* (3), 453.

[144] Y. Lu, L. Li, Q. Zhang, Z. Niu, J. Chen. Electrolyte and interface engineering for solid-state sodium batteries. *Joule* 2018, *2* (9), 1747.

[145] J.-J. Kim, K. Yoon, I. Park, K. Kang. Progress in the development of sodium-ion solid electrolytes. *Small Methods* 2017, *1* (10), 1700219.

[146] J.A.S. Oh, L. He, B. Chua, K. Zeng, L. Lu. Inorganic sodium solid-state electrolyte and interface with sodium metal for room-temperature metal solid-state batteries. *Energy Storage Materials* 2021, *34*, 28.

[147] S. Wenzel, T. Leichtweiss, D.A. Weber, J. Sann, W.G. Zeier, J. Janek. Interfacial Reactivity benchmarking of the sodium ion conductors $Na_3PS_4$ and sodium β-alumina for protected sodium metal anodes and sodium all-solid-state batteries. *ACS Appl. Mater. Interfaces* 2016, *8* (41), 28216.

[148] Z. Zhu, I.-H. Chu, Z. Deng, S.P. Ong. Role of $Na^+$ interstitials and dopants in enhancing the $Na^+$ conductivity of the cubic $Na_3PS_4$ superionic conductor. *Chem. Mater.* 2015, *27* (24), 8318.

[149] N.J.J. de Klerk, M. Wagemaker. Diffusion mechanism of the sodium-ion solid electrolyte $Na_3PS_4$ and potential improvements of halogen doping. *Chem. Mater.* 2016, *28* (9), 3122.

[150] D.W. McOwen, S. Xu; Y. Gong, Y. Wen, G.L. Godbey, J.E. Gritton, T.R. Hamann, J. Dai, G.T. Hitz, L. Hu, E.D. Wachsman. 3D-printing electrolytes for solid-state batteries. *Adv. Mater.* 2018, *30* (18), 1707132.

[151] Y. Kato, S. Hori, T. Saito, K. Suzuki, M. Hirayama, A. Mitsui, M. Yonemura, H. Iba, R. Kanno. High-power all-solid-state batteries using sulfide superionic conductors. *Nat. Energy* 2016, *1* (4), 16030.

[152] T. Wiley, J. Whitacre, E. Weber, M. Eshoo, J. Noland, D. Blackwood, W. Campbell, E. Sheen, C. Spears, C. Smith. Recovery act–demonstration of sodium ion battery for grid level applications, 2012, https://digital.library.unt.edu/ark:/67531/metad c838317.

[153] J.A. Dowling, K.Z. Rinaldi, T.H. Ruggles, S.J. Davis, M. Yuan, F. Tong, N.S. Lewis, K. Caldeira. Role of long-duration energy storage in variable renewable electricity systems. *Joule* 2020, *4* (9), 1907.

[154] D. Kundu, E. Talaie, V. Duffort, L.F. Nazar. The emerging chemistry of sodium ion batteries for electrochemical energy storage. *Angew. Chem, Int. Ed.* 2015, *54* (11), 3431.

[155] L. Fan, X. Li. Recent advances in effective protection of sodium metal anode. *Nano Energy* 2018, *53*, 630.

[156] J. Peters, A. Peña Cruz, M. Weil. Exploring the economic potential of sodium-ion batteries. *Batteries* 2019, *5* (1), 10.

[157] H. Hou, X. Qiu, W. Wei, Y. Zhang, X. Ji. Carbon anode materials for advanced sodium-ion batteries. *Adv. Energy Mater.* 2017, *7* (24), 1602898.

[158] B. Zhang, C.M. Ghimbeu, C. Laberty, C. Vix-Guterl, J.-M. Tarascon. Correlation between microstructure and Na storage behavior in hard carbon. *Adv. Energy Mater.* 2016, *6* (1), 1501588.

[159] C. Bommier, T.W. Surta, M. Dolgos, X. Ji. New mechanistic insights on Na-ion storage in nongraphitizable carbon. *Nano Lett.* 2015, *15* (9), 5888.

[160] X. Yang, A.L. Rogach. Electrochemical techniques in battery research: A tutorial for nonelectrochemists. *Adv. Energy Mater.* 2019, *9* (25), 1900747.

# 3 Potassium-Ion Batteries
## *Future of Metal Ion Batteries*

*Peddinti V.R.L. Phanendra, Anoopkumar V.,
Sumol V. Gopinadh, Bibin John and Mercy T.D.*

## 3.1 INTRODUCTION

Among the various energy storage systems, those based on electrical energy are the most promising ones. As electrical energy can't be stored in its basic form, fossil fuels are widely used to meet the demand of most of the electrical energy requirements, but global warming is a concern. The energy generated from renewable sources such as sunlight, tides and wind is intermittent in nature, that is, it changes with weather conditions. It is in this context that the need for energy storage systems arises. Energy storage systems can be divided into electrochemical, chemical, mechanical and electrical on the basis of storage of energy. Among these, electrochemical energy storage is more energy efficient and is the mechanism in rechargeable batteries. Examples of various rechargeable batteries are lead-acid, Li-ion, Ni-metal hydride (Ni-MH), Ni-Cd etc. [1].

Nowadays, many rechargeable batteries such as lead-acid and metal hydride are replaced by lithium-ion batteries (LIBs). The characteristics such as higher voltage (>3.6 V), specific energy (100–265 Wh kg$^{-1}$), superior rate performance, good cycle life and shelf life made LIBs more popular than other battery technologies [2]. LIB is one of the sustainable and clean energy storage systems and has great commercial potential as it is extensively used in portable electronics, large-scale energy storage devices, electric vehicles (EV), etc. [3]. For example, in the 1990s, for hybrid vehicles (HVs), high-power Ni-MH cells were used and since 2011, for large HVs, LIBs have been largely used. For the LIB, Li and Co are essential, which are expensive and not abundant on the earth [4]. However, due to the steady increase in demand, lithium production has increased from 31,700 tons to 77,000 tons from 2014 to 2019, which increased the material cost [5].

Due to the inadequate lithium reserves, uneven geographical distribution and the increasing cost of lithium extraction, long-term sustainability of lithium-based energy storage systems comes under scrutiny. So, it is time to focus on alternative energy storage technologies [6], which are viable, sustainable and have comparable electrochemical performance to that of LIBs [7,8]. Potassium-ion batteries (KIBs) have received much attention recently as an alternative to LIBs [9, 10]. This chapter deals with various aspects of KIBs, viz. their advantages and limitations, different materials used and recent advances.

DOI: 10.1201/9781003208198-3

## 3.2 COMPARATIVE STUDY OF Li, Na AND K

A comparison of the characteristics of lithium, sodium and potassium is summarized in Table 3.1.

A comparison of KIB with Li-ion and Na-ion batteries is given below:

a.  For the insertion (deinsertion) of K-ion, the cathode must possess a large inter-layer gap as ionic radius of $K^+$ (1.38 Å) is higher than that of $Li^+$ (0.76 Å) and $Na^+$ (1.02 Å). In comparison with $Li^+$ and $Na^+$, the Lewis acidic nature of $K^+$ is low. As a result, during solvation, the ionic size of $K^+$ formed is too smaller [18], which leads to improved ionic conductivity and good mobility. For $FSA^-$ [FSA: bis(fluorosulfonyl)amide] based alkali metal salt, conductivity increases with increasing size of alkali metal, and larger ions are expected to lower desolvation energy in metal insertion/deinsertion process in alkali metal ion batteries. As the alkali metal size increases, viscosity decreases with concentration due to lower desolvation energy, that is, weak interactions of larger size ions with the solvent [16].

b.  The output capacity of KIB is more beneficial than sodium ion battery (NIB) due to lower potential of $K^+/K$ (–2.93 V vs. NHE) compared to $Na^+/Na$ (–2.71

**TABLE 3.1**
**Comparison of the Characteristics of Lithium, Sodium and Potassium**

|  | Lithium | Sodium | Potassium | Reference |
|---|---|---|---|---|
| Atomic mass (amu) | 6.941 | 22.989 | 39.098 | [11] |
| Ionic radius (Å) | 0.76 | 1.02 | 1.38 | [11] |
| Melting Point (°C) | 180.54 | 97.72 | 63.38 | [11] |
| Abundance (%) | 0.0017 | 2.36 | 2.09 | [12] |
| Electrode potential $(A^+_{aq}/A)$ vs SHE (V) | -3.04 | -2.71 | -2.93 | [13] |
| Cost of industrial- grade metal (US $ $ton^{-1}$) | 100000 | 3000 | 13000 | [14] |
| Desolvation energy (kJ $mol^{-1}$) in Propylene carbonate | 215.8 | 158.2 | 119.2 | [15] |
| In $AMO_2$,preference of coordination of $A^+$ | Octahedral and tetrahedral | Octahedral and prismatic | Prismatic | [15] |
| Specific capacity (mAh $g^{-1}$) | 3860 | 1166 | 685 | [15] |
| Specific capacity for Graphite (mAh $g^{-1}$) | 372 ($LiC_6$) | <35 (Negligible insertion) | 279 ($KC_8$) | [16] |
| Theoretical $ACoO_2$ capacity (mAh $g^{-1}$) | 274 | 235 | 206 | [17] |
| Theoretical $ACoO_2$ capacity (mAh $cm^{-3}$) | 1378 | 1193 | 906 | [17] |

Note: A = Li, Na, K

V vs. NHE) [19]. The electrode potential of K$^+$/K is lower in some non-aqueous and two-component electrolytes (e.g., EC:DEC, EC=ethylene carbonate, DEC=diethyl carbonate) than Li$^+$/Li. Thus, KIB delivers high specific energy due to elevated electrochemical window. For example, the electrode potentials of K$^+$/K, Na$^+$/Na and Li$^+$/Li electrodes are −2.88 V, −2.79 V, −2.56 V, respectively, in propylene carbonate (PC) [20]. The K$^+$/K potential is 15 mV lower than Li$^+$/Li in EC:DEC. The standard electrode potential of K$^+$/K vs SHE is −2.93 V, so KIBs may exhibit elevated cell voltage, which is comparable to LIBs. Due to large size and high mass of potassium, the gravimetric energy density and volumetric energy density are lower than LIBs. The low charge density of K$^+$ increases the ionic mobility within the electrolyte and the electrode and hence improves the electrochemical performance.

c.  During the first stage intercalation of K$^+$ in graphite, KC$_8$ is the compound formed and the capacity achieved is nearly the theoretical capacity (279 mAh g$^{-1}$) at 0.23V (vs K$^+$/K) whereas for Li$^+$ in graphite, Li insertion takes place at 0.1V. Due to this reason, the possibility for K dendrite formation is less, so it is safer in the cycling process. The KIB has better high-rate performance than NIB as intercalation of K$^+$ is faster in carbon. However, because of high ionic size of K$^+$, graphite loses its structure resulting in sudden capacity fade [21].

d.  A major problem in LIB and lithium metal battery is dendrite formation. In the KIB, the formed K dendrites can be minimized by controlling temperature due to its much weaker mechanical properties and less boiling point. Lithium cannot form alloy with Na or K whereas the Na and K mixture can form and changes to liquid near RT [22]. Na–K alloy as anode delivers elevated specific capacity (579 mAh g$^{-1}$) and it doesn't form dendrite as the formed metal becomes liquid during charging in KIB. In KIBs, cheaper aluminium can be used as a current collector for anode due to its inertness toward the formation of Al-K alloy at lower voltages [23]; this makes battery cheaper and light weight.

Due to the safety issues of potassium metal [24], studies on KIBs were limited for years and now more focus is given to KIBs as can be seen from the increase in number of publication in the last few years (Figure 3.1).

## 3.3  BASIC COMPONENTS OF KIB

Similar to LIB, KIB is also composed of cathode, anode, electrolyte, separator and similar hardware components. The capacity of KIB depends on the electrode materials (cathode and anode) employed. The binder in the electrode helps to improve the cohesive forces between electrode constituents (active material, conducting agent and binder) and adhesive forces between electrode constituents and current collector. The separator used should be a good ionic conductor of K$^+$ and electronic insulator. Electrolyte plays a vital role in the safety, cycle life and the specific charge/discharge capacity. Presently, the studies on aqueous vs non-aqueous electrolytes are one of the hot topics in KIB [25]. For aqueous solution with the same concentration and same anion, K$^+$ is more ionic conductive than Li$^+$ and Na$^+$ due to weaker solvation. When compared to LIBs and NIBs with aqueous medium, KIB has advantages but

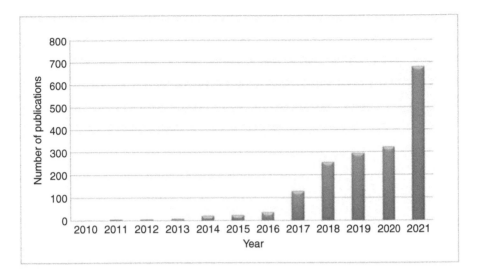

**FIGURE 3.1**   Number of publications on KIBs from 2010 to 2021 (generated by SciFinder using the keyword "potassium ion batteries").

commonly used aqueous electrolytes have limitations such as small voltage windows and most of the electrode materials may dissolve in aqueous medium. So, non-aqueous solvents are extensively used in most of the KIBs [26].

### 3.3.1   Cathode Materials

The various types of cathode materials used in KIB are discussed in this section.

#### 3.3.1.1   Layered Transition Metal Oxides (TMOs)

In the recent years, layered transition metal oxides (TMOs) with high theoretical specific capacity have been greatly used as cathode active material in LIBs and NIBs. The basic formula of layered TMO is $A_xMO_2$ [A= alkali metals—Li, Na, K; M= 1st transition series metals—Fe, Co, Mn, Ni, V, Cr or mixture of two or three elements with +2 or a mixture of +4 and +3 or +2 oxidation states]. As they have high theoretical capacity and good stability, they are used as a cathode in LIB and NIB. The electrochemical intercalation nature of Li and Na in $LiCoO_2$ (LCO) and $NaCoO_2$ (NCO), respectively, were first reported in 1980s [27].

In layered metal oxides, the accepted mechanism can be expressed as

$$AMO_2 \leftrightarrow A_{1-x}MO_2 + xA^+ + xe^- \ (A=K)$$

##### 3.3.1.1.1   Mn-Based Layered TMOs

Mn-based layered TMO materials are less expensive and less toxic in nature, so they have attracted attention as cathode active materials in KIBs, but current rate and cycling performance cannot satisfy the demand of high energy and high power

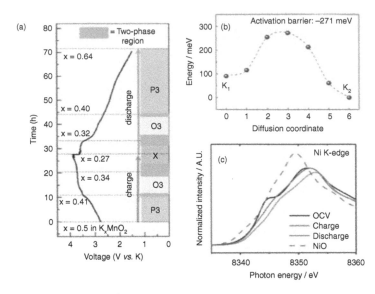

**FIGURE 3.2** (a) Structural change of P3-$K_{0.5}MnO_2$ during cycling. Reproduced with permission from Ref. 29, Copyright 2017 WILEY-VCH Verlag GmbH & Co. KGaA. (b) Activation barrier in P'2-$K_x[Ni_{0.05}Mn_{0.95}]O_2$. Reproduced with permission from Ref.30, Copyright 2020 Elsevier B.V. (c) XANES Ni K-edge for P2-$K_{0.75}[Ni_{1/3}Mn_{2/3}]O_2$ in the voltage window 1.5–4.3 V. Reproduced with permission from Ref. 31, Copyright 2020 WILEY-VCH Verlag GmbH & Co. KGaA.

requirements. The performance of layered Mn oxides can be improved by doping with various ions and surface coating.

$K_{0.3}MnO_2$ compound of P2-type is successfully prepared and characterized by Vaalma et al. [28], which has capacity $\approx$ 136 mAh g$^{-1}$ between 1.5 and 4.0 V, but observed intermediate stages at 3.7 V and 3.9 V causing huge irreversible expansion of oxygen layers. So, capacity decay is minimized by operating in the 1.5–3.5 V voltage range. Under this voltage range, the material exhibited a reversible capacity of 65 mAh g$^{-1}$with capacity retention of 57% after 685 cycles.

By using high-temperature process, the compound $K_{0.5}MnO_2$ of the type P3 (space group: R-3m) is synthesized [29]. During charging, P3 phase is converted to O3 and then finally reaches an unidentified phase X and discharge takes place in the same opposite direction (Figure 3.2(a)). The complete reversible cycling process is shown by in-situ X-ray diffraction (XRD) and additionally first-principle calculations also confirm that the whole process is reversible in nature. This phase change depends on the relative stability of oxygen stacking with respect to the content of potassium. To overcome the capacity fading at high voltage ($\approx$4.3 V) and to keep the layered structure stable and reversible, Choi et al. [30] reported a layered TMO of P'2 type– $K_{0.83}[Ni_{0.05}Mn_{0.95}]O_2$—which is synthesized by electrochemical ion displacement of P'2-$Na_{0.67}$ [$Ni_{0.05}Mn_{0.95}$] $O_2$. Its prominent power capability is confirmed by nudged elastic band (NEB) method and also by first-principle calculations by finding activation barrier (Figure 3.2(b)).

So far, a synthetic approach for a stable P2 structure hasn't been established due to large size of $K^+$, which favors crystallization in the P3 type structure. Jo et al. [31] designed a P2-class $K_{0.75}[Ni_{1/3}Mn_{2/3}]O_2$ cathode by the electrochemical ion-exchange method from the P2 type $Na_{0.75}[Ni_{1/3}Mn_{2/3}]O_2$, which can allow repeated de/insertion of huge $K^+$ and maintains more structural stability due to its single-phase mechanism with the redox couple $Ni^{+4/+2}$, which is concluded by operando X-ray diffraction (o-XRD) and ex situ X-ray absorption near edge structure spectroscopy (XANES) (Figure 3.2(c)).

### 3.3.1.1.2 Cobalt-Based Layered TMOs

In LIBs, cobalt-based transition metal layered oxides are commonly used due to steady discharge voltage and good rate performance. So, cobalt-based layered compounds are extensively studied as cathode materials for KIBs also. Yu et al. [32] designed a porous nano frame (PNF) layered oxide, $K_{0.6}CoO_{2-x}N_x$ (KCO PNF). Density functional theory (DFT) calculations and experimental observations showed that partial substitution of oxygen by nitrogen accelerates the electronic conductivity and increases the interlayer gap, which enables accommodation of more number of $K^+$ during intercalation and accelerate the $K^+$ migration (Figure 3.3).

Masese et al. [33] prepared a layered P2-class honeycomb $K_2NiCoTeO_6$ as a lofty voltage ($\approx 4.3V$ vs $K^+/K$) cathode material. The layers of $K_2NiCoTeO_6$ consist of edge-sharing octahedra $Co^{+2}/Ni^{+2}O_6$ and $Te^{+6}O_6$ itself in the $ab$ axis, which arrange regularly to form honeycomb lattice. The layers are connected by triangular prisms of $KO_6$ (K-ions are present in intermediate layer through the $c$-axis). In these layers, $O_2^{2-}$ ions are substituted by highly electronegative $[TeO_6]^{-6}$ species and its high voltage can be due to the lowering of electron density in Co/Ni-O bonds by intense O 2p-Te 5P hybridization, which improves the ionic nature of Co/Ni-O bonds and thus more energy is required for the redox process of transition metals (Co/Ni).

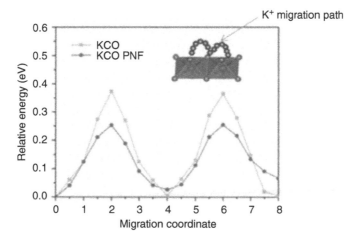

**FIGURE 3.3** Migration energy barrier comparison of KCO with KCO PNF. Reproduced with permission from Ref. 32, Copyright 2020 Elsevier B.V.

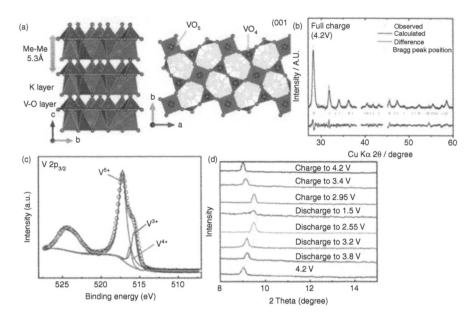

**FIGURE 3.4** (a) Crystal structure of $K_2V_3O_8$ from the refinement result. (b) Rietveld refinement analysis of fully charged $K_2V_3O_8$/C. Reproduced with permission from Ref. 34, Copyright 2019 Elsevier B.V. (c) $K_{0.486}V_2O_5$: XPS spectrum of $V2p_{3/2}$ region. (d) Ex-situ XRD of $K_{0.486}V_2O_5$ taken at different voltages. Reproduced with permission from Ref. 35, Copyright 2019 WILEY-VCH Verlag GmbH & Co. KGaA.

### 3.3.1.1.3 Vanadium-Based Layered TMOs

Jo et al. [34] reported a new cathode material, $K_2V_3O_8$ for KIBs. In its crystal structure $VO_4$, $KO_{11}$ and $VO_5$ are coordinated with 'four' hendecahedra in such a way that four oxygens of $VO_5$ share with four units of $VO_4$ tetrahedra to form four penta-shaped empty spaces in which the de/insertion of $K^+$ ions take place (Figure 3.4(a)). Vanadium in $K_2V_3O_8$ crystallographic site is represented by V1 in $V^{+4}O_5$ and V2 in $V^{+5}O_4$. XANES showed that $V1^{+4}$ oxidizes to $V1^{+5}$ while charging. Rietveld refinement of XRD pattern (Figure 3.4(b)) showed that when $K_2V_3O_8$/C is charged to 4.2 V, V1–O bond reduction in $VO_5$ is observed. It showed that capacity delivery is due to the redox pair $V^{+4/+5}$ in $VO_5$ and $VO_4$ supports the crystal structure.

Yuan et al. [35] synthesized a high-voltage $K_{0.486}V_2O_5$ nanobelt as a cathode material for KIBs. X-ray photoelectron spectroscopy (XPS) showed the splitting of V $2p_{3/2}$ peak into three sub-categories, which indicates that the atomic amounts of $V^{+3}$, $V^{+4}$ and $V^{+5}$ are 21.7%, 7.7% and 70.6%, respectively (Figure 3.4(c) ). Ex-situ XRD revealed that during complete charge–discharge cycles no additional peaks are noticed, which suggests the superior structural reversibility of $K_{0.486}V_2O_5$ (Figure 3.4(d)) .

### 3.3.1.1.4 Chromium-Based Layered TMOs

Kim et al. [36] showed that for cathode in KIBs, O3-class stoichiometric layered $KCrO_2$ can be used and presented a practical de/intercalation mechanism (rocking chair). The layered structure becomes unstable as the concentration of K increase (K/M=1) due to huge electrostatic repulsions between $K^+$-$K^+$ ions; but this repulsion is compensated by the ligand field selection of $Cr^{+3}$ for the octahedral environment and stabilizes the layered structure of $KCrO_2$. Inductively coupled plasma-mass spectrometry (ICP-MS) showed that K/Cr ratio is 1.02. From scanning electron microscopy (SEM), the approximate particle size observed is 1 μm and energy-dispersive spectroscopy (EDS) revealed that K, Cr and O are uniformly distributed in $KCrO_2$.

The incessant capacity fading and lower charge efficiency (CE) in the first charge/discharge are observed in the stoichiometric O3-class $KCrO_2$. So, Naveen et al. [37] synthesized a stable non-stoichiometric P'3-class $K_{0.8}CrO_2$ from the industrially available $K_2CrO_4$ as a superior cathode in KIBs. In the stoichiometric $KCrO_2$, multiphase changes take place while charging (O3 changes to O'3-P'3-P3-P'3-P3-O3) with the incomplete recovery during the discharge (O3–P3–P'3-P3-P'3-O'3). But $K_{0.8}CrO_2$ at the first charge almost retains its basic P'3 phase along with a small region of P3 and due to limited phase changes and almost no change of volume during cycling makes the P'3-class $K_{0.8}CrO_2$ more stable.

Hwang et al. [38] proposed a layered oxide of P3-class $K_{0.69}CrO_2$ synthesized by electrochemical ion-exchange process. Nudged elastic band (NEB) calculations showed that its high-power capability (superior $K^+$ movement) is due to the low $E_a$ (activation energy) barrier (≈230 meV). The Galvanostatic intermittent titration technique (GITT) confirmed that P3-$K_xCrO_2$ can accommodate ≈0.69 mol of 'K' up to 1.5 V (potassiation) and ≈0.47 mol of K up to 3.8 V (depotassiation).

### 3.3.1.2 Polyanion-Based Compounds

The general structure of polyanionic compounds is composed of $MO_x$ (M is the transition metal, x ≥ 0) and polyhedra of $(XO_4)_n$ (X = P, S, As, Si, Mo or W) with the 3D open structural network and possess good structural stability along with thermal stability. During charge and discharge, M forms redox pair and contributes to specific capacity with the de/intercalation of K-ions. The $(PO_4)^{3-}$ anion is in the tetrahedral or in the octahedral coordination in the crystal structure. The inductive effect of the polyanion boosts the transition metal redox potential.

Zheng et al. [39] synthesized Rb-doped composite $K_{2.95}Rb_{0.05}V_2(PO_4)_3$/C by an easy sol-gel process. Rb ($Rb^{+1}$) doping doesn't affect the V valence ($V^{+3}$), but few of K sites are occupied by Rb and improves the diffusion coefficient ($D_K$) of K-ion, which is determined by CV and calculated by Randles–Sevick expression (Figure 3.5(a)).

Park et al. [40] introduced a novel $K_4Fe_3(PO_4)_2(P_2O_7)$ cathode with excellent electrochemical performance. Its crystal structure possesses plenty of sheets of $Fe_3P_2O_{13}$, arranged through '*bc*' plane and within the $Fe_3P_2O_{13}$ layer, $FeO_6$ octahedra and $P_2O_7$ are connected through *a*-axis which provides a huge 3D pathway for the $K^+$ ion diffusion (Figure 3.5(b)). The superior power-capability of the cathode is due to small activation barrier for the $K^+$ ion diffusion (huge empty space and 3D pathways), which is verified by first-principle computation method (Figure 3.5(c)).

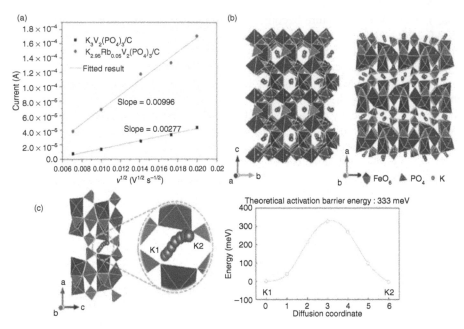

**FIGURE 3.5** (a) Peak current ($i_p$) vs $V^{1/2}$ (scan rate) [$i_p$=2.69 × $10^5n^{3/2}$ $AD_k^{1/2}C^0V^{1/2}$, where $i_p$= peak current in ampere, $n$=number of exchanged electrons, A=electrode area (cm²), V= sweep rate (V/s) and $C^0$=$K^+$ concentration (mol/cm³)]. Reproduced with permission from Ref. 39, Copyright 2019 Elsevier B.V. (b) Crystal structure of $K_4Fe_3(PO_4)_2(P_2O_7)$. (c) Diffusion pathway and energy landscape in $K_4Fe_3(PO_4)_2(P_2O_7)$. Reproduced with permission from Ref. 40, Copyright © 2019 Elsevier B.V.

Liao et al. [41] synthesized $KVPO_4F$ (KVPF) with carbon coating as a cathode for KIBs. Among the different kinds of samples, viz. KVPF-P/B/F (P-platelet, B-ball and F-flower-like), the carbon content is high in the KVPF-F, it partially diminishes the V oxidation and F volatilization. These three KVPF samples undergo redox reaction through four redox pair: $KVPO_4$ F ↔ $K_{0.75}VPO_4$ F ↔ $K_{0.625}VPO_4$ F ↔ $K_{0.5}VPO_4F$ ↔ $VPO_4F$. KVPF-F peaks are almost overlapped and sharpened, indicating superior reaction rate with better reversibility (Figure 3.6(a) and(b) ).

### 3.3.1.3 Prussian Blue and Its Analogs (PBAs)

The huge potassium insertion ability of Prussian blue and its analogs has been greatly studied for the past several years. Around the year 1936, Prussian blue's (PB's) face-centered cubic structure was first proposed by Keggin and Miles [42]. Further in 1962, its complete structural arrangement and electronic configuration were determined by Melvin B. Robin [43]. The general chemical formula of PBA can be represented by $K_xM_A[M_B(CN)_6]_y.zH_2O$ (0 <$x$< 2, $y$< 1), $M_A$=Ti, Mn, Fe, Co, Ni, Cu, Zn, Ba; $M_B$=Fe. $M_A$ and $M_B$ are the transition metal ions, which are exchangeable, coordinated to N and C, respectively, and interconnected by the ligand CN. In PBAs, each repeating unit possesses eight subunits associated

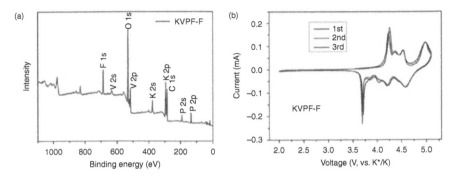

**FIGURE 3.6** (a) XPS pattern and (b) Cyclic voltammogram of KVPF-F at 0.1 mV s$^{-1.}$ Reproduced with permission from Ref. 41, Copyright 2019 Elsevier B.V.

with eight interstitial sites which can occupy transition metal ions and neutral species. PBAs possess a few advantages, viz. open framework, superior cycling performance, low cost, structural stability and easy preparation. Due to the open framework, PBAs possess superior diffusion of various ions and improve the rate stability. PB is a superior de/intercalating component even in the solid state without having a liquid electrolyte.

In PBAs, the possible chemical reaction can be represented by

$$K_xM_A^{II}[M_B^{II}(CN)_6].nH_2O \leftrightarrow K_{x-y}M_A^{III}[M_B^{III}(CN)_6].nH_2O + yK^+ + ye$$

Eftekhari et al. [44] for the first time designed a secondary cell with PB as a cathode and K as anode. Figure 3.7 (a) and (b) show the cyclic voltammogram and charge–discharge curve, respectively of the cell. The material retained 88% of intial capacity after 500 complete charge–discharge cycles at C/10 in the electrolyte 1 M KBF$_4$ in EC/EMC (3:7 by weight, EMC= ethyl methyl carbonate) (Figure 3.7(b)). The CV curves are associated with two redox pairs as follows:

$$KFe^{III}[Fe^{II}(CN)_6]\text{“PB”} = Fe^{III}[Fe^{III}(CN)_6]\text{“BG”} + K^+ + e^- \text{(Oxidation)}$$

$$KFe^{III}[Fe^{II}(CN)_6]\text{“PB”} + K^+ + e^- = K_2Fe^{II}[Fe^{II}(CN)_6]\text{“PW”} \text{(Reduction)} \quad (3.1)$$

At 0.86 V$_{SCE}$, PB oxidation takes place and becomes BG (Berlin green), reduction takes place and becomes PW (Prussian white).

Zhang et al. [45] first time disclosed a potential cathode-K$_{0.22}$Fe[Fe(CN)$_6$]$_{0.805}$.4.01H$_2$O nanoparticles (KPBNPs) for KIBs. The first cycle coulombic efficiency of 44% is due to the presence of interstitial water but in further cycles residual water decreases and coulombic efficiency increases (Figure 3.7(c)). XPS (during charging Fe$^{3+}$ 2p increases and decreases with discharging), ex situ X-ray diffraction (before and after cycling shows same pattern) and Raman spectroscopy (stretching vibration of CN with Fe) conclude that C-coordinated Fe$^{III}$/Fe$^{II}$ couple is responsible for the K de/intercalation and structural stability.

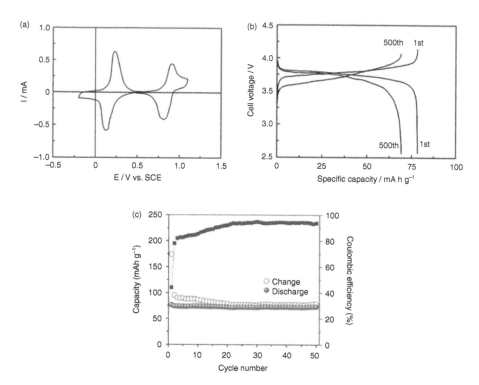

**FIGURE 3.7**  (a) CV of PB at 10 mV s$^{-1}$ scan rate. (b) Charge–discharge curve of K (anode) and PB (cathode) cell at C/10. Reproduced with permission from Ref. 44, Copyright 2003 Elsevier B.V. (c) Cycling performance of KPBNPs/K half-cell at 50 mA g$^{-1}$. Reproduced with permission from Ref. 45, Copyright 2016 WILEY-VCH Verlag GmbH & Co. KGaA.

### 3.3.1.4  Organic Cathode Materials

Organic electrodes possess distinct merits than inorganic electrodes such as different chemical and flexible structures, environment friendliness, lower cost and electrochemical stability. Due to weak intermolecular interactions, K$^+$ ions easily de/intercalate and achieve excellent specific capacity. According to various reaction mechanisms in organic electrodes, carbonyl bond (C=O), C=N and doping reactions are the redox reaction centers in NIB and LIB. In LIBs and NIBs, quinone compounds, anhydrides, conductive polymers, radical and carboxylate compounds have been extensively studied.

Chen et al. [46] first reported 3,4,9,10-perylene–tetracarboxylic acid–dianhydride (PTCDA) as an organic and low-cost cathode in non-aqueous KIBs. The electrochemical reactions that take place during cycling can be represented as in Figure 3.8(a). It exhibits a specific capacity of 131 mAhg$^{-1}$ between 1.5 and 3.5V vs. K$^+$/K (2 mol K$^+$ insertion per mole of PTCDA) and superior cycling stability with 66.1% capacity retention after 200 complete charge–discharge cycles (Figure 3.8(b)).

Flipp et al. [47] investigated poly(N-phenyl-5,10-dihydrophenazine) [p-DPPZ] as a high-voltage cathode which delivers 162 mAhg$^{-1}$ at 200 mAg$^{-1}$ and can operate at

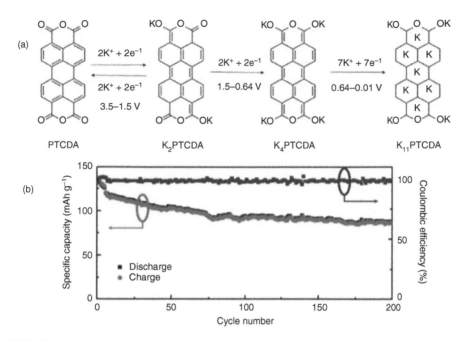

**FIGURE 3.8** (a) Proposed electrochemical mechanism during cycling. (b) Cycling performance at 50 mA g⁻¹of PTCDA cathode. Reproduced with permission from Ref. 46, Copyright 2015 Elsevier.

high current densities (2–10 Ag⁻¹) and delivers better capacity with 96% and 79% capacity retention over 100 and 1,000 complete charge–discharge cycles, respectively. For the superior performance of p-DPPZ, the chosen electrolyte was 2.2 M $KPF_6$ in diglyme (DG).

### 3.3.1.5 Other Cathode Materials

Due to the large ionic size of $K^+$, the traditional cathode materials have limited application in KIBs. So it is relevant to explore other cathode materials which possess stable structure during de/intercalation process such as layered $TiS_2$ [48], $KCrS_2$ [49] and $V_2O_5 \cdot 0.6H_2O$ [50].

The performances of various cathode materials reported for KIB are consolidated in Table 3.2.

### 3.3.2 ANODE MATERIALS

For the alkali metal-ion batteries, the main requirements of anode are low reduction potential, structural stability, elevated capacity and superior electrochemical reversibility. On the basis of material type and potassium storage mechanism, anode materials of KIBs can be categorized into intercalation type, conversion type, alloying and organic materials.

**TABLE 3.2**
**The Performance of Various Cathode Materials Used in KIBs**

| Cathode Material | Electrolyte | Specific Capacity (mAh g$^{-1}$)@ rate | Rate Performance | First Cycle Coulombic Efficiency (%) | Capacity Retention @cycles @rate | Voltage Window (V) | Ref. |
|---|---|---|---|---|---|---|---|
| K$_{0.3}$MnO$_2$ | 1.5 M KFSI in EC:DMC (1:1, v/v) | 70@ 0.1C | 40 mAh g$^{-1}$ @ 5C | ≈99 | 57% @ 685@ 0.1C | 1.5–3.5 | [28] |
| K$_{0.5}$MnO$_2$ | 0.7 M KPF$_6$ in EC/DEC (1:1, v/v). | 94@20 mA/g | 38 mAh g$^{-1}$ @ 300 mA g$^{-1}$ | ≈95 | 70%@50 @ 20 mA g$^{-1}$ | 1.5–3.9 | [29] |
| P'2 type- K$_{0.83}$[Ni$_{0.05}$Mn$_{0.95}$]O$_2$ | 0.5 M KPF$_6$ in EC:DEC (1:1, v/v) | 155@ 52 mA/g | 78 mAh g$^{-1}$ @ 2600 mA g$^{-1}$ | ≈96 | 77%@ 500@ 520 mA g$^{-1}$ | 1.5–4.3 | [30] |
| P2 type K$_{0.75}$[Ni$_{1/3}$Mn$_{2/3}$]O$_2$ | 0.5 M KPF$_6$ in EC:DEC (1:1, v/v) | 110@ 20 mA/g | 91 mAh g$^{-1}$ @ 1.4 A g$^{-1}$ | ≈80 | 83%@ 500@ 1400 mA g$^{-1}$ | 1.5–4.3 | [31] |
| Porous nano frame K$_{0.6}$CoO$_{2-x}$N$_x$ | 1 M KPF$_6$ in EC:DEC (1:1, v/v) | 86 @ 50 mA/g | 30 mAh g$^{-1}$ @ 500 mA g$^{-1}$ | 81.7 | 77.3% @ 400 @ 50 mA g$^{-1}$ | 1.5–4.2 | [32] |
| Honeycomb K$_2$NiCoTeO$_6$ | 0.5 M KTFSI in Pyr13 TFSI | 30 @ C/20 | 15 mAh g$^{-1}$ @ C/10 | – | >90% @ 25@ C/20 | 1.3–4.7 | [33] |
| K$_2$V$_3$O$_8$/C | 0.5 M KPF$_6$ in EC:DEC (1:1, v/v). | 75@ 20 mA/g | 60%@ 200 mA g$^{-1}$ | 90 | 80% @ 200@ 20 mA g$^{-1}$ | 1.0–4.2 | [34] |
| K$_{0.486}$V$_2$O$_5$ nanobelt | 0.8 M KPF$_6$ in EC:DEC (1:1, v/v). | 159 @ 20 mA/g | 78.8 mAh g$^{-1}$@ 100 mA g$^{-1}$ | 99 | 67.4% @ 100 @ 100 mA g$^{-1}$ | 1.5–4.2 | [35] |
| KCrO$_2$ | 0.7 M KPF$_6$ in EC:DEC (1:1, v/v). | 65 @ 10 mA/g | 31 mAh g$^{-1}$@ 0.5 A g$^{-1}$ | – | 67%@ 100@ 10 mA g$^{-1}$ | 1.5–4.0 | [36] |

| | | | | | | | |
|---|---|---|---|---|---|---|---|
| P'3 $K_{0.8}CrO_2$ | 0.5 M KPF$_6$ in EC:DEC (1:1, v/v). | 60@ 1C | 58% @ 2C | 104 | 99% @ 300@ 1C | 1.5–3.9 | [37] |
| P3 $K_{0.69}CrO_2$ | 0.5 M KPF$_6$ in EC:DEC (1:1, v/v) | 100@ 1C | 65 mAh g$^{-1}$ @10C | 96 | 65% @ 1000 @ 1C | 1.5–3.8 | [38] |
| $K_{2.95}Rb_{0.05}V_2(PO_4)_3$/C | 0.8 M KPF$_6$ in EC/DMC (1:1, v/v) | 55.7 @ 20 mA/g | 34.5 mAh g$^{-1}$ @ 200 mA g$^{-1}$ | - | 94.43% @ 50 @ 20 mA g$^{-1}$ | 2.5–4.6 | [39] |
| $K_4Fe_3(PO_4)_2(P_2O_7)$ | 0.5 M KPF$_6$ in EC/PC/FEC (20:20:1, v/v) | 118@ C/20 | 70% @ 5C | - | 82% @ 500@ 5C | 2.1–4.1 | [40] |
| $KVPO_4F$ | 1 M KPF$_6$ EC/PC (1:1, v/v) | 103@ 20 mA/g | 87.6 mAh g$^{-1}$ @ 5 mA g$^{-1}$ | 69 | 78.57% @ 900@ 1 A g$^{-1}$ | 2.0–4.8 | [41] |
| $K_{0.220}Fe[Fe(CN)]_{6-0.805}\cdot 4.01H_2O$ | 0.8 M KPF$_6$ EC: DEC (1:1, v/v) | 76.7@ 50 mA/g | 90% @ 100 mA g$^{-1}$ | 44 | 96.3% @50@ 50 mA g$^{-1}$ | 2.0–4.0 | [45] |
| PTCDA (3,4,9,10-perylene–tetracarboxylicacid–dianhydride) | 0.5 M KPF$_6$ in EC/DEC (1:1, v/v) | 131 @ 10 mA/g | 73 mAh g$^{-1}$ @ 500 mA g$^{-1}$ | - | 66.1% @ 200@ 50 mA g$^{-1}$ | 1.5–3.5 | [46] |
| p-DPPZ (poly(N-phenyl-5,10-dihydrophenazine)) | 2.2M KPF$_6$ DG | 162 @ 200 mAg | 127 mAh g$^{-1}$ @ 1 A g$^{-1}$ | 90@ 2A/g | 59% @ 2000@ 2 A g$^{-1}$ | 2.5–4.5 | [47] |
| Layered $TiS_2$ | 1M KPF$_6$ in DME | 134@ 0.48 A/g | 80 mAh g$^{-1}$ @ 4.8 A g$^{-1}$ | 96 | 78.75% @ 600@ 4.8 A g$^{-1}$ | 1.5–3.0 | [48] |
| $KCrS_2$ | 1M KFSI in EC/DEC (1:1, v/v) | 71@ 0.05C | 68% @ 5C | 98 | 90% @ 1000 @ 1C | 1.8–3.0 | [49] |
| $V_2O_5\cdot 0.6H_2O$ | 0.8 M KPF$_6$ EC/DEC (1:1, v/v) | 224@ 50 mA/g | ≈25 mAh g$^{-1}$ @ 1 A g$^{-1}$ | 82 | 46% @ 500 @ 50 mA g$^{-1}$ | 1.5–4.0 | [50] |

### 3.3.2.1   Intercalation-Based Anodes

In these materials, K-ion intercalates into the empty space present in their structure. Generally, these materials possess superior cycling stability, good safety and rate capability. The materials are mainly carbon and titanium based.

#### 3.3.2.1.1   *Carbon-Based Anodes*

Carbon-based anodes are widely used in rechargeable alkali metal-ion (Li and K) batteries due to their fantastic electronic conductivity, high crystallinity and good stability of chemical structure for the reversible de/insertion of the guest metal ion in between the carbon layers. Graphite is one of the carbon-based anode materials which is widely used in LIBs. Graphite with expanded layer is more conducive for the de/intercalation of K (due to its larger size than Li). Besides graphite, other materials such as hard carbon, amorphous carbon, soft carbon, carbon nanotube (CNT), graphene, carbon nanofibers etc. have been studied as potential anode materials in KIBs.

*3.3.2.1.1.1 Graphite-Based Anodes*   It was observed that alkali metal ions intercalate into graphite layers and form a lamellar pattern, mainly with $Li^+$. During charging, $Li^+$ intercalates into the layers of graphite through various intermediate stages and finally forms $LiC_6$. Similar to Li, K with graphite also forms various intermediate stages during de/intercalation process. Both ex-situ and in-situ XRD analysis showed that during K intercalation into graphite, it first forms $KC_{36}$ between 0.3 and 0.2 V, then $KC_{24}$ from 0.2 to 0.1 V and finally becomes $KC_8$ at 0.01 V. During deintercalation, $KC_8$ directly becomes $KC_{36}$ at 0.3 V and then returns as low crystalline graphite above 0.5 V. Other studies suggested that the K intercalation into graphite takes place through $C \rightarrow KC_{24} \rightarrow KC_{16} \rightarrow KC_8$ [51]. To study the $K^+$ de/insertion mechanism in graphite, DFT calculations are also used. The simulation output confirms that K intercalates more easily into graphite than Li due to the fact that enthalpy of formation of $KC_8$ ($-27.5$ kJmol$^{-1}$) is lower than $LiC_6$ ($-16.5$ kJmol$^{-1}$) and also $KC_8$ ($2.0 \times 10^{-10}$ m$^2$s$^{-1}$) possesses higher diffusion coefficient than $LiC_6$ ($1.5 \times 10^{-15}$ m$^2$s$^{-1}$), which suggests a kinetically more favorable reaction [52]. The theoretical capacity of graphite in KIB is 279 mAh g$^{-1}$.

An et al. [53] for the first time used a graphite-derivative which consists of big particles of expanded graphite possessing superior conductivity and larger interlayer space to improve $K^+$ diffusion during de/intercalation. Li et al. [54] reported $FeCl_3$-intercalated expanded graphite (EG) in which $FeCl_3$ is sandwiched between graphene layers and delivers high first cycle reversible capacity of 269.5 mAh g$^{-1}$ at 50 mAg$^{-1}$. The mechanism involves $K^+$ de/intercalation in between the graphite sheets and adsorption/desorption on the graphite surface with the conversion reaction between $FeCl_3$ and $K^+$ to form nano Fe and KCl (Figure 3.9(a) and (b)).

Cao et al. [55] reported highly graphitized carbon nanocage (CNC) as a negative electrode in KIB. It possesses superior cycle stability and rate performance than mesophase graphite (MG) due to its interconnected cage arrangement (which avoids slipping of interlayers and reduces K-ion diffusion distance) and it exhibits hybrid K storage mechanism (adsorption and intercalation) (Figure 3.10 (a) and (b)).

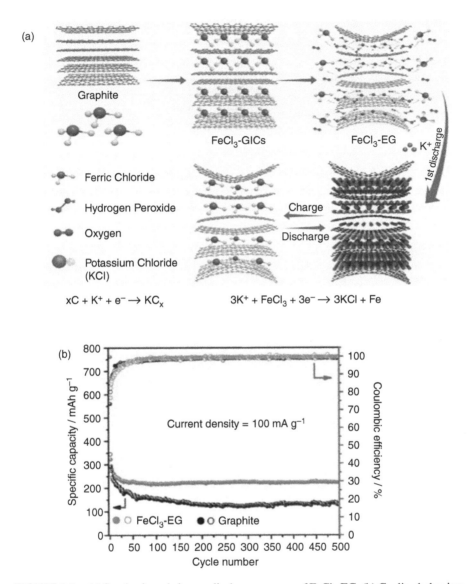

**FIGURE 3.9** (a) Synthesis and charge–discharge process of $FeCl_3$-EG. (b) Cycling behaviors of $FeCl_3$-EG and graphite at 100 mA g$^{-1}$. Reproduced with permission from Ref. 54, Copyright 2018 WILEY-VCH Verlag GmbH & Co. KGaA.

Zhang et al. [56] demonstrated a new class of N-doped, more efficient graphitic nanocarbons (GNCs) as anode. Its performance is better than nanocage and nanotubes due to the availability of defects (for de/intercalation), high N-doping and short range-order of graphite structure. Zhao et al. [57] synthesized F-doped enlarged graphite by the purification of cheap microcrystalline graphite (MG) immersed in HF solution with the composition, 98.59% MG and 1.02% fluorine. The F incorporation

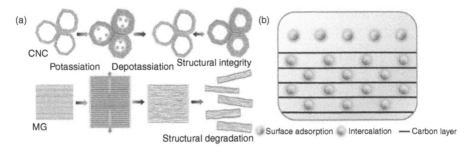

**FIGURE 3.10** (a) Structural change of carbon nanocage and mesophase graphite during cycling & (b) Schematic illustration of hybrid K storage nature of CNC. Reproduced with permission from Ref. 55, Copyright2018 WILEY-VCH Verlag GmbH & Co. KGaA.

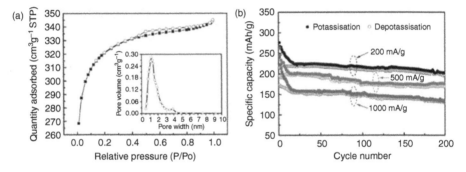

**FIGURE 3.11** (a) $N_2$ adsorption–desorption isotherm with pore volume distribution. (b) Cycling performance at different rates of PCMs. Reproduced with permission from Ref. 58, Copyright 2018 Wiley-VCH Verlag GmbH & Co. KGaA.

results in semi-ionic bond (70.53%), which improves the distance ($d_{002}$) between the MG layers from 3.356 Å to 3.461 Å and $K^+$ diffusion kinetics.

*3.3.2.1.1.2 Non-graphitic Carbon-Based Anodes* The irregularly located graphitized domains with partially or fully disordered structure possess electronic transport and ionic diffusion, which are more advantageous to improve capacity. So non-graphitic carbon materials are also well studied as negative electrode in KIBs. Non-graphitic carbon materials generally include several amorphous carbons like hard and soft carbons, and non-graphitic carbon with heteroatom doping.

Chen et al.[58] reported sulfur (S)/oxygen (O) co-doped porous hard carbon microspheres (PCM) as a superior negative electrode for boosting the performance of KIBs. It was found that the as prepared PCMs exhibit amorphous carbon property, enlarged interlayer distance of 0.393 nm (crystalline graphite ≈0.334 nm), high Brunauer–Emmett–Teller (BET) surface area of 983.2 $m^2g^{-1}$ (Figure 3.11 (a)) and unusual structural defects; these characteristics improved the K adsorption and also

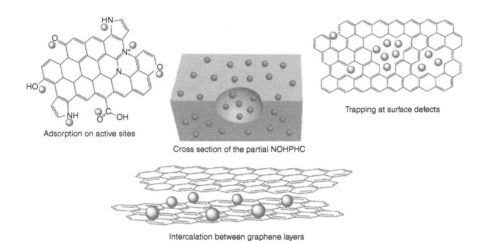

Adsorption on active sites

Cross section of the partial NOHPHC

Trapping at surface defects

Intercalation between graphene layers

**FIGURE 3.12** Mixed mechanism of K storage in NOHPHC. Reproduced with permission from Ref. 59, Copyright 2017 Wiley-VCH Verlag GmbH & Co. KGaA.

minimize the structural deformation of PCMs during K-ion intercalation process, leading to improved performance (Figure 3.11(b)).

Yang et al. [59] introduced nitrogen (N) /oxygen (O) dual-doped hierarchical porous hard carbon (NOHPHC) as anode by carbonizing and acidic treatment of $NH_2$-MIL-101(Al) precursor (aluminium-based MOF component with N and O). It possesses high reversible capacity, excellent rate performance and long cycle life due to improved interlayer distance of 0.39 nm, large surface area ($\approx 1030$ $m^2g^{-1}$) and abundant active sites/surface defects/nanovoids (Figure 3.12).

In the KIB, hard carbon is more stable than soft carbon for long cycle life applications and soft carbon possesses excellent high-rate capability than hard carbon (at 10C, soft carbon delivered 121 $mAhg^{-1}$ vs. 45 $mAhg^{-1}$ by hard carbon). So, by merging both the advantages, Jian et al. [60] designed hard carbon spheres–soft carbon (HCS-SC) composite with 20% soft carbon by ball milling (soft carbon spread into the pores of hard carbon). It delivered $\approx 200$ $mAhg^{-1}$ with 93% capacity retention after 200 charge–discharge cycles between 0.01 and 2.0 V (vs $K^+/K$) with >90% coulombic efficiency. Figure 3.13(a) and (b) show the XRD pattern and cycling performance of the composite, respectively.

### 3.3.2.1.2 Other Intercalation-Based Materials

Kishore et al. [61] synthesized potassium tetratitanate ($K_2Ti_4O_9$) by a solid-state route using potassium carbonate and $TiO_2$. It is capable of cycling up to 15 C (high rate) with a small discharge capacity, but its discharge capacity is maintained at low rates. While charging, two $Ti^{4+}$ becomes $Ti^{3+}$, which facilitates two $K^+$ insertion/formula unit ($K_2Ti_4O_9 + 2K^+ + 2\ e^- \leftrightarrow K_4Ti_4O_9$).

Li et al. [62] synthesized fluff-like HNTO (hydrogenated $Na_2Ti_3O_7$) nanowires developed on nitrogen (N)-doped CS (carbon sponge), which is a flexible negative electrode without current-collector and binder for KIBs. First-principle calculations

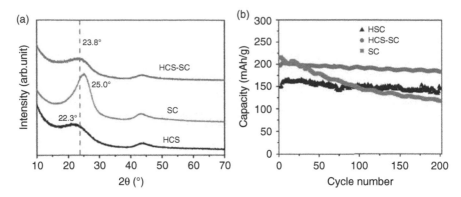

**FIGURE 3.13** (a) XRD.(b) Cycling performance at 1 C of HCS, SC and HCS-SC. Reproduced with permission from Ref. 60, Copyright 2017 Wiley-VCH Verlag GmbH & Co. KGaA.

showed that hydrogenation improves the electronic conductivity by shifting Fermi level to conduction band because Ti–OHs and oxygen vacancies are almost same as n-type doping. In addition to improving conductivity, N-doped CS also controls the aggregation of HNTO during cycling.

### 3.3.2.2   Conversion-Based Anodes

These types of anode materials possess higher theoretical capacity than the intercalation materials due to the formation of new compound with the reduction of transition metal ion. The overall reversible reaction in the conversion-type can be represented by

$$M_mA_a + (an)\ K^+ + (an)\ e^- \leftrightarrow mM + aK_nA \qquad (3.2)$$

where M = Co, Mo, Sn, Fe, Sb, etc and A = N, P, O, S, Se etc. The materials coming under this category are transition metal oxides, sulfides, selenides and phosphides. Transition metal oxides exhibit good reversibility during the conversion process; CuO nanoparticles of $\approx$20 nm delivered 342.5 mAh g$^{-1}$ at 200 mAg$^{-1}$ through conversion reactions,

$$CuO + K \rightarrow KCuO,$$

$$KCuO + K \rightarrow Cu + K_2O,$$

$$2Cu + K_2O \leftrightarrow Cu_2O + 2K\ [63] \qquad (3.3)$$

Another conversion type material, $Co_3O_4@C@MoS_2$ delivered 256 mAhg$^{-1}$ at 500 mAg$^{-1}$ with 88.3% capacity retention after 500$^{th}$ cycle mainly due to

$$MoS_2 + xK^+ + xe^- \leftrightarrow K_xMoS_2\ (>0.54V\ vs\ K/K^+),$$

$$K_xMoS_2 + (4 - x) K^+ + (4 - x) e^- \leftrightarrow Mo + K_2S \; (< 0.54V \; vs \; K/K^+). \; [64] \qquad (3.4)$$

Transition metal selenides show superior conductivity and $K^+$ diffusion kinetics than TMO and TMS, but the volume expansion during cycling need to be reduced. $Co_{0.85}Se$@carbon nanoboxes delivered 299 mAhg$^{-1}$ after 400$^{th}$ cycle at 1 Ag$^{-1}$ through the formation of potassium selenide and cobalt [65]. Transition metal phosphides have a semiconductor nature and good first cycle discharge capacity but have shortcomings such as poor electrochemical reversibility, lower conductivity and severe capacity fading. $Sn_4P_3$/C composite delivered 384.8 mAh g$^{-1}$ at 50 mA g$^{-1}$ and good rate performance of 221.9 mAh g$^{-1}$ at 1 A g$^{-1}$ through the formation of phases such as $K_4Sn_{23}$, KSn and $K_{3-x}P$ during discharge [66].

### 3.3.2.3 Alloy-Based Anodes

The common alloying mechanism in KIBs can be expressed as

$$xA + yK^+ + ye^- \leftrightarrow K_yA_x \qquad (3.5)$$

where A=alloying element (Sn, P, Sb and Bi), $K_yA_x$= final alloy product. It has some disadvantages such as fast capacity fade due to huge volume change and consumes electrolyte for the solid electrolyte interphase (SEI) reformation. Recently, new strategies have been reported to overcome these shortcomings, such as anodic protection and designing specific nanostructures. Examples of alloy-based anodes are P, Sb, Sn, Bi etc. Phosphorous forms $K_3P$ alloy and exhibits a theoretical capacity of 2596 mAh g$^{-1}$. For the practical KIBs, both red and black P are potential anodes in their composite forms. Pure forms limit the battery life as they are reactive in nature and appreciably minimize the first cycle coulombic efficiency [67]. Antimony-based anodes possess low potassiation potential and relatively high theoretical capacity ($K_3Sb$: 660 mAhg$^{-1}$), so it is fit as the most favorable anode for KIBs [68]. Tin-based anodes due to low price and high theoretical capacity ($Li_{22}Sn_4$-990 mAh g$^{-1}$; $Na_{15}Sn_4$-867 mAh g$^{-1}$) are important alloy-based anode for both LIBs and NIBs, but it showed high volume expansion ($\approx$420% in LIBs and $\approx$260% in NIBs), resulting in huge structural damage and rapid capacity decay [69]. Bismuth can also form $K_3Bi$ alloy with K and delivered 385 mAh g$^{-1}$ but possesses low coulombic efficiency and fast capacity fading [70].

Electrochemical performances of various anode materials explored for KIB are summarized in Table 3.3.

### 3.3.3 Electrolytes

Electrolytes are the essential components which act as medium for the movement of ions between the electrodes and play an important role in the electrochemical process. SEI is a passive layer formed on the surface of electrodes due to the reduction of electrolyte (irreversible process) and due to its high thickness in the KIBs, it resists the flow of ions. To improve the SEI, ionic conductivity needs to be enhanced and certain

TABLE 3.3
Electrochemical Performance of Various Anode Materials

| Anode Material | Electrolyte | Capacity (mAh/g) @ rate | Rate Performance (mAh g⁻¹) @ rate | First Cycle Coulombic Efficiency (%) | Capacity Retention @cycles @rate | Voltage Window (V) | Ref. |
|---|---|---|---|---|---|---|---|
| Expanded graphite | 1 M KFSI in EC/DEC (1:1, v/v) | 263 @ 10 mA/g | 175 @ 0.2 A g⁻¹ | 81.56 | ~97% @500 @ 0.2 A g⁻¹ | 0.01–3 | [53] |
| FeCl₃ intercalated expanded graphite | 0.8 M KPF₆ in EC/DEC (1:1, v/v) | 259.5@ 100 mA/g | 133.1@ 5 A g⁻¹ | 34.06 | 70.38% @ 1300 @ 2 A g⁻¹ | 0.01–3 | [54] |
| Graphitic C hollow Nanocage | 1 M KFSI in EC:PC (1:1, v/v) | 212 @ 0.2 C | 40@5C | 40 | 92% @ 100 @ 0.2 C | 0.01–3 | [55] |
| N-doped, GNC (graphitic nanocarbons) | 0.8 M KPF₆ in EC/DEC (1:1, v/v) | 280@ 50 mA/g | 152@ 1 A g⁻¹ | - | 75.6% @ 200 @ 200m A g⁻¹ | 0.8–2.0 | [56] |
| F-doped graphite | 0.8 M KPF₆ in EC/DEC (1:1, v/v) | 303 @ 100 mA/g | 126 @ 600 mA g⁻¹ | 44.1 | 74.6% @100 @0.1 A g⁻¹ | 0.01–3 | [57] |
| S/O-doped Hard Carbon Microspheres | 0.8 M KPF₆ in EC/DEC (1:1, v/v) | 226.6 @ 50 mA/g | 158 @ 1 A g⁻¹ | 61.7 | ~68% @ 2000 @1 A g⁻¹ | 0.01–2.5 | [58] |
| N/O dual doped porous hard carbon | 1.0 M KPF₆ in EC/DMC (1:1, v/v) | 365 @ 25 mA/g | 118 @3000 mA g⁻¹ | 25 | 69.5% @ 1100@ 1050 mA g⁻¹ | 0.001–3.0 | [59] |
| Hard-soft Composite Carbon | 0.8 M KPF₆ in EC/DEC (1:1, v/v) | 230 @ 0.5 C | 45 @ 10 C | 67 | 93% @200 @1 C | 0.01–2 | [60] |
| K₂Ti₄O₉ | 1 M KPF₆ in PC (1:1, v/v) | 80@ 100 mA/g | 97@ 30 mA g⁻¹ | 20 | 43.75% @ 30@ 100 mA g⁻¹ | 0.01–2.5 | [61] |
| HNTO/CS (hydrogenated Na₂Ti₃O₇/ carbon sponge) | 1 M KPF₆ in EC/DEC (1:1, v/v) | 101 @ 100 mA/g | 25 @ 1 A g⁻¹ | 20 | 82.5% @ 1555@ 100 mA g⁻¹ | 0.01–2.6 | [62] |

additives are added to form stable SEI. New electrolytes should minimize dendrite formation and unwanted reactions. Among the various K salts, KFSI possesses more conductivity than $KPF_6$ but the uniformity of composition is a concern. Ether-based solvents (DME, diglyme, etc.) can form stable SEI and exhibit high initial coulombic efficiency due to powerful chemical adsorption and superior charge-transfer rate than carbonate-based solvents (PC, DEC etc). Polymer electrolytes, K-IL (ionic liquid) and blending with few binary/ternary solvent support stable SEI formation and minimize dendrite formation. Additionally, SEI formation with the electrolyte additive (such as FEC) also minimizes the unwanted reactions and makes it more stable.

### 3.3.4 BINDER AND SEPARATOR

The electrode construction technology is a major challenge for alkali metal-ion batteries. In LIBs, polyvinylidene fluoride (PVDF) is widely used as the binder, but it can't sustain the volume change during de/potassiation, which can result in the separation of electrode materials from the current collector (exfoliation), leading to capacity fading and inferior electronic conductivity. So, future works are to be focused on the binders which possess superior elastic nature and improve the mechanical strength of the electrodes. Separators also have to be developed by modifying the sides which face the electrodes for the optimization of their performance by minimizing the dendrite development and side reactions.

## 3.4   STUDIES IN FULL CELL

The full cells are reported in literature in the form of coin cell (CR2025 and CR2032) and used to power compact portable devices such as LED. From the literature, the commonly reported positive electrodes for KIB full-cell are organic-based, polyanionic compounds, PB and layered TMO, which are matched with negative electrodes, containing carbon, metallic alloy-based and few other materials. For the better performance of full cell, most of the cathodes and anodes must be prepotassiated with K in the half-cells. The KIB possesses low operating potential, high price of cathodes and poor cycle life, restricting its practical applications [71]. Due to various technical issues, cathode development is slower than anode [72]. The performance and the development of the full-cell KIBs for commercial applications are directly determined by the cathode materials [71]. The electrochemical performance reported for full cell is consolidated in Table 3.4.

## 3.5   SAFETY ASPECTS

KIBs have a lower susceptibility toward thermal runway and liberate less heat than LIBs. Inflammable organic electrolytes and thermally less stable SEI are responsible for the safety issues. By using thermally stable separators, solid-state (SS) electrolytes, flame retarding agents and polymer electrolytes, the safety of KIBs can be improved. By studying the practical and theoretical aspects, an efficient model needs to be constructed. This model should minimize the complex thermal runaway and accelerate the construction of superior-safety KIBs.

**TABLE 3.4**
**Electrochemical Performance of KIB Full Cells**

| Cathode | Anode | Electrolyte | Initial Capacity (mAh g⁻¹)@ rate | Capacity Retention@ cycles@ rate | Voltage Window (V) | Ref. |
|---|---|---|---|---|---|---|
| KPB (pottassiated Prussian blue) | Graphite | 1 M $KPF_6$ in Diglyme | 110@1C | 60% @2000 @2 A g⁻¹ | 1.5-4 | [73] |
| $K_{1.69}Mn[Fe(CN)]_{6-0.85} \cdot 0.4H_2O$ | Hard Carbon | 1.5 M $KPF_6$ in diglyme (G2) | 121 @0.6 C | 86%@300 @0.6 C | 1.5-4.5 | [74] |
| $K_{1.92}Fe[Fe(CN)]_{6-0.94} \cdot 0.5H_2O$ | $K_2TP$ (dipotassium terephthalate) @CNT | 1 M $KPF_6$ in DME | 99 @ 0.5 C | 90%@ 60 @ 0.5 C | 1.5-3.8 | [75] |
| P2-$K_{0.44}Ni_{0.22}Mn_{0.78}O_2$ | Soft Carbon | 0.8 M $KPF_6$ in EC:DEC (1:1) | 78.9 @50 mAg⁻¹ | 90%@ 500 @50 mA g⁻¹ | 0.5-3.5 | [76] |
| $K_{0.77}MnO_2 \cdot 0.23H_2O$ | Hard-Soft Carbon | 0.8 M $KPF_6$ in EC:DEC (1:1) | 121 @ 100 mA g⁻¹ | 80%@ 500@ 1000 mAh g⁻¹ | 0.5-3.5 | [77] |
| KVO (Potassium vanadate)@rGO (reduced graphene oxide) | $V_2O_{3-x}$ @rGO | 2M KFSI in EC:DEC (1:1) | 68 @ 20 mAg⁻¹ | 75%@250 cycles @ (0.1 A g⁻¹) | 1.5-3.1 | [78] |
| Na2AQ26DS (sodium anthraquinone-2,6-disulfonate) | Graphite | 1 M $KPF_6$ in DME | 82 @50 mA g⁻¹ | 55%@ 250 @ 0.1 A g⁻¹ | 0.5-3.0 | [79] |
| PTCDI-DAQ (N,N'-bis(2-anthraquinone)]-perylene-3,4,9,10-tetracarboxydiimide) | $K_4TP$ (potassium terephthalate) | 1 M $KPF_6$ in DME | 213 @ 100 mA g⁻¹ | 50%@ 10000 @ 3 A g⁻¹ | 0.2-3.2 | [80] |
| PTCDA (3,4,9,10-perylene-tetracarboxylic acid-dianhydride) | Soft Carbon | 3 M KFSI in DME | 116 @ 50 mA g⁻¹ | 65%@ 3000 @ 0.6 A g⁻¹ | 0.5-3.5 | [81] |

## 3.6 AQUEOUS K-ION BATTERY

Aqueous alkali metal ion batteries possess some favorable characteristics than non-aqueous batteries, such as environmentally friendly components and more safe systems. The present studies on aqueous LIBs and NIBs are mainly focused toward their intercalation components. For aqueous LIBs, natural abundance of Li is a concern and for aqueous NIBs capacity is a concern. K is relatively 1,000 times more abundant than Li and the minimum Stokes radius in water is responsible for the superior transportation properties in the aqueous systems. So far, very few studies are reported on aqueous KIBs due to the giant ionic size and huge ionization potential of K; it has limited electrode materials and possess low energy density because of its low operating voltage (<1.2V)

Pasta et al.[82] studied Prussian blue with its analogous (PBAs) of open frameworks such as copper hexacyanoferrate (CuHCF) and nickel hexacyanoferrate ($K_{0.6}Ni_{1.2}Fe(CN)_6 \cdot 3.6H_2O$) with noticeable performance. But these components have less K content and from each formula one electron transfer through $Fe^{3+}/Fe^{2+}$ redox pair takes place and then the capacity becomes <70 mAhg$^{-1}$ [83]. Some PBAs such as potassium iron (II) hexacyanoferrate ($K_2Fe^{II}[Fe^{II}(CN)_6] \cdot 2H_2O$) can store two K atoms, so multi-electron transport takes place [84], which could deliver ≈120 mAhg$^{-1}$. During cycling, it maintained the structural stability and improved the kinetics. Nickel ferrocyanide (II) ($K_2Ni^{II}[Fe^{II}(CN)_6] \cdot 1.2H_2O$) maintained ≈59 mAhg$^{-1}$ (98.6% capacity retention) after 5000 cycles at 30C with ≈100% coulombic efficiency and no noticeable volume change (≈7.7%) during the K de/intercalation at the sites of C-$Fe^{3+}$ /$Fe^{2+}$ [85].

## 3.7 ALL SOLID-STATE KIBS

In most of the traditional secondary alkali metal-ion batteries, liquid electrolyte connects the electrodes and improves the connected area, but inflammable organic electrolytes (liquid) can elevate the risk factor. Hence attention has been paid toward solid-state secondary alkali metal-ion batteries. All solid electrolytes can be divided into organic-based (solid polymer electrolyte-SPE), inorganic-based (inorganic solid electrolyte-ISE) and hybrid or composite (organic and inorganic) electrolytes. All SPEs possess superior performance in LIBs due to broad electrochemical stability range, flexibility, excellent safety, good thermal stability and no leakage. The SEI formed with the liquid electrolyte could break and reform but with solid electrolytes the formed SEI is stable, difficult to break, minimize the unwanted reactions and improve the cycling performance. Currently, there is a need to improve ionic conductivity and reduce the resistance of charge-transfer between the electrode and the electrolyte.

Fei et al. [86] assembled 3,4,9,10-perylene-tetracarboxylicacid-dianhydride (PTCDA) organic cathode with K anode and poly (propylene carbonate) based solid polymer electrolyte with cellulose non-woven backbone (PPCB-SPE) having ionic conductivity of 1.36 × 10$^{-5}$ Scm$^{-1}$ at 20 °C. It delivered 118 mAhg$^{-1}$ at 20 mAg$^{-1}$ and maintained 91.17 mAhg$^{-1}$ over 40$^{th}$ cycle.

Fie et al. [87] demonstrated $Ni_3S_2$@Ni/poly(ethylene oxide)-potassium bis(flourosulfonyl)imide (PEO—50% KFSI)/K cell, which delivered 312 mAhg$^{-1}$ at

25 mAg$^{-1}$ at 55°C with 97% coulombic efficiency. The ionic conductivity of the SPE is $1.14 \times 10^{-5}$ Scm$^{-1}$ at 40°C and at 60°C the ionic conductivity is $2.7 \times 10^{-4}$ Scm$^{-1}$.

## 3.8 POTASSIUM DUAL-ION BATTERY

Dual-ion batteries (DIBs) present stunning advantages such as cost effectiveness, eco-friendliness, excellent safety and high operating voltage [88]. DIB generally consists of cathode and anode of same material, usually graphite, thus eliminating the necessity for high cost and non-green transition metal compounds. Unlike the other secondary alkali metal ions batteries, the working mechanism of DIB is quite different. Here both the ions of the electrolyte must be electrochemically active so as to get inserted simultaneously into the respective electrodes. During charging, the cations (such as Li$^+$, Na$^+$, K$^+$) and anions such as BF$_4^-$ and PF$_6^-$ in the electrolyte get intercalated into the anode and cathode respectively and during discharging, the intercalated anions and cations leave their host matrix and get dissolved in the electrolyte [89]. Recently, as a part of improvising KIB performance, researchers have extended the concept of dual-ion battery to KIBs and designated such combination as KDIBs.

### 3.8.1 CATHODE MATERIALS FOR KDIBs

Graphite is the widely known cathode material for KDIBs. However, recent results suggest that KDIBs made from graphite cathode display very poor capacity and cycling performance. Therefore, it is being replaced by some other materials. Organic compounds are nowadays identified as a suitable candidate for KDIBs. They possess advantages like low cost, renewability, large-scale production and controllable synthesis process. Polytriphenylamine (PTPA) and poly(N-vinylcarbazole) (PVK) are few such examples. As this compound contains positively charged N atom in carbazole ligand, it facilitates fast insertion of PF$_6^-$ ions from the electrolyte [90].

### 3.8.2 ANODE MATERIALS FOR KDIBs

From the earlier period onward, carbonaceous materials have been used as an anode for the dual ion system. However, in pristine form, it is not recommended for KDIBs, because the insertion of heavy K$^+$ causes the exfoliation of host materials and lead to sudden capacity decay. Doping is one of the effective ways to improve the electrochemical performance of the carbonaceous materials. Doping can not only create enough free space to accommodate the volume change upon cycling but also supply intrinsic electronic pathways and well-organized ion transport channels [90]. Nowadays, organic compounds are also identified to exhibit sufficient active centers and flexible structural designability, which enable them to show excellent cycling stability and high theoretical specific capacity. Many organic materials such as potassium Prussian blue, 3,4,9,10-perylene-tetracarboxylic acid-dianhydride, dyes, poly(anthraquinonyl sulfide), sodium rhodizonate dibasic, metal–organic frameworks and terephthalate compounds have been studied for cationic storage [91].

## 3.9  CONCLUSION

To fulfil the energy storage demand of modern society, it is necessary to widen the search for an innovative energy storage system. KIBs are one among the advanced technologies. Nonetheless, in order to realize commercialization, KIBs have to conquer some challenges, which include poor gravimetric/volumetric energy density and K$^+$-ion induced electrode exfoliation. Hence, it is vital to explore the appropriate electrode as well as electrolyte materials to meet their demand in the market. In light of previous studies, organic electrode materials are recommended to be the most appropriate candidate, as it contains flexible molecular structures for the fast and faultless reversible insertion/extraction of K$^+$ ion. KIBs can cope with all-solid state as well as aqueous based electrolytes. It can also experience the benefit of dual ion system.

## ACKNOWLEDGMENT

The authors thank the Director, Vikram Sarabhai Space Centre, Thiruvananthapuram, for granting permission to publish the chapter.

## REFERENCES

1.  M. Armand, J. M. Tarascon. Building better batteries. *Nature* 2008, *451*, 652.
2.  J. M. Tarascon, M. Armand. Issues and challenges facing rechargeable lithium batteries. *Nature* 2001, *414*, 359.
3.  B. Scrosati, J. Hassoun, Y-K. Sun. Lithium-ion batteries. A look into the future. *Energy Environ Sci.* 2011, *4*, 3287.
4.  C. Vaalma, D. Buchholz, M. Weil, S. Passerini. A cost and resource analysis of sodium-ion batteries. *Nat. Rev. Mater.* 2018, *3*, 18013.
5.  M. Gregory LaRocca. Global value chains: Lithium in lithium-ion batteries for electric vehicles2020, www.usitc.gov/publications/332/working_papers/no_id_069_gvc_lith ium-ion_batteries_electric_vehicles_final_compliant.pdf
6.  A. Eftekhari. Lithium batteries for electric vehicles: from economy to research strategy. *ACS Sustainable Chem Eng.* 2019, *7*(6), 5602.
7.  T. C. Wanger. The lithium future—resources, recycling, and the environment. *Conserv. Lett.* 2011, *4*(3), 202.
8.  V. Anoopkumar, B. John, T.D. Mercy. Potassium ion batteries: Key to future large scale energy storage *ACS Appl. Ener. Mat.* 2020, *3*(10), 9478.
9.  Wen Li, Zimo Bi, Wenxin Zhang, Jian wang, Ranjusha Rajagopalan, Qiujun Wang, Di Zhang, Zhaojin Li, Haiyan Wang and Bo Wang. Advanced cathodes for potassioum-ion batteries with layeted transition metal oxides: A Review. *J. Mater. Chem.* A. 2021, *9*, 8221–8247.
10. Ranjusha Rajagopalan, Yougen Tang, Xiaobo Ji, Chuankun Jia, Haiyan Wang. Advancements and challenges in potassium ion batteries: A comprehensive review. *Adv. Funct. Mat.* 2020, *30*(12), 1909486.
11. J. Zhang, T. Liu, X. Cheng, M. Xia, R. Zheng, N. Peng, H. Yu, M. Shui, J. Shu. Development status and future prospect of non-aqueous potassium ion batteries for large scale energy storage. *Nano Energy.* 2019, *60*, 340–61.
12. K. K. Turekian, K. H. Wedepohl. Distribution of the elements in some major units of the earth'scrust. *GeolSocAmBull.* 1961, *72*, 175–192

13. Y. H. Zhu, X. Yang, T. Sun, S. Wang, Y. L. Zhao, J. M. Yan, X. B. Zhang. Recent progresses and prospects of cathode materials for non-aqueous potassium ion batteries. *Electrochem Energy Rev.* 2018, *1*, 548–66.
14. G. Martin, L. Rentsch, M. H€ock, M. Bertau. Lithium market research—global supply, future demand and price development. *Energy Storage Mater.* 2017, *6*, 171–9.
15. Kei Kubota, Mouad Dahbi, Tomooki Hosaka, Shinichi Kumakura, Shinichi Komaba. Towards K-ion and Na-ion batteries as "beyond Li-ion". *Chem Rec.* 2018, *18* (4),459–79.
16. Jian, Z, Luo, W., Ji, X., Carbon electrodes for K-ion batteries *J. Am. Chem. Soc.* 2015, *137*, 11566–11569.
17. T. Hosaka, K. Kubota, A. S. Hameed, S. Komaba. Research development on potassium-ion batteries. *Chem. Rev.* 2020, *120* (14), 6358–6466.
18. S. Komaba, T. Hasegawa, M. Dahbi, K. Kubota. Potassium intercalation into graphite to realize high-voltage/high-power potassium-ion batteries and potassium-ion capacitors. *Electrochem Commun.* 2015, *60*, 172.
19. M. Niro, U. Kisaburo, T. Zen'ichi. Standard potential of alkali metals, dilver, thallium metal/ion couples in DMF, DMSO and PC. *Bull. Chem. Soc. Japan* 1974, *47*(4), 813.
20. Y. Marcus. Thermodynamic functions of transfer of single ions from water to non-aqueous and mixed solvents. *Pure Appl. Chem.* 1985, *57*, 1129.
21. J. Zhao, X. Zou, Y. Zhu, Y. Xu, C. Wang. Electrochemical intercalation of potassium into graphite. *Adv. Funct. Mater.* 2016, *26*, 8103.
22. L. Xue, H. Gao, W. Zhou, S. Xin, K. Park, Y. Li, J. B. Goodenough. Liquid K–Na alloy anode enables dendrite-free potassium batteries. *Adv Mater.* 2016, *28*, 9608–12.
23. J. C. Pramudita, D. Sehrawat, D. Goonetilleke, N. Sharma. An initial review of the status of electrode materials for potassium-ion batteries, *Adv. Energy Mater.* 2017, *7*(24), 1602911.
24. S. M. Ahmed, G. Suo, W. A. Wang, K. Xi, S. B. Iqbal. Improvement in KIBs electrodes: Recent developments and efficient approaches. *J. Ener. Chem.* 2021, *62*, 307.
25. R. Zhang, J. Bao, Y. Wang, C.F. Sun. Concentrated electrolytes stabilize bismuth potassium batteries. *Chem. Sci.* 2018, *9*, 6193.
26. Y. Chen, W. Luo, M. Carter, L. Zhou, J. Dai, K. Fu, S. Lacey, T. Li, J.Wan, X. Han, Y. Bao, L. Hu. Organic electrode for non-aqueous KIBs. *Nano Energy.* 2015, *18*, 205.
27. K. Mizushima, P. Jones, P. Wiseman, J. B. Goodenough. LixCoO$_2$ (0<x<–1): A new cathode material for batteries of high energy density. *Mater. Res. Bull.* 1980, *15*, 783.
28. C. Vaalma, G. A. Giffin, D. Buchholz, S. Passerini. Non-aqueous K-ion battery based on layered K$_{0.3}$MnO$_2$ cathode. *J. Electrochem. Soc.* 2016, *163*, A1295.
29. H. Kim, D. H. Seo, J. C. Kim, S. H. Bo, L. Liu, T. Shi, G. Ceder . Investigation of potassium storage in layered P3-Type K$_{0.5}$MnO$_2$ cathode. *Adv. Mater.* 2017, *29*, 1702480.
30. J. U. Choi, Y. J. Park, J. H. Jo, Y. H. Jung, D. C. Ahn, T. Y. Jeon, K. S. Lee, H. Kim, S. Lee, J. Kim, S. T. Myung. An optimized approach towards high energy density cathode materials in K-ion batteries. *Ener. Stor. Mater.* 2020, *27*, 342.
31. J. H. Jo, J. U. Choi, Y. J. Park, Y. H. Jung, D. Ahn, T. Y. Jeon, H. Kim, J. Kim S. T. Myung. P2-K$_{0.75}$[Ni$_{1/3}$Mn$_{2/3}$]O$_2$cathode material for high power and long life KIBs. *Adv. Energy Mater.* 2020, *10*, 1903605.
32. Q. Y. Yu, J. Hu, W. Wang, Y. Li, G. Q. Suo, L. P. Zhang, K. Xi, F. L. Lai, D. N. Fang. K$_{0.6}$CoO$_{2-x}$N$_x$ porous nanoframe: A co-enhanced ionic and electronic transmission for K-ion batteries. *Chem. Eng. J.* 2020, *396*, 125218.

33. T. Masese, K. Yoshii, M. Kato, K. Kubota, Z. D. Huang, H. Senoh, M. Shikano. A high voltage honeycomb layered cathode framework for rechargeable K-ion battery: P2-type $K_{2/3}Ni_{1/3}Co_{1/3}Te_{1/3}O_2$. *Chem. Commun. (Camb).* 2019, *55*, 985.

34. J. H. Jo, J. Y. Hwang, J. U. Choi, H. J. Kim, Y. K. Sun, S. T. Myung. Potassium vanadate as a new cathode material for K-ion batteries. *J. Power Sources.* 2019, *432*, 24.

35. K. Yuan, R. Ning, M. Bai, N. Hu, K. Zhang, J. Gu, Q. Li, Y. Huang, C. Shen, K. Xie. Prepotassiated $V_2O_5$ as the cathode material for high-voltage K-ion batteries. *Energy Technology.* 2020, *8*, 1900796.

36. H. Kim, D. H. Seo, A. Urban, J. Lee, D. H. Kwon, S. H. Bo, T. Shi, J. K. Papp, B. D. McCloskey, G. Ceder. Stoichiometric layered potassium transition metal oxide for rechargeable K-ion batteries. *Chem. Mater.* 2018, *30*, 6532.

37. N. Naveen, S. C. Han, S. P. Singh, D. Alan, K. S. Sohn, M. Pyo. Highly stable P'3-$K_{0.8}CrO_2$ cathode with limited dimensional changes for K-ion batteries. *J. Power Sources* 2019, *430*, 137.

38. J. –Y. Hwang, J. Kim, T.-Y. Yu, S.-T. Myung, Y.-K. Sun. Development of P3- $K_{0.6}9CrO_2$ as an ultra-high-performance cathode material for K-ion batteries. *Energy Environ. Sci.* 2018, *11*, 2821.

39. S. Zheng, S. Cheng, S. Xiao, L. Hu, Z. Chen, B. Huang, Q. Liu, J. Yang, Q. Chen. Partial replacement of K by Rb to improve electrochemical performance of $K_3V_2(PO_4)_3$ cathode material for K-ion batteries. *J. Alloys Comp.* 2020, *815*, 152379.

40. H. Park, H. Kim, W. Ko, J. H. Jo, Y. Lee, J. Kang, I. Park, S.-T. Myung, J. Kim. Development of $K_4Fe_3(PO_4)_2(P_2O_7)$ as a novel Fe-based cathode with high energy densities and excellent cyclability in rechargeable K-ion batteries. *Ener. Stor. Mat.* 2020, *28*, 47.

41. J. Liao, Q. Hu, X. He, J. Mu, J. Wang, C. Chen. A long lifespan of K-ion full battery based on $KVPO_4F$ cathode and $VPO_4$ anode. *J Power Sources.* 2020, *451*, 227739.

42. J. Keggin, F. Miles. Structure and formulae of the Prussian blues and related compounds. *Nature*, 1936, *137*, 577.

43. M. B. Robin. The color and electronic configurations of Prussian blue. *Inorg. Chem.* 1962, *1*(2), 337.

44. A. Eftekhari, Potassium secondary cell based on Prussian blue cathode. *J. Power Sources.* 2004, *126*, 221.

45. C. Zhang, Y. Xu, M. Zhou, L. Liang, H. Dong, M. Wu, Y. Yang, Y. Lei, Potassium Prussian blue nano particles: A low-cost cathode material for potassium-ion batteries. *Adv. Funct. Mater.* 2017, *27*, 1604307.

46. Y. Chen, W. Luo, M. Carter, L. Zhou, J. Dai, K. Fu, S. Lacey, T. Li, J. Wan, X. Han, Y. Bao, L. Hu. Organic electrode for non-aqueous K-ion batteries. *Nano Energy.* 2015, *18*, 205.

47. A. F. Obrezkov, V. Ramezankhani, I. Zhidkov, V. F. Traven, E. Z. Kurmaev, K. J. Stevenson, P. A. Troshin. High-energy and high-power-density K-ion batteries using dihydrophenazine-based polymer as active cathode material. *J. Phys. Chem. Lett.* 2019, *10*(18), 5440.

48. L. Wang, J. Zou, S. Chen, G. Zhou, J. Bai, P. Gao, Y. Wang, X. Yu, J. Li, Y-S. Hu, H. Li. $TiS_2$ as a high performance KIB cathode in ether-based electrolyte. *Energy Storage Mater.* 2018, *12*, 216.

49. N. Naveen, W. B. Park, S. P. Singh, S. C. Han, D. Ahn, K-S. Sohn, M. Pyo. $KCrS_2$ cathode with considerable cyclability and high rate performance: The first $K^+$ stoichiometric layered compound for KIBs. *Small.* 2018, *14*(49),e1803495.

50. B. Tian, W. Tang, C. Su, Y. Li. Reticular $V_2O_5.0.6H_2O$ xerogel as cathode for rechargeable KIBs. *ACS Appl Mater Interfaces.* 2018, *10*, 642.

51. W. Luo, J. Wan, B. Ozdemir, W. Bao, Y. Chen, J. Dai, H. Lin, Y. Xu, F. Gu, V. Barone, L. Hu. Potassium ion batteries with graphitic materials. *Nano Lett.* 2015, *15*, 7671.

52. Z. Wang, A. P. Ratvik, T. Grande, S. M. Selbach. Diffusion of alkali metals in the first stage graphite intercalation compounds by vdW-DFT calculations. *RSC Adv.* 2015, *5*, 15985.

53. Y. An, H. Fei, G. Zeng, L. Ci, B. Xi, S. Xiong, J. Feng. Commercial expanded graphite as a low-cost, long-cycling life anode for KIBs with conventional carbonate electrolyte. *J Power Sources.* 2018, *378*, 66.

54. D. Li, M. Zhu, L. Chen, L. Chen, W. Zhai, Q. Ai, G. Hou, Q. Sun, Y. Liu, Z. Liang, S. Guo, J. Lou, P. Si, J. Feng, L. Zhang, L. Ci. Sandwich-like FeCl3@C as high-performance anode materials for KIBs. *Adv Mater Interfaces.* 2018, *5*, 1800606.

55. B. Cao, Q. Zhang, H. Liu, B. Xu, S. Zhang, T. Zhou, J. Mao, W. K. Pang, Z. Guo, A. Li, J. Zhou, X. Chen, H. Song. Graphitic carbon nanocage as a stable and high power anode for KIBs. *Adv. Energy, Mater.* 2018, *8*, 1801149.

56. W. Zhang, J. Ming, W. Zhao, X. Dong, M. N. Hedhili, P. M. F. J. Costa, H. N. Alshareef. Graphitic nanocarbon with engineered defects or high-performance K-ion battery anodes. *Adv. Funct. Mater.* 2019, *29*(35), 1903641.

57. Y. Zhao, L. Yang, C. Ma, G. Han. One-step fabrication of fluorine-doped graphite deriving from low-grade microcrystalline graphite ore for KIBs. *Energy & Fuels.* 2020, *34*(7), 8993.

58. M. Chen, W. Wang, X. Liang, S. Gong, J. Liu, Q. Wang, S. Guo, H. Yang. Sulfur/oxygen codoped porous hard carbon microspheres for high-performance KIBs. *Adv Energy Mater.* 2018, *8*(19), 1800171.

59. J. Yang, Z. Ju, Y. Jiang, Z. Xing, B. Xi, J. Feng, S. Xiong. Enhanced capacity and rate capability of nitrogen/oxygen dual-doped hard carbon in capacitive potassium-ion storage. *Adv Mater.* 2018, *30*, 1700104.

60. Z. Jian, S. Hwang, Z. Li, A. S. Hernandez, X. Wang, Z. Xing, D. Su, X. Ji. Hard-soft composite carbon as a long-cycling and high-rate anode for KIBs. *AdvFunct Mater.* 2017, *27*(26), 1700324.

61. B. Kishore, G. Venkatesh, N. Munichandraiah. $K_2Ti_4O_9$: A promising anode material for KIBs. *J. Electrochem. Soc.* 2016, *163*(13), A2551.

62. P. Li, W. Wang, S. Gong, F. Lv, H. Huang, M. Luo, Y. Yang, C. Yang, J. Zhou, C. Qian, B. Wang, Q. Wang, S. Guo. Hydrogenated $Na_2Ti_3O_7$ epitaxially grown on flexible N-doped carbon sponge for KIBs. *ACS Appl. Mater. Interfaces* 2018, *10*, 37974.

63. K. Cao, H. Liu, W. Li, Q. Han, Z. Zhang, K. Huang, Q. Jing, L. Jiao. CuO nanoplates for high-performance KIBs. *Small.* 2019, *15*(36), e1901775.

64. G. Qin, Y. Liu, P. Han, F. Liu, Q. Yang, C. Wang. Dispersed $MoS_2$ nanosheets in core shell $Co_3O_4$@C nanocubes for superior K-ion storage. *Appl. Surface Sci.* 2020, *514*, 145946.

65. C. A. Etogo, H. Huang, H. Hong, G. Liu, L. Zhang. Metal-organic-frameworks-engaged formation of $Co_{0.85}Se$@C nanoboxes embedded in carbon nanofibers film for enhanced K-ion storage. *Energy Storage Mater.* 2020, *24*, 167.

66. W. Zhang, J. Mao, S. Li, Z. Chen, Z. Guo. Phosphorus-based alloy materials for advanced potassium-ion battery anode. *J. Am. Chem. Soc.* 2017, *139*(9), 3316.

67. C. Liu, X. Han, Y. Cao, S. Zhang, Y. Zhang, J. Sun. Topological construction of phosphorus and carbon composite and its application in energy storage. *Energy Storage Mater.* 2019, *20*, 343.

68. J. Zheng, Y. Yang, X. Fan, G. Ji, X. Ji, H. Wang, S. Hou, M.R. Zachariah, C. Wang. Extremely stable antimony-carbon composite for KIBs. *Energy Environ. Sci.* 2019, *12*, 615.

69. I. Sultana, T. Ramireddy, M. M. Rahman, Y. Chen, A. M. Glushenkov. Tin-based composite anodes for KIBs. *Chem. Commun. (Camb)*. 2016, *52*, 9279.

70. A. Wang, W. Hong, L. Yang, Y. Tian, X. Qiu, G. Zou, H. Hou, X. Ji. Bi-based electrode materials for alkali metal-ion batteries. *Small*. 2020, *16*(48), 2004022.

71. V. Etacheri, R. Marom, R. Elazari, G. Salitra, D. Aurbach. Challenges in the development of advanced LIBs. *Energ Environ Sci*. 2011, *4*, 3243.

72. M. Li, J. Lu, Z. Chen, K. Amine. 30 years of LIBs. *Adv Mater*. 2018, *30*(33), 1800561.

73. K. Moyer, J. Donohue, N. Ramanna, A. P. Cohn, N. Muralidharan, J. Eaves, C. L. Pint. High-rate K-ion and Na-ion batteries by co-intercalation anodes and open framework cathodes. *Nanoscale*, 2018, *10*, 13335.

74. N. S. Katorova, S. S. Fedotov, D. P. Rupasov, N. D. Luchinin, B. Delattre, Y.-M. Chiang, A. M. Abakumov, K. J. Stevenson. Effect of concentrated diglyme-based electrolytes on the electrochemical performance of KIBs. *ACS Appl. Energ. Mater.* 2019, *2*(8), 6051.

75. J. Liao, Q. Hu, Y. Yu, H. Wang, Z. Tang, Z. Wen, C. Chen. A potassium-rich iron hexacyanoferrate/dipotassium terephthalate@carbon nanotube compositeused for K-ion full-cells with an optimized electrolyte. *J Mater Chem A*. 2017, *5* 19017.

76. X. Zhang, Y. Yang, X. Qu, Z. Wei, G. Sun, K. Zheng, H. Yu, F. Du. Layered P2-Type $K_{0.44}Ni_{0.22}Mn_{0.78}O_2$ as a high-performance cathode or KIBs. *Adv. Funct. Mater.* 2019, *29*(49), 1905679.

77. B. Lin, X. Zhu, L. Fang, X. Liu, S. Li, T. Zhai, L. Xue, Q. Guo, J. Xu, H. Xia. Birnessite nanosheet arrays with high K content as a high-capacity and ultrastable cathode for KIBs. *Adv. Mat.* 2019, *31*(24), 1900060.

78. Z. Tong, R. Yang, S. Wu, D. Shen, T. Jiao, K. Zhang, W. Zhang, C.S. Lee. Defect-engineered vanadium trioxide nanofiber bundle @ graphene hybrids for high-performance all vanadate NIBs and KIBs. *J. Mater. Chem. A*. 2019, *7*, 19581.

79. D. Li, W. Tang, C. Wang, C. Fan. A polyanionic organic cathode for highly efficient K-ion full batteries. *Electrochem. Commun*. 2019, *105*, 106509.

80. Y. Hu, W. Tang, Q. Yu, X. Wang, W. Liu, J. Hu, C. Fan. Novel insoluble organic cathodes for advanced organic KIBs. *Adv. Funct. Mater*. 2020, *30*(17), 2000675.

81. L. Fan, R. Ma, J. Wang, H. Yang, B. Lu. An ultrafast and highly stable potassium-organic battery. *Adv. Mat*. 2018, *30*(51), 1805486.

82. M. Pasta, C. D. Wessells, R.A. Huggins, Y. Cui. A high-rate and long cycle life aqueous electrolyte battery for grid-scale energy storage. *Nat. Commun*. 2012, *3*, 1149.

83. X. Wu, Y. Cao, X. Ai, J. Qian, H. Yang. A low-cost and environmentally benign aqueous rechargeable sodium-ion battery based on $NaTi_2(PO_4)_3$ -$Na_2NiFe(CN)_6$ intercalation chemistry. *Electrochem. Commun*. 2013, *31*, 145.

84. D. Su, A. McDonagh, S. Z. Qiao, G. Wang. High-capacity aqueous potassium-ion batteries for large-scale energy storage. *Adv. Mater.* 2017, *29*, 1604007.

85. W. Ren, X. Chen, C. Zhao, Ultrafast aqueous potassium-ion batteries cathode for stable intermittent grid-scale energy storage. *Adv. Energy Mater.* 2018, 1801413.

86. H. Fei, Y. Liu, Y. An, X. Xu, G. Zeng, Y. Tian, L. Ci, B. Xi, S. Xiong, J. Feng. Stable all-solid-state potassium battery operating at room temperature with a composite polymer electrolyte and a sustainable organic cathode. *J. Power Sources*. 2018, *399*, 294.

87. H. Fei, Y. Liu, Y. An, X. Xu, J. Zhang, B. Xi, S. Xiong, J. Feng. Safe all-solid-state potassium batteries with 3D, flexible and binder-free metal sulfide array electrode. *J. Power Sources*. 2019, *433*, 226697.

88. B. Ji, F. Zhang, X. Song, Y. Tang. A novel potassium-ion-based dual-ion battery. *Adv. Mater*. 2017, *29*(19), 1700519.

89. L. Fan, Q. Liu, S. Chen, K. Lin, Z. Xu, B. Lu. Potassium-based dual ion battery with dual-graphite electrode. *Small* 2017, *13*(30), 1701011.
90. C. Li, J. Xue, A. Huang, J. Ma, F. Qing, A. Zhou, Z. Wang, Y. Wang, J. Li. Poly (N-vinyl carbazole) as an advanced organic cathode for potassium-ion-based dual-ion battery. *Electrochimica Acta*. 2019, *297*, 850.
91. A. Yu, Q. Pan, M. Zhang, D. Xie, Y. Tang. Fast rate and long life potassium-ion based dual-ion battery through 3D porous organic negative electrode. *Adv. Funct. Mat.* 2020, *30*(24), 2001440.

# 4 Zinc-Ion Batteries
## A Promising Solution for Future Energy Demands

*Ranjusha Rajagopalan, Haiyan Wang and Yougen Tang*

## 4.1 INTRODUCTION

In the present scenario, the advancements of energy storage and generation technologies are critical to meet the day-to-day requirements. Further, the demand for portable electronic gadgets and electric vehicles is huge. These technologies depend on the performance and availability of energy storage devices. For instance, the mobile phone technology primarily depends on the performance of the lithium-ion batteries (LIBs). For example, weight, cost and charge storage capacity of the devices depend on the performance of the associated LIBs. Even though the LIB is a world leader in the area of portable devices, the lithium resources are limited, which makes it unsustainable. Further, its electrochemical performance is nearing the theoretical limit, which led the scientists across the world to think of an alternative energy storage systems. In addition, there are a number of concerns regarding load levelling of renewable energy sources as well as the smart grid and the limited availability of lithium resources resulting in the cost increase. Therefore, alternate energy storage technologies should be brought into the market to meet the increasing energy demands. Hence, the constant search for an alternate battery system resulted in the emergence of different metal ($Na^+$, $K^+$ and $Mg^+$) ion, metal sulfur and metal air batteries.

The path to commercial success of any battery technology is strewn with challenges. Even though, the research on different battery chemistries such as sodium ion battery (SIB) have been investigated since the early 1980s. But these technologies are not yet commercialized due to the lack of further follow up in the respective field due to the high popularity and commercialization of LIBs. However, recently researchers have started focusing more on these alternative battery technologies due to the limited lithium resources and the increasing demand for energy storage systems. Apart from the successful development of anodes and cathodes, appropriate electrolytes, binders and additives have also been researched widely to develop fully functional batteries. Despite all these developments in different active materials and components, there remain numerous challenges, which include full cell design, electrode material balancing, etc., in the practical applications aspect of battery technologies.

In this regard, this chapter will mainly focus on the current research on materials for zinc-ion battery (ZIB) technologies and propose future directions. ZIBs are considered as a potential candidate in the post LIB era due to their high volumetric

DOI: 10.1201/9781003208198-4

energy density, functional safety, cost effectiveness and environment-friendly nature. The high redox potential of –0.76V (vs standard hydrogen electrode (SHE)) is also accounted as the positive attributes. In addition to this, we can translate most of the fabrication and processing technologies of the established LIBs into ZIBs, which could expedite the commercialization of ZIBs. One of the main challenges of the ZIB is its instability and rapid capacity deterioration during the prolonged cycling [1, 2]. Nonetheless, companies like Salient Energy are working explicitly on ZIBs to bring this technology into market [3, 4].

## 4.2   DIFFERENT ZIB TECHNOLOGIES

### 4.2.1   AQUEOUS ZIBs

Many researchers have started working on aqueous ZIBs due to their potential properties such as inexpensive and safer nature as compared to organic electrolyte based ZIBs [5, 6]. However, these batteries suffer from dendrite formation (especially with Zn anode), unwanted side-reactions and low full-cell voltage [7]. The dendrite formation can cause uneven deposition of zinc on the anode, resulting in a change in morphology and loss of active sites. The dendrite formation could also cause low Coulombic efficiency and electric short-circuit during the prolonged cycling [8]. Zinc corrosion can result in unwanted zinc ion deposition on the anode and the $H_2$ evolution can cause the battery gassing, which can result a deteriorated cycling and rate capabilities [7, 9]. However, the low cost, better safety and relatively long cycle of the aqueous battery system as compared to the organic electrolyte based batteries have motivated many researches to work on aqueous ZIBs.

Moreover, the aqueous electrolytes provide much higher (~two magnitude) ionic conductivity than the non-aqueous electrolytes. A study demonstrated that the aqueous (with neutral or slightly acidic pH electrolyte) ZIB can be used in the grid-scale energy storage. In this study, they employed α-$MnO_2$ as cathode, a zinc as anode and a mild $ZnSO_4$ or $Zn(NO_3)_2$ as aqueous electrolyte [10].

#### 4.2.1.1   Energy Storage Mechanism in Aqueous ZIBs

Unlike the LIBs and SIBs, the charge storage mechanism of ZIB in aqueous electrolytes is complex and controversial. Hence, the exact energy storage mechanism is still unclear and still remain a topic of discussion [11]. The three common reaction mechanisms in aqueous electrolyte system are: (1) $Zn^{2+}$ insertion/extraction; (2) chemical conversion reaction and (3) $H^+$ /$Zn^{2+}$ insertion/extraction [6].

*4.2.1.1.1   $Zn^{2+}$ Insertion/Extraction*

Most of the tunnel and layer structured compounds aid the $Zn^{2+}$ insertion and extraction into and from their hosts due to the small ionic radius (0.74 Å) of the zinc ion. For instance, a recent study conducted by Xu et al. demonstrated that the energy storage mechanism in an aqueous ZIB with α-$MnO_2$ as a cathode, a zinc as an anode and a mild $ZnSO_4$ or $Zn(NO_3)_2$ as an aqueous electrolyte is resulted of the migration of $Zn^{2+}$

between the tunnels of α-MnO$_2$ and a zinc anode. The Zn$^{2+}$ insertion/extraction reaction is given below [10].

At cathode:

$$Zn^{2+} + 2e^- + 2\alpha MnO_2 \leftrightarrow ZnMn_2O_4 \tag{1}$$

At anode:

$$Zn \leftrightarrow Zn^{2+} + 2e^- \tag{2}$$

In this study, it is revealed that the structure of ZIB is close to that of the alkaline Zn/MnO2 battery, which normally uses γ-MnO$_2$ or δ-MnO$_2$ as the cathode and KOH or NaOH alkaline solutions as the electrolyte. Nonetheless, their reaction chemistries are quite different [10]. The open circuit potential of ZIB was reported to be ~1.5 V, which is similar to that of Zn/MnO$_2$ primary battery. Further, at 0.5 C and 126 C, the ZIB could deliver a capacity of 210 and 68 mAhg$^{-1}$, respectively [10], indicating the high capacity and superior rate capability of ZIBs.

### 4.2.1.1.2 Chemical Conversion Reaction Mechanism

The common charge storage mechanism in ZIB is the Zn$^{2+}$ intercalation/de-intercalation. However, studies have demonstrated that chemical conversion reaction (between α-MnO$_2$ and H$^+$) can also be a viable option to store the charge in ZIB [12]. The MnOOH state here is the fully discharged state, which arises due to the reaction between α-MnO$_2$ and the H$^+$ from the water. To balance the charge, the OH$^-$ interact with the ZnSO$_4$ and H$_2$O to produce ZnSO$_4$[Zn(OH)$_2$]$_3$.$x$H$_2$O [12].

At cathode:

$$H_2O \leftrightarrow H^+ + OH^- \tag{3}$$

$$\alpha MnO_2 + H^+ + e^- + \leftrightarrow MnOOH \tag{4}$$

$$\frac{1}{2}Zn^{2+} + OH^- + \frac{1}{6}ZnSO_4 + \frac{x}{6}H_2O \leftrightarrow \frac{1}{6}ZnSO_4\left[Zn(OH)_2\right]_3.xH_2O \tag{5}$$

At anode:

$$\frac{1}{2}Zn \leftrightarrow \frac{1}{2}Zn^{2+} + e^- \tag{6}$$

In this study, they utilized the α-MnO$_2$ nanowires as the cathode, metallic Zn as the anode and mild ZnSO$_4$ as the aqueous electrolyte. The operating potential of this aqueous ZIB was reported to be 1.44 V and could achieve a capacity of 285 mAh g$^{-1}$ with a capacity retention of 92% after 5,000 charge/discharge cycles.

### 4.2.1.1.3  H$^+$ and Zn$^{2+}$ Insertion/Extraction Mechanism

Open tunnel and layered structures can facilitate the co-insertion of H$^+$ and Zn$^{2+}$. However, the insertion kinetics and mechanism are different for both these ions. So, studies have reported a consequent insertion reaction of H$^+$ and Zn$^{2+}$ at two different regions in aqueous ZIBs [13]. In this study, the electrodeposited MnO$_2$ on carbon fiber paper was used as the cathode, Zn was utilized as the anode and mild acidic ZnSO$_4$+MnSO$_4$ was used as the aqueous electrolyte. This ZIB showed a specific capacity of 50–70 mAh g$^{-1}$ at 6.5 C for 10,000 cycles with a coulombic efficiency of ~100% and a capacity decay rate of 0.007% per cycle. At lower C-rates of 1.3 C and 0.3 C, this electrode could deliver an initial discharge capacity of 260 and 308 mAh g$^{-1}$, respectively [13]. A study conducted by Chen et al. proposed that the H$^+$ and Zn$^{2+}$ co-insertion mechanism can be observed in vanadium-based electrodes [14]. In this study, sodium vanadate hydrate nanobelts were used as the cathode and aqueous zinc sulfate with sodium acetate additive was utilized as the electrolyte. Here also the H$^+$ and Zn$^{2+}$ co-insertion was considered as the charge storage mechanism and it could provide a high capacity of ~380mAh g$^{-1}$ with a retention of 82% even after 1,000 cycles [14].

First charge:

At cathode:

$$3.9\ H_2O \leftrightarrow 3.9\ H^+ + 3.9\ OH^- \tag{7}$$

$$1.95\ Zn^{2+} + 0.65\ ZnSO_4 + 3.9\ OH^- + 2.6\ H_2O \leftrightarrow 0.65Zn_4SO_4(OH)_6.4H_2O \tag{8}$$

$$NaV_3O_8.1.5H_2O + 3.9H^+ + 0.5Zn^{2+} + 4.9e^- \rightarrow H_{3.9}NaZn_{0.5}V_3O_8.1.5H_2O \tag{9}$$

At anode:

$$2.45\ Zn \leftrightarrow 2.45\ Zn^{2+} + 4.9e^- \tag{10}$$

Overall:

$$NaV_3O_8.1.5H_2O + 0.65ZnSO_4 + 6.5H_2O + 2.45Zn \rightarrow 0.65Zn_4SO_4 \\ (OH)_6.4H_2O + H_{3.9}NaZn_{0.5}V_3O_8.1.5H_2O \tag{11}$$

Subsequent cycle:

At cathode:

$$H_{3.9}NaZn_{0.5}V_3O_8.1.5H_2O \leftrightarrow NaZn_{0.1}V_3O_8.1.5H_2O + 3.9H^+ + 0.4Zn^{2+} + 4.7e^- \tag{12}$$

$$0.65\ Zn_4SO_4(OH)_6.4H_2O \leftrightarrow 1.95Zn^{2+} + 0.65ZnSO_4 + 3.9OH^- + 2.6H_2O \tag{13}$$

$$3.9 \, H^+ + 3.9OH^- \leftrightarrow 3.9H_2O \tag{14}$$

At anode:

$$2.35 \, Zn^{2+} + 4.7e^- \leftrightarrow 2.35Zn \tag{15}$$

Overall:

$$0.65Zn_4SO_4(OH)_6 \cdot 4H_2O + H_{3.9}NaZn_{0.5}V_3O_8 \cdot 1.5H_2O \leftrightarrow \tag{16}$$
$$NaZn_{0.1}V_3O_8 \cdot 1.5H_2O + 0.65ZnSO_4 + 6.5H_2O + 2.35Zn$$

During the initial discharge, both the ions simultaneously inserted into the $NaV_3O_8 \cdot 1.5H_2O$ to generate $H_{3.9}NaZn_{0.5}V_3O_8 \cdot 1.5H_2O$. This reaction is not fully reversible. In the charging process (to 1.25 V), the proton and partial zinc ion are extracted at the same time and forms $NaZn_{0.1}V_3O_8 \cdot 1.5H_2O$, which is fully reversible and the reaction continues in the subsequent cycles. At the end of fully charged state to 1.25 V, the proton and partial zinc ion get extracted at the same time to produce $NaZn_{0.1}V_3O_8 \cdot 1.5H_2O$. This reaction is reversible and continues in the subsequent cycles. The capacity contribution of proton and zinc ion was calculated to be ~315 mAhg$^{-1}$ (which is ~83% of total capacity) and ~65 mAhg$^{-1}$ (which is ~17% of total capacity), respectively [14]. This reaction is different from the other reported reaction where either proton or zinc ion participate in the reaction. In the present reaction, both $ZnSO_4$ and $H_2O$ participate in the reaction, which provide an energy and power density of 180 and 2160 W kg$^{-1}$, respectively [12, 13, 15].

### 4.2.2 NON-AQUEOUS ZIBS

A study conducted by Manthiram et.al. demonstrated that diffusion of $Zn^{2+}$ can be considerably affected within the NASICON-based $V_2(PO_4)_3$ host during the $Zn^{2+}$ insertion process. Further, the insertion process requires an elevated temperature of 80°C. The loss of structural integrity can also be observed in the host structure due to the strong interaction between the host and zinc ions, which can lead to a poor cycling stability [16, 17]. In another study conducted by the same group revealed that unlike aqueous ZIB (where both $H^+$ and $Zn^{2+}$ participate in the charge storage mechanism) in non-aqueous ZIB, the $Zn^{2+}$ acts as the guest ion and participates in the charge storage mechanism. Study conducted by Nazar et.al. utilized $V_3O_7 \cdot H_2O$ as the cathode, Zn metal as the anode and 0.25 M $Zn(CF_3SO_3)_2$ (ZnOTf) in acetonitrile (AN) as the electrolyte. This combination could exhibit a potential window of ~3.6 V (vs $Zn^{2+}/Zn$) [18], It was also reported that the diffusion of $Zn^{2+}$ into the $V_3O_7 \cdot H_2O$ cathode from the surface forms an ion pair with the AN solvent, resulting in a higher de-solvation energy at the surface, while demonstrating similar overall electrochemical performance in both aqueous and non-aqueous ZIBs [18]. This study revealed that ZnOTf/ AN is one of the most suitable systems for reversible cycling of the zinc metal anode, with almost no oxidative degradation. A recent study demonstrated that the common issues with $H_2$ evolution, Zn dendrite formation, corrosion and passivation persist and can be addressed by utilizing non-aqueous ZIBs [19].

### 4.2.3 Solid State ZIBs

It is widely known that solid-state electrolytes can improve the safety of the battery. In addition, the Zn metal anode can be employed to enable high operation potential and superior energy density [20]. During charge/discharge, the electrolytes in solid state ZIB facilitates the transport of $Zn^{2+}$ between the anode and cathode. To achieve this, the solid-state electrolyte must have (1) a high ionic conductivity at room temperature ($\sim$0.1mS cm$^{-1}$); (2) low electronic conductivity; (3) high thermal/chemical stability; (4) wide electrochemical stability window and (5) high mechanical strength to minimize the Zn dendrite formation [8]. Zhi et al. fabricated a flexible/wearable solid state ZIB, which demonstrated a high specific capacity of 306 mA h g$^{-1}$ with a superior cycling stability (97% capacity retention) even after 1,000 cycles of charge discharge [21]. A rechargeable solid-state ZIB fabricated using Zn/MnO$_2$ fiber was integrated in the electronic textiles due to the superior cyclic stability of more than 500 hours (1,000 cycles) with 98.0% capacity retention [22].

## 4.3 CHALLENGES AND OPPORTUNITIES

ZIBs have gained attention because of their low cost, high capacity and lower flammability than LIBs. Although sodium-ion and potassium-ion batteries are seen as alternatives to LIBs, they also have flammability issues, like LIBs. This makes ZIBs a next-generation battery technology. Unfortunately, there have been numerous claims in the open literature that have been mistakenly overestimated, and that ZIBs still face many issues.

For instance, a zinc anode brings a high level of safety and low-cost solution for aqueous ZIBs because it is stable with most of the water-based electrolytes. However, during charge–discharge, zinc tends to grow as random dendrites, which damages the separator and causes short-circuit. Although many strategies have been proposed to solve the dendrite issue, still they are enough not to meet the requirements of practical applications. There are challenges also on the cathode side. For instance, the water molecules in the aqueous-based electrolyte can naturally split into OH$^-$ and H$^+$, where H$^+$ competes with the zinc ions in the process of intercalation into the cathode. This results in the OH$^-$ to combine with the $Zn^{2+}$, causing layered precipitates of Zn(OH)$_2$ on the cathode surface. This causes deficit of Zn and creates an insulating layer on the surface, which can lead to undesirable side reaction.

In order to address these issues, researchers have been investigating non-aqueous/aqueous hybrid electrolyte that can help to prevent water splitting. These electrolytes have been shown to be quite effective while ensuring that $Zn^{2+}$ intercalation dominates the chemistry. Another strategy that has been proposed is to reduce the cycling rates, which tend to favour H$^+$ insertion more than $Zn^{2+}$ intercalation at the cathode. Further, high C-rates lead to lowering of capacity in ZIBs. Hence it becomes imperative that C-rates need to be lowered, while engineering the electrolyte to avoid water splitting.

Most of the reported ZIB cathodes are based on oxides, which in aqueous media claim notable cycle lives at high C rates; however, their relatively poor cyclability at lower C-rates indicates that a significant contribution of storage comes from capacitive behaviour rather than to reported capacities, that is, true $Zn^{2+}$ storage. High C-rate testing procedures can be a misrepresentation, leading to yet another misconception that high C-rate cyclability will translate to good low-rate performance. Although standardizing onto specific testing protocols is difficult at these stages of ZIB development, however real capacity, current density and electrode–electrode distance need to be investigated in the literature. These parameters are very important in the context of cycling, dendrite growth and cathode life. Modifying testing protocols can identify some of the important issues. For instance, providing a pause time after discharge cycle influences the charge capacity that can be recovered and also provides a signal that $H^+$ /$Zn^{2+}$ co-intercalation has occurred. Interval pauses in (dis)charge during cycling can help determine if good performance can be obtained at slow rates. Such procedures are of particular importance for practical applications of ZIBs, where the need for more advanced test conditions may become evident as the research in the field progresses.

Ultimately, the realistic commercialization of ZIBs will depend on how low the cost for Li-ion batteries will evolve, and whether their recycling becomes a viable technology for vehicular e-mobility. However, from the safety standpoint, ZIB technology will likely find applications where the low-voltage platform defined by the Zn anode (not offering the high energy densities) are needed for mobility applications (i.e., electric mobility and drones) and getting high capacities (within a required voltage range) are unattainable. The stationary applications, such as replacements for car batteries, inverter batteries, mobile robotic applications or distributed power for telecommunication stations, can be realistic applications (i.e., anywhere that lead acid is currently used and where safety is a prime concern) for ZIBs. Further, in order to assess progress in the ZIBs, it is important to conduct cost analyses, corrosion performance and volumetric capacity, as it moves into the product development phase. These are important metrics for future development for a wide variety of ZIB systems. From research standpoint, an in-depth understanding of cathode–electrolyte and anode–electrolyte compatibility, particularly concerning $H^+$ intercalation and oxygen evolution, is essential for material selection and design in different ZIB systems. Limiting high C-rates during charging that only benefits from capacitive contributions, performing careful electrochemical studies to differentiate between actual $Zn^{2+}$ and $H^+$ intercalation (i.e., ensure that contributions to capacity arises primarily from $Zn^{2+}$ intercalation), investigating differences between $Zn^{2+}$ intercalation in aqueous and non-aqueous media and deep dive into the fundamental science to understand the role of de-solvation of electrolyte at the electrode–electrolyte interface are needed.

Identifying host chemistries and functionalities that trigger parasitic side reactions in aqueous electrolytes can help to develop surface modifications and electrolyte optimizations toward a better performing ZIBs, which are likely to be low-cost, eco-friendly with elemental-abundant inorganic hosts that minimizes or eliminates $H^+$ co-intercalation. Researchers have started focusing on organic cathode materials that have already demonstrated impressive performance metrics such as cycle life (particularly at low C-rates exhibiting reasonable capacity). This provides new avenues for future research and also for investigating these electrodes in non-aqueous cells.

The advantage of elemental abundance of Zn for anodes also holds true for non-aqueous systems. Although restricted diffusion of $Zn^{2+}$ in most structures can limit the selection of active host materials, it has been shown that $Zn[Ni_{1/2}Mn_{1/2}Co]O_4$ can deliver energy density as high as 305 Wh kg$^{-1}$ at an open circuit voltage of 2.05 V, suggesting that new materials involving spinel oxides may hold much promise. Nevertheless, complex non-aqueous Zn electrochemistry adds a layer of complexity in identifying practical ZIBs. Although one solution to poor solid-state diffusion of $Zn^{2+}$ is rooted in implementing open framework 3D electrode architectures with interconnected pathways and chemistries with weaker electrostatics, an approach to counter slow charge transfer kinetics in the non-aqueous system is yet to be researched. Strong solvation and cation–anion pairing in non-aqueous electrolytes hinder charge transfer at the electrode–electrolyte interface, which limits the overall ZIBs performance. Remedies include (1) electrolyte modulation to ensure better anion solvation that can further hinder ion-pairing; (2) architecting layered hosts with large interplanar spacing to lodge $Zn^{2+}$ together with its solvation sheath or utilizing open frameworks of organic materials and (3) altering the surface chemistries of cathode to minimize the charge transfer barrier at the electrode–electrolyte interface by facilitating strong bonds to intercalating $Zn^{2+}$.

As the energy storage research continues to thrive for innovative solutions, ZIBs can provide an alternate solution that will change the dynamics of LIBs as the leader in the category. It is important that research in new materials and electrolytes continues to enhance the success of the ZIBs. Although ZIBs are relatively new, it has been demonstrated that ZIBs can show substantial improvements in supply chain security and safety. Combining these advantages with a scalable and safe manufacturing process will ensure that ZIBs have an important role in the future clean technologies. As the world continues to combat carbon emissions, the need for better energy storage systems from easier-to-source materials will become more important. ZIBs are well-positioned to meet this growing demand through research and increasing credibility within the industry, making them the default choice for stationary energy storage.

## ACKNOWLEDGMENTS

This work was financially supported by the National Nature Science Foundation of China (no.21975289), the Hunan Provincial Science and Technology Plan Projects of China (no.2022RC3050, no.2017TP1001), degree and postgraduate Education Reform Research Project of Hunan Province (2020JGYB026). This work was also financially supported by Postdoctoral International Exchange Program Funding of China (No. 115) and China Postdoctoral Science Foundation (2019M652802).

## REFERENCES

[1] Zhu, Y.; Yin, J.; Zheng, X.; Emwas, A.-H.; Lei, Y.; Mohammed, O. F.; Cui, Y.; Alshareef, H. N. Concentrated dual-cation electrolyte strategy for aqueous zinc-ion batteries. *Energy & Environmental Science* 2021, *14* (8), 4463.
[2] Shin, J.; Lee, J.; Park, Y.; Choi, J. W. Aqueous zinc ion batteries: focus on zinc metal anodes. *Chemical Science* 2020, *11* (8), 2028.

[3] Brown, R., Zinc-ion: A competitive alternative to lithium-ion for stationary energy storage, *Solar power world* 2021.

[4] Bellini, E., Salient Energy develops Zinc-ion battery for residential applications, PV magazine global . 2022.

[5] Tang, B.; Shan, L.; Liang, S.; Zhou, J. Issues and opportunities facing aqueous zinc-ion batteries. *Energy & Environmental Science* 2019, *12* (11), 3288.

[6] Fang, G.; Zhou, J.; Pan, A.; Liang, S. Recent Advances in Aqueous Zinc-Ion Batteries. *ACS Energy Letters* 2018, *3* (10), 2480.

[7] Zhang, N.; Chen, X.; Yu, M.; Niu, Z.; Cheng, F.; Chen, J. Materials chemistry for rechargeable zinc-ion batteries. *Chemical Society Reviews* 2020, *49* (13), 4203.

[8] Hansen, E. J.; Liu, J. Materials and structure design for solid-state zinc-ion batteries: A mini-review. *Frontiers in Energy Research* 2021, *8*, 616665.

[9] Ma, L.; Chen, S.; Li, N.; Liu, Z.; Tang, Z.; Zapien, J. A.; Chen, S.; Fan, J.; Zhi, C. Hydrogen-free and dendrite-free all-solid-state Zn-ion batteries. *Advanced Materials* 2020, *32* (14), 1908121.

[10] Xu, C.; Li, B.; Du, H.; Kang, F. Energetic zinc ion chemistry: The rechargeable zinc ion battery. *Angewandte Chemie International Edition* 2012, *51* (4), 933.

[11] Palacín, M. R. Recent advances in rechargeable battery materials: A chemist's perspective. *Chemical Society Reviews* 2009, *38* (9), 2565.

[12] Pan, H.; Shao, Y.; Yan, P.; Cheng, Y.; Han, K. S.; Nie, Z.; Wang, C.; Yang, J.; Li, X.; Bhattacharya, P.et al. Reversible aqueous zinc/manganese oxide energy storage from conversion reactions. *Nature Energy* 2016, *1* (5), 16039.

[13] Sun, W.; Wang, F.; Hou, S.; Yang, C.; Fan, X.; Ma, Z.; Gao, T.; Han, F.; Hu, R.; Zhu, M.et al. Zn/MnO$_2$ battery chemistry with H+ and Zn$^{2+}$ coinsertion. *Journal of the American Chemical Society* 2017, *139* (29), 9775.

[14] Wan, F.; Zhang, L.; Dai, X.; Wang, X.; Niu, Z.; Chen, J. Aqueous rechargeable zinc/sodium vanadate batteries with enhanced performance from simultaneous insertion of dual carriers. *Nature Communications* 2018, *9* (1), 1656.

[15] Kundu, D.; Adams, B. D.; Duffort, V.; Vajargah, S. H.; Nazar, L. F. A high-capacity and long-life aqueous rechargeable zinc battery using a metal oxide intercalation cathode. *Nature Energy* 2016, *1* (10), 16119.

[16] Park, M. J.; Yaghoobnejad Asl, H.; Therese, S.; Manthiram, A. Structural impact of Zn-insertion into monoclinic V2(PO4)3: implications for Zn-ion batteries. *Journal of Materials Chemistry A* 2019, *7* (12), 7159.

[17] Park, M. J.; Manthiram, A. Unveiling the charge storage mechanism in nonaqueous and aqueous Zn/Na3V2(PO4)2F3 batteries. *ACS Applied Energy Materials* 2020, *3* (5), 5015.

[18] Kundu, D.; Hosseini Vajargah, S.; Wan, L.; Adams, B.; Prendergast, D.; Nazar, L. F. Aqueous vs. nonaqueous Zn-ion batteries: Consequences of the desolvation penalty at the interface. *Energy & Environmental Science* 2018, *11* (4), 881.

[19] Kao-ian, W.; Mohamad, A. A.; Liu, W.-R.; Pornprasertsuk, R.; Siwamogsatham, S.; Kheawhom, S. Stability enhancement of zinc-ion batteries using non-aqueous electrolytes. *Batteries & Supercaps* 2022, *5*, e202100361.

[20] Wang, L.; Li, J.; Lu, G.; Li, W.; Tao, Q.; Shi, C.; Jin, H.; Chen, G.; Wang, S. Fundamentals of Electrolytes for Solid-State Batteries: Challenges and Perspectives. *Frontiers in Materials* 2020, *7*, 1.

[21] Li, H.; Han, C.; Huang, Y.; Huang, Y.; Zhu, M.; Pei, Z.; Xue, Q.; Wang, Z.; Liu, Z.; Tang, Z. et al. An extremely safe and wearable solid-state zinc ion battery based on a

hierarchical structured polymer electrolyte. *Energy & Environmental Science* 2018, *11* (4), 941

[22] Xiao, X.; Xiao, X.; Zhou, Y.; Zhao, X.; Chen, G.; Liu, Z.; Wang, Z.; Lu, C.; Hu, M.; Nashalian, A. et al. An ultrathin rechargeable solid-state zinc ion fiber battery for electronic textiles. *Science Advances* 2021, *7* (49), eabl3742.

# 5 Magnesium Ion Batteries

## Promising Application for Large-Scale Energy Storage

Yuan Yuan, Dachong Gu, Xingwang Zheng,
Ligang Zhang, Liang Wu, Jingfeng Wang,
Dajian Li and Fusheng Pan

## 5.1 INTRODUCTION

Recently, battery energy storage has received widespread attention, of which the most representative lithium-ion batteries (LIBs) are widely applied in electric vehicles, electronic devices and grid energy storage fields [1, 2]. However, LIBs are still difficult to meet the growing demand for safety and energy density requirements [3]. In addition, the lithium reserves available on the earth's crust are limited, cost-effective lithium recovery technology is still lacking and the continuous depletion of lithium resources will inevitably lead to the rising cost of LIBs. The reserves of more abundant metals are used as alternative batteries, such as sodium (Na), potassium (K), magnesium (Mg), zinc (Zn) and aluminum (Al) batteries. Among these alternative batteries, magnesium ion batteries (MIBs) have received much attention due to their unique advantages and performance [4, 5].

Magnesium reserves on the earth's crust (2.9 wt%) are much higher than lithium (0.002 wt%), making it a low-cost raw material. With a melting point of 660°C, magnesium is more stable in the atmosphere and easier to handle than lithium. Magnesium has a low redox potential ($-2.37$ V vs. $H^+/H_2$) and two charges, which makes the volume theoretical capacity of 3,833 mAh cm$^{-3}$ of magnesium ion battery close to the twice that of the LIBs. Additionally, the radius of magnesium ion (0.86 Å) is similar to that of lithium ion (0.9 Å). Unlike lithium metals exhibiting low safety due to the formation of dendrites, alkaline earth metal ions have lower diffusion energy and weaker bonding and the formed deposition layers is generally smoother [6–8]. To conclude, MIBs are promising for the large-scale applications.

Despite the mentioned advantages of MIBs, currently, there are still two important challenges which hinder their commercial application. One is that the strong electrostatic interaction of divalent $Mg^{2+}$ with the electrode matrix makes it diffuse slowly, which will lead to increased polarization and reduced cycle life. The second is the incompatibility of anode–electrolyte–cathode. The Mg anode is incompatible with traditional electrolytes due to the passivation reaction between the Mg surface and electrolytes. Specifically, the active material in the electrolyte will be reduced for the formation of an insulating passivation film on the electrode surface [9–11]. The LIBs can also form the solid-electrolyte interphase (SEI), but the insulating film in LIBs

DOI: 10.1201/9781003208198-5

can still conduct Li$^+$. The passivation film on the surface of a magnesium anode can hardly conduct Mg$^{2+}$ [12]. Within the developed electrolytes, ether-based electrolytes are found to have good compatibility with magnesium electrodes. However, ether electrolytes are poorly compatible with many high-voltage cathode materials. Thus, finding suitable electrolytes to match the promising cathode materials is still in study.

In 1990, Gregory et al. [13] developed ether solutions containing magnesium salts with organoboranes or organoaluminates as anions and first reported Mg//0.25 M Mg(B(Bu$_2$RPh$_2$))$_2$/THF-DMF//Co$_3$O$_4$ rechargeable magnesium batteries. In 2000, Aurbach et al. [14] developed an ether solution containing a magnesium halide aluminate complex electrolyte, which can effectively circumvent the problem of passivation of Mg electrodes and achieve reversible magnesium plating and stripping. Their group developed a prototype magnesium battery, the chevrel phase Mo$_6$S$_8$ as the positive electrode and Mg metal as the negative electrode in the electrolyte 0.25 M Mg(AlCl$_2$BuEt)$_2$/THF, cycling for more than 2,000 cycles with an actual energy density of 60 Wh kg$^{-1}$. This groundbreaking discovery is considered a milestone in the development of magnesium batteries. In 2007, Aurbach et al. [15] developed an all-phenyl complex electrolyte (APC) with enhanced anodic stability. Notably, Wang et al. [16] discovered the spontaneous degradation behavior of metal Mg and Mg$_2$Sn in APC electrolytes. To date, electrolyte discoveries are still limited and suffer from low operating voltage windows, corrosive electrolytes and high volatility. For electrodes (cathode and anode), there are two main strategies to obtain applicable MIBs. The first one is the rational design of high-voltage transition metal compounds to obtain superior electrochemical performance. The second one is that a suitable alternative anode compatible with conventional electrolytes may bring a breakthrough in MIBs. The electrode materials can only work if it is matched with a compatible electrolyte. The exploration of achieving the compatibility of the alternative negative electrode with the conventional electrolyte may be also the key to the practical application. In this chapter, we will review the research progress of electrolytes and electrode materials in MIBs and focus on their performance and potential solutions.

## 5.2 ELECTROLYTE

One key challenge with an Mg metal anode is its propensity to be easily passivated by electrolytes consisting of conventional polar solvents (such as carbonate) and common commercial magnesium salts (such as Mg(ClO$_4$)$_2$), which are detrimental to reversible magnesium deposition/stripping. Unlike lithium batteries, the solid-electrolyte interface (SEI) phase formed on the surface of the metal lithium can conduct lithium ions, while the passivation film on the metal magnesium cannot conduct magnesium ions. As such, the development of Mg electrolytes is strongly tied to salts and solvents compatible with the Mg metal, which limited electrolyte designs and the development of MIBs. Over the past decade, there have been two breakthroughs in the electrolyte field that realized reversible cycle in magnesium ion batteries with Grignard reagents in 1990 [13] and 0.25 M Mg (AlCl$_2$BuEt)$_2$/THF electrolyte in 2000 [14]. Due to the above findings, the study of magnesium battery electrolytes shifted mainly to the non-aqueous electrolytes of magnesium. For magnesium ion batteries, criteria of noncorrosive and non-toxic are also critical for Mg electrolytes besides the properties of fast magnesium conductivity and chemical stability toward electrodes. Halide anions, especially chloride, have been historically strongly embedded in Mg electrolyte development mainly because Grignard reagents were readily available precursors in the early days of Mg electrolyte development. The current trend shifts away from chloride anions, which develop the noncorrosive electrolytes to reduce corrosion of current collectors and battery shells, etc. In this section, Mg electrolytes can be classified into halide-ion containing electrolytes, chloride-free electrolytes and solid-state electrolytes.

## 5.2.1 Halide-Ion Containing Electrolyte

### 5.2.1.1 Organic Electrolytes

The electrolytes of Mg rechargeable batteries date back to the 1920s with the first observation of reversible deposition/stripping of Mg in Grignard solutions [17, 18]. Grignard reagent, with chemical formula as RMg-X (R may be alkyl or aryl; X is Cl, Br, or other halides), can prevent the formation of the passivation film. However, due to the strong reducing character of Grignard reagents, they show limited anodic stability, a narrow potential window (1.9 V vs. $Mg^{2+}/Mg$) and low ionic conductivity (0.398 mS cm$^{-1}$). Early work on the reversible electrodeposition of magnesium promoted the birth of magnesium batteries, of which the pioneering work of Gregory et al. [13] in 1990 was representative. Gregory et al. [13] first showed the elaboration of solutions containing Mg organo-borate moieties as improvements to organomagnesium reagents such as $Mg(BBu_2Ph_2)_2$ in THF, in which magnesium is reversibly deposited/dissolved. In addition, the reported methodology – adding appropriate Lewis acid (such as $AlCl_3$) to Grignard reagents to yield stable complex anions – is widely adopted to improve electrolyte performance. Gregory's work showed higher anodic stability compared to Grignard reagents, but the coulombic efficiency (CE) was observed to be low. Gregory's electrolytes cannot exhibit impressive performance in cell of transition metal sulfides and transition metal oxides as the counterelectrode.

In 2000, based on the concept that Gregory outlined, Aurbach et al. [14] reported the electrolytes composed of ether solutions with Mg organo-borate or organo-aluminate complexes and ethereal solutions of magnesium halo-alkylaluminate complex. These electrolytes have demonstrated superior reversibility of $Mg^{2+}$ intercalation in particular with Chevrel-type cathodes and improved conductivity. The prototype magnesium batteries proposed by Aurbach and his group [14] in 2000 using 0.25 M $Mg(AlCl_2BuEt)_2$/tetrahydrofuran (THF) electrolyte solution with a $Mo_6S_8$ cathode and Mg metal anode. The dichloro complex (DCC) electrolytes were practically products of the reaction between the $Bu_2Mg$ Lewis base and $EtAlCl_2$ Lewis acid. The magnesium battery prototype demonstrated over 2,000 reversible cycles with little capacity fade. However, the electrochemical stability window (2.2 V vs. $Mg^{2+}/Mg$) of this electrolyte is still too narrow to exert the high energy density of magnesium ion batteries. Therefore, Aurbach's group developed a high-voltage stable all-phenyl complex (APC) electrolyte in 2007 [15], which was composed of the reaction product of PhMgCl and $AlCl_3$ in an ether solution. The substitution of the alkyl groups in DCC with aromatics led to the synthesis of APC electrolyte allowing for the increase in the oxidative stability to 3.0 V vs. $Mg^{2+}/Mg$. The obtained APC electrolyte exhibited a high electrochemical window, specific conductivity of about 2 mS $cm^{-1}$, low overpotential and approached 100% cycling efficiency. The APC electrolyte is widely used in the basic research of MIBs. However, due to the nucleophilicity of APC electrolytes, it is difficult to be compatible with the sulfur cathode.

To develop the non-nucleophilic electrolytes and match sulfur as a cathode, Kim et al. of Toyota [19] introduced the hexa-methyl-disilazide ($HMDS^-$) into magnesium electrolytes. The $HMDSMgCl-AlCl_3/THF$ could be used for an Mg/S system and the first rechargeable Mg/S batteries were prepared. However, this system shows apparent capacity fading with increasing cycling. Zhao-Karger et al. [20] replaced $HMDSMgCl$ with $(HMDS)_2Mg$ and obtained $(HMDS)_2Mg-2AlCl_3/THF$, which exhibited the high anodic stability of 3.5 V vs. $Mg^{2+}/Mg$ and an overpotential of less than 150 mV. In addition, the $(HMDS)_2Mg-AlCl_3$ electrolytes dissolving in the mixed solvents displayed a discharge potential of ~ 1.65 V in the Mg/S system for the first time. To avoid the $Al^{3+}$ co-deposit with $Mg^{2+}$, Liao et al. [21] developed an Al-free concentrated $(HMDS)_2Mg-MgCl_2$ electrolyte, which displayed the anodic stability of 2.8 V vs. $Mg^{2+}/Mg$ and Mg stripping/plating with a coulombic efficiency of 99%.

### 5.2.1.2   MACC

The magnesium–aluminum chloride complex (MACC) electrolyte is obtained by mixing a certain proportion of $MgCl_2$ and $AlCl_3$ in ether solvents. MACC electrolyte, as a non-Grignard-based electrolyte, has the characteristics of simple synthesis, low cost and high efficiency of magnesium deposition and dissolution. Moreover, the non-nucleophilicity of MACC electrolytes is a concern. Viestfrid et al. [22] prepared the MACC electrolytes and obtained a cycling efficiency of 37%. However, it cannot be used as an electrolyte in practice due to its poor electrochemical behavior. In 2014, Doe et al. [23] showed the MACC electrolyte overpotential below 200 mV and the CE above 99% by simply adding more $MgCl_2$. The MACC electrolyte shows that overpotential can be lowered and experienced some cycles, which is called the electrochemical conditioning process. At present, the research on MACC electrolytes is in the preliminary stage. There are few studies on magnesium full battery with MACC electrolyte.

### 5.2.2   CHLORIDE-FREE ELECTROLYTE

### 5.2.2.1   $Mg(TFSI)_2$ Electrolytes

$Mg(TFSI)_2$ electrolytes have good ionic conductivity and high solubility in various solvents. However, Ha et al. [24] found in 2014 that $Mg(TFSI)_2$ electrolytes displayed a low CE (<50%) and a large overpotential (>2 V). Although the interface between $Mg(TFSI)_2$ electrolytes and the Mg anode has not been fully investigated, $TFSI^-$ anions are believed to be electrochemically unstable with the Mg anode. The recent study by Jay et al. [25] also suggests the passivation layer from $Mg(TFSI)_2$ consisted of Mg fluoride, sulfide and oxide, etc. from the decomposition of $TFSI^-$ anions. The introduction of $Cl^-$ such as $MgCl_2$ into $Mg(TFSI)_2$ electrolytes was an effective strategy to reduce the overpotential during the Mg plating/stripping process. Cheng et al. [26] reported the $Mg(TFSI)_2-MgCl_2$ electrolytes dissolving in DME for the first time exhibited an improved CE (80%) and reduced overpotential (400 mV). It is noted that the active cations of $Mg(TFSI)_2$-$MgCl_2$ electrolytes are Mg-Cl monomers or dimers; so $Mg(TFSI)_2$-$MgCl_2$ electrolytes are not simple salt electrolytes. $Cl^-$ can protect the Mg surface layer from passivating by a trace amount of moisture [27]. In addition, the presence of the $MgCl_2$-rich surface layer may protect against the further decomposition of $TFSI^-$ on the Mg surface [28, 29]. The electrochemical performance of $Mg(TFSI)_2$-based electrolytes is highly dependent on the purity of Mg salt and solvents, which caused increased costs. To develop Cl-free, noncorrosive $Mg(TFSI)_2$-based electrolytes, Ma et al. [30] added a trace amount of $Mg(BH_4)_2$ to remove moisture. The electrolytes showed the initial CE up to 84% and 75% for 500 cycles on a Pt substrate.

### 5.2.2.2  Boron-Based Electrolytes

$Mg(BH_4)_2$ has been known as a hydrogen storage material due to its high hydrogen capacity (14.9 wt%) and low hydrogen-release temperature. As early as 1957, Conner et al. [31] reported Mg deposition from $Mg(BH_4)_2$ in ether solvents. The $Mg(BH_4)_2$ electrolyte has excellent compatibility with the Mg anode due to the strong reducibility of $[BH_4]^-$. However, the strong reductive $Mg(BH_4)_2$ results in a narrow electrochemical window (1.7 V vs. $Mg^{2+}/Mg$) and the strong ionic bond between $Mg^{2+}$ and $[BH_4]^-$ leads to low ionic conductivity (< 0.1 mS $cm^{-1}$) [32, 33]. $Mg(BH_4)_2$/DME electrolytes were observed to be electrochemically active for reversible Mg deposition/stripping by Mohtadi et al. in 2012 [34]. Mohtadi et al. [34] also found that adding a certain amount of $LiBH_4$ into $Mg(BH_4)_2$/DME electrolytes can significantly improve the kinetics and CE (from 70% to 94%) of Mg plating/stripping, which could attribute to the increase of $[BH_4]^-$ concentration. As reported by Su et al., [35] the electrochemical stability could be improved by dissolving $Mg(BH_4)_2$ and $LiBH_4$ in hybrid TG-DME-$PP_{14}TFSI$ solvent, which suggests the significance of solvent in $[BH4]^-$-based electrolytes. Moreover, Seh et al. [36] suggested that $[BH_4]^-$ could be used as the moisture scavenger to alleviate the passivation of impurities such as $H_2O$ on the Mg surface. Incorporating a larger boron cluster is another method to widen the electrochemical windows. In 2015, Tutusaus et al. [37] reported $Mg(CB_{11}H_{12})_2$/tetraglyme (MMC/G4) electrolytes, which exhibited electrochemical behavior such as ionic conductivities of about 1.8 mS $cm^{-1}$ and the high anodic stability around 3.8 V vs. $Mg^{2+}/Mg$. However, the configuration of ion species in the MMC/G4 electrolytes was complicated and the difficult preparation process also limits its application. The $Mg[B(hfip)_4]_2$-based electrolytes [38–40] prepared by redissolution in the microcrystal or in-situ synthesis are studied and for some cases, the high anodic stability (>3 V even 4 V) and high cycle efficiency (>98%) were presented.

### 5.2.3  SOLID-STATE ELECTROLYTES

#### 5.2.3.1  Inorganic Solid Electrolytes

Solid-state electrolytes have attracted much attention due to their advantages over liquid electrolytes such as non-flammability, non-volatility, good thermal stability, reduced leakage and reactivity. However, solid-state electrolytes suffer from fixed ions and show low ionic conductivity. The high charge density of divalent $Mg^{2+}$ makes it migration more difficult than monovalent $Li^+$ in solid-state electrolytes.

In 1987, Ikeda et al. [41] observed $Mg^{2+}$ ion conductivity in the composition $MgZr_4(PO_4)_6$ to be 0.029 mS cm$^{-1}$ at 400°C and 6.1 mS cm$^{-1}$ at 800°C. Imanaka et al. [42–44] reported improved $Mg^{2+}$ ion conductivity in the magnesium zirconium phosphate electrolytes through doping Nb and the ratio adjustment of the materials to stoichiometric and nonstoichiometric, while the ionic conductivity of the electrolytes is still limited. In 2014, Higashi et al. [45] studied $Mg(BH_4)(NH_2)$ solid-state electrolytes. They found that $Mg(BH_4)(NH_2)$ can support an ionic conductivity of $10^{-3}$ mS cm$^{-1}$ at a relatively low temperature of 150°C. From the results they calculated, the favorable $Mg^{2+}$ ion conduction could be attached to the ionic bonding character. A spinel $MgX_2Z_4$ (where X is In, Y and Sc; Z is S and Se) was reported by Canepa et al., which could be considered as a significant development in Mg solid-state electrolyte research [46]. Specifically, $MgSc_2Se_4$ exhibited a high ionic conductivity of 0.1 mS cm$^{-1}$ at 25°C, but its electronic conductivity needs to be reduced. Wang et al. [47] tried to synthesize Se-rich phases by doping an aliovalent element but failed to reduce the electronic conductivity sufficiently.

### 5.2.3.2 Polymeric Electrolytes

The organic solid-state electrolytes can be called polymeric electrolytes, which can be classified into three different categories: solid polymer electrolyte, gel polymer electrolyte and composite polymer electrolyte. The solid polymer electrolyte system generally consists of the polymer matrix and Mg salt. The common organic polymer substrates include poly(ethylene oxide) (PEO), poly(ethylene oxide)-polymethyl acrylate (PEO-PMA), ploy(vinylidenefluoride) (PVdF), poly(vinylidenefluoride-hexafluoro propylene) P(VdF-HFP) and metal–organic frameworks (MOFs). For the gel polymer electrolyte, the liquid plasticizer is generally added to increase ionic conductivity, which includes dimethyl carbonate (DMC), propylene carbonate (PC), polyethyleneglycol (PEG), acetonitrile, etc. In the case of composite polymer electrolytes, inorganic fillers such as $Al_2O_3$, $SiO_2$ and MgO are also added to improve ionic conductivity and enhance mechanical stability. In 2004, Chusid et al. [48] prepared PVdF or PEO polymer-based electrolytes by coupling $Mg(AlCl_{4-n}R_n)_2$/ether solution. The best one of these combinations was the PVdF-$Mg(AlEtBuCl_2)_2$-tetraglyme system, which showed a high electrochemical window of 2.5 V vs. $Mg^{2+}$/Mg, the high ionic conductivity of 3.7 mS cm$^{-1}$ at 25°C and the high discharge capacity of 115 mAh g$^{-1}$ using $Mo_6S_8$ cathode. In 2005, Morita et al. [49] synthesized a gel polymer electrolyte consisting of a PEO–PMA matrix with $Mg(CF_3SO_3)N_2$-based electrolyte using EMITFSI as solvent. The ionic conductivity of the electrolyte, composed of 50 wt% of EMITFSI dissolving in 20 mol% $Mg(CF_3SO_3)N_2$, was 0.11 mS cm$^{-1}$ at 20°C. Du et al. [50] prepared a gel polymer electrolyte (PTB@GF-GPE) by a cross-linking reaction of hydroxyl-terminated polytetrahydrofuran with $Mg(BH_4)_2$. The PTB@GF-GPE demonstrated a high ionic conductivity of 0.476 mS cm$^{-1}$ and showed the adaptation of −20 to 60°C in the assembled $Mo_6S_8$/Mg battery.

## 5.3 CATHODE

MIBs have become the subject of high attention due to the high specific volume capacity (3833 mAh cm$^{-3}$), comparable high safety (no obvious magnesium dendrite formation) and natural abundance of Mg resources. The energy density of MIBs depends, to a large extent, on the capacity and operating potential of the cathode. However, the high-voltage cathodes compatible with current Mg electrolytes are still not found. The slow kinetics of Mg$^{2+}$ in cathode hosts is still a challenge. Many efforts have been made to develop a high-energy-density Mg battery prototype with a high voltage/capacity cathode and appropriate electrolyte since the first rechargeable magnesium batteries prototype was reported in 2000. Currently, there are four primary groups of cathode materials for MIBs that have been extensively studied: transition metal oxides, transition metal sulfides, polyanionic compounds and Prussian blue analogs.

### 5.3.1 TRANSITION METAL OXIDE

#### 5.3.1.1 VO$_2$

The strong oxygen–metal bond in the oxide has high ionic characteristics, which provides transition metal oxide with a high anodic oxidation potential. Only a few transition metal oxides can be used as cathode materials for rechargeable magnesium batteries. The existing research mainly includes V$_2$O$_5$, MoO$_3$ and spinel compounds with the general formula A$_x$O$_y$ (VO$_2$, Co$_3$O$_4$, Mn$_3$O$_4$, Pb$_3$O$_4$). At present, transition metal oxides as electrode materials have attracted extensive attention owing to their advantages of high capacity, low cost and suitability for commercial development. However, the slow diffusion of Mg$^{2+}$ ions in the host hinders the development of the transition metal oxide cathodes. In addition, the huge volume change in transition metal oxides, during the charging and discharging process, could lead to electrode pulverization and damage its good contact with the current collector, significantly reducing the battery capacity.

Vanadium dioxide, a transition metal oxide with phase-change properties, has received extensive research as a cathode material for LIBs due to its layered structure, high capacity and high electrical conductivity. Therefore, monoclinic VO$_2$ (B) has also attracted increasing attention in magnesium batteries.

Gao et al. [52] used the DFT+U calculation method for the first time to study systematically the effects of magnetism and van der Waals (vdW) forces on the electronic structure and diffusion kinetics of magnesium in bulk VO$_2$ (B). At 300 K, the diffusion coefficient of magnesium can reach $1.62 \times 10^{-7}$ cm$^2$ s$^{-1}$, which is comparable to that of Li$^+$. In addition, the lowest diffusion energy barrier of Mg$^{2+}$ is calculated to be only 0.26 eV. These results demonstrate that bulk VO$_2$ (B) is a cathode material for magnesium secondary batteries with fast kinetics. Luo et al. [51] synthesized nanosheet-like and nanorod-like materials of VO$_2$ (B) and tested their electrochemical performance as a cathode for magnesium batteries. The VO$_2$ (B) nanorod electrode exhibited a high discharge voltage plateau of 2.17 V (vs. Mg$^{2+}$/Mg) at a current density of 25 mA g$^{-1}$, and the first discharge specific capacity is 391 mAh g$^{-1}$ (Figure 5.1).

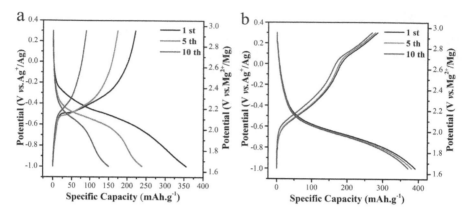

**FIGURE 5.1**   The discharge/charge curves of (a) VO₂ (B) nanosheets; (b) VO₂ (B) nanorods at a current density of 25 mA g⁻¹. Reprinted with permission from [51], Copyright 2018, Elsevier.

### 5.3.1.2   V₂O₅

V₂O₅ can provide high operating voltage and excellent electrochemical performance when used as a cathode material of LIBs. Motivated by the successful application in LIBs, V₂O₅ is also widely investigated in MIBs.

V₂O₅ crystal is composed of V₂O₅ polyhedral layers [53], as shown in Figure 5.2, whose layered structure is beneficial to the de-intercalation of Mg²⁺. However, most layered oxide materials have poor stability in the air, which is not only detrimental to electrochemical performance but also has certain limitations on its manufacture, transportation and application. The main reason for poor air stability is that water and carbonate ions in the air will be embedded in the TMO₂ layer of layered materials, where TM refers to one or more transition metal elements. The interlayer spacing of layered materials with such layered structure is the main factor affecting the normal de-intercalation of Mg²⁺.

Magnesium ions can reversibly intercalatedeintercalated into a V₂O₅ crystal lattice and the reactions are as follows [54]:

$$\text{Discharge: } x\text{Mg}^{2+} + \text{V}_2\text{O}_5 + 2xe^- \rightarrow \text{Mg}_x\text{V}_2\text{O}_5$$

$$\text{Charge: } \text{Mg}_x\text{V}_2\text{O}_5 \rightarrow x\text{Mg}^{2+} + \text{V}_2\text{O}_5 + 2xe$$

The gap between V₂O₅ layers can be used for embedding magnesium ions, and the theoretical specific capacity can reach 589 mAh g⁻¹ [55]. However, the current experimental capacity is less than one-third of the theoretical value, which is mainly due to the low conductivity of V₂O₅ and the extremely slow diffusion process of Mg²⁺ in its crystal lattice [56]. A nanostructure can effectively increase the specific surface area of materials and shorten the diffusion distance of magnesium ions [57]. Some research teams have obtained V₂O₅ with different morphologies such as nanowires [58] and nanosheets [59] by various methods to improve the electrochemical performance of the materials. Gershinsky et al. [60] prepared a high-purity thin film

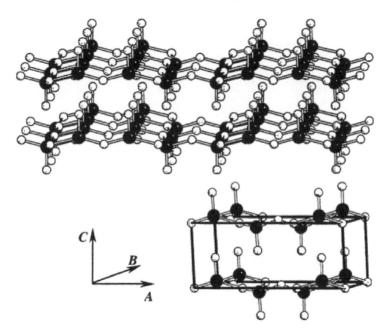

**FIGURE 5.2** Schematic diagram of the layered structure of $V_2O_5$ (the V atom is black and the O atom is white). Reprinted with permission from [53]. Copyright 2001, American Chemical Society.

$V_2O_5$ electrode with a thickness and particle size of the nanometer scale and studied their interaction with magnesium ions. Their results showed that highly reversible Mg insertion/de-insertion is possible using electrochemical, spectroscopic, microscopic and X-ray diffraction analysis. $V_2O_5$ thin-film electrodes can be cycled over a potential range of 2.2–3.0 V vs $Mg^{2+}/Mg$ with a specific capacity of 150 mAh $g^{-1}$ corresponding to one electron transfer per unit and these electrodes can be cycled at ~100% Coulombic efficiency, which demonstrates a very stable capacity upon cycling (Figure 5.3).

### 5.3.1.3　MnO$_2$

$MnO_2$ has been studied as the cathode material for MIBs by many research teams due to its inherent high capacity, low cost, environmental friendliness and natural abundance. Manganese dioxide used as cathode materials for MIBs can be traced back to 2001 with a discharge capacity of 85 mAh $g^{-1}$ [61]. $MnO_2$ has various crystal forms, which can be roughly divided into three types: 1D tunnel structures, 2D layered materials and 3D spinel. The research of $MnO_2$ as cathode material of magnesium batteries mainly focuses on the different crystal structures formed by manganese atoms and oxygen atoms.

As shown in Figure 5.4, there are three kinds of square tunnel structures of $MnO_2$, namely pyrolusite (1×1), Hollandite (2×2) and Todorokite (3×3) [62–64]. It is preferable to use tunnel $MnO_2$ as the cathode material of MIBs than other structures

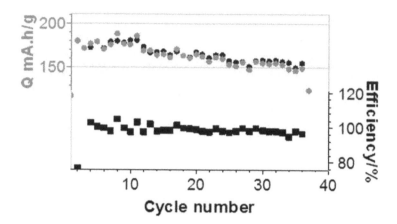

**FIGURE 5.3** The cycle life of the $V_2O_5$ thin-film electrodes. Reprinted with permission from [60]. Copyright 2013, American Chemical Society.

because its tunnel structure is more conducive to the diffusion of $Mg^{2+}$ than that of other crystal forms. Kumagai [62] prepared Todorokite-type $MnO_2$ and carried out a discharge/charge test on Todorokite-type $Mg_{0.21}MnO_{2.03} \cdot 0.24H_2O$, which exhibited higher capacity compared to Hollandite. The initial discharge capacity of 85 mAh $g^{-1}$ was observed in 1 mol $L^{-1}$ magnesium perchlorate electrolyte (Figure 5.4).

The insertion of $Mg^{2+}$ into oxides is more difficult than $Li^+$ in oxides and almost irreversible due to higher charge density and stronger electrostatic interaction with the host. Todorokite delivered a larger discharge capacity than Hollandite, which can be attributed to its larger (3×3) tunnel size to weaken the electrostatic effect between $Mg^{2+}$ and the host. To enlarge the pore diameter size and improve the conductivity, compound Hollandite $MnO_2$ with acetylene black (AB) has been proposed [65, 66]. When the material is cycled at 20°C with a current density of 100 mA $g^{-1}$, it has delivered a discharge capacity as high as 310 mAh $g^{-1}$. In this study, $K^+$ was used to stabilize the Hollandite tunnel. Although the capacity of Hollandite $MnO_2$/AB composite gradually decreased with the cycle, its capacity retention rate was improved.

Nam et al. [67] discovered that $Mn_3O_4$ of spinel was converted into layered Birnessite-type $\delta$-$MnO_2$ in cycling tested in $MgSO_4$ aqueous solution, which is due to the dissolution of $Mn^{2+}$ and oxidation of $Mn^{3+}$ driven electrochemically. During this period, the electrostatic interaction between $Mg^{2+}$ and the host anions was weakened by the insertion of water molecules between layers, so $Mg^{2+}$ was more easily intercalated between $\delta$-$MnO_2$ layers. Kim et al. [68] used scanning transmission electron microscopy (STEM) to observe directly the transformation of $Mn_3O_4$ of spinel into layered $\delta$-$MnO_2$ induced by water molecules. Although $H_2O$ molecules shield the strong Coulomb force of $Mg^{2+}$ and lattice oxygen anions inside the lattice, and effectively enhance the diffusion ability of $Mg^{2+}$, passivation layers of MgO and $Mg(OH)_2$ will be formed on the surface of the Mg anode. Zhao et al. [69] successfully synthesized Birnessite-type $MnO_2$ nanoflowers assembled by multi-layer nanoplatelets by a simple hydrothermal method. Zhou's team has reported the rapid

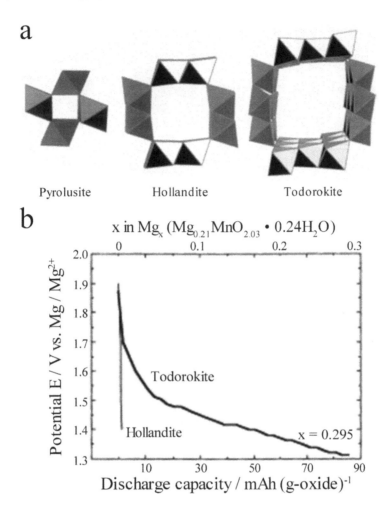

**FIGURE 5.4**   (a) Crystal structure models of three manganese oxides with a tunnel structure (Pyrolusite, Hollandite and Todorokite) and (b) charging and discharging curves of Todorokite and Hollandite $MnO_2$ in magnesium perchlorate electrolyte. Reprinted with permission from [62]. Copyright 2001, Elsevier.

(1 h) synthesis of Birnessite-type $MnO_2$ nanostructures via a polyol-reflux process [70]. Through oxidizing $MnCl_2$ with $H_2O_2$ under basic conditions in the presence of polyvinylpyrrolidone (PVP), the obtained flower-like Birnessite-type $MnO_2$ nanostructure is composed of nanosheets with an average diameter of 300–500 nm and shows mesoporous characteristics with a pore diameter of 20 nm (Figure 5.5) [70].

Recently, Wang et al. [71] expanded the interlayer spacing of Birnessite $MnO_2$ from 0.70 nm to 0.97 nm by introducing $K^+$ and $H^+$ plasma and increased its capacity from 58.6 to 110.8 mAh $g^{-1}$. The results have demonstrated that the diffusion channel of $Mg^{2+}$ is enlarged by the expansion of interlayer spacing, and its rate performance is also improved. This method provides a new idea for improving the reactivity of $Mg^{2+}$.

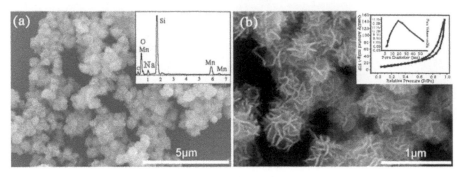

**FIGURE 5.5**   SEM images of the birnessite-type $MnO_2$. Reprinted with permission from [70]. Copyright 2013, American Chemical Society.

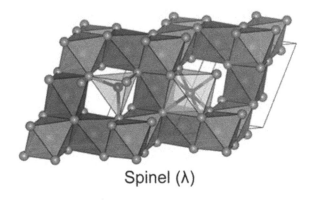

Spinel (λ)

**FIGURE 5.6**   The crystal structure of $\lambda$-$MnO_2$. Reprinted with permission from [72]. Copyright 2017, American Chemical Society.

$\Lambda$-$MnO_2$ is a typical spinel structure with an Fd3m space symmetry group (Figure 5.6). As shown in Figure 5.6, tetrahedral lattice points and octahedral lattice points are coplanar, forming an interconnected three-dimensional tunnel structure [72]. Spinel-structured $\lambda$-$MnO_2$ materials have been synthesized as cathode materials for MIBs by microwave method [73, 74]. However, their electrochemical results showed that the discharge capacity of the $\lambda$-$MnO_2$ electrode is about 80 mAh g$^{-1}$ in $Mg(ClO_4)_2$-$H_2O$/THF electrolyte solution with poor cycling performance.

The characteristics of $Mg^{2+}$, such as high charge density, strong polarization effect and slow diffusion kinetics, seriously restrict the development of cathode materials for MIBs. Therefore, it is necessary to develop new nanomaterials or design suitable structures to resolve these bottlenecks. Increasing the energy density of spinel $MnO_2$ materials is an important direction for the development of cathode materials for MIBs in the future [75, 76]. In future studies, the research on the compatibility and cycling performance of $\lambda$-$MnO_2$ with high initial energy and non-aqueous electrolyte can be intensified. The particle size of the material needs to be further reduced or the

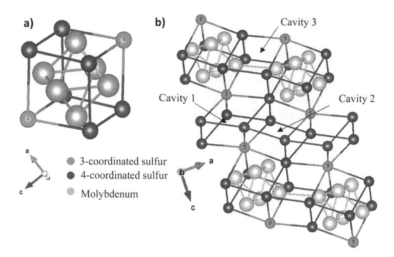

**FIGURE 5.7** (a) Structure of the Chevrel phase $Mo_6S_8$; (b) the three types of cavities (1, 2 and 3). Reprinted with permission from [79]. Copyright 2017, American Chemical Society.

diffusion path of $Mg^{2+}$ should be shortened, and auxiliary materials can also be added to improve the mobility of $Mg^{2+}$.

## 5.3.2 Transition Metal Sulfide

### 5.3.2.1 Chevrel Phase

Sluggish diffusion of magnesium ions in most electrode materials, caused by a strong interaction between divalent $Mg^{2+}$ ions and host matrixes, inhibits the development of cathode for MIBs. Chevrel sulfide ($Mo_6S_8$, $Mo_6S_6Se_2$, $Cu_2MoS_8$, etc.) can be capable of embedding monovalent and multivalent cations as a unique insertion-type material.

Saha et al. [77] first synthesized $Cu_2Mo_6S_8$ through a high-temperature solid-state reaction, then leached Cu atoms from $Cu_2Mo_6S_8$ by acid, [78] leaving $Mo_6S_8$ with the cavity, so $Mo_6S_8$ was thermodynamically metastable. The crystal structure of the Chevrel phase can be viewed as a stack of $Mo_6S_8$ blocks (Figure 5.7). Six molybdenum atoms form an octahedron in a cube composed of eight sulfur atoms, and $Mg^{2+}$ is embedded between these $Mo_6S_8$ blocks and can occupy two main gap positions [79].

Aurbach et al. [80] used a $Mo_6S_8$ cathode to match Mg negative electrode in $Mg(AlCl_2BuEt)_2/THF$ electrolyte, and there were two reversible intercalation reactions:

$$Mg^{2+} + Mo_6S_8 + 2e^- \rightarrow MgMo_6S_8 \tag{1}$$

$$Mg^{2+} + MgMo_6S_8 + 2e^- \rightarrow Mg_2Mo_6S_8 \tag{2}$$

The specific capacity of 122 mAh $g^{-1}$ can be achieved, and the capacity barely decreases after 2,500 cycles.

Although Chevrel $Mo_6S_8$ showed excellent thermodynamic reversibility of Mg intercalation, its limited specific capacity and poor kinetics limit its practical application. It is proposed to further improve the discharge voltage and capacity of the cathode material to obtain magnesium ion batteries with high working voltage and high energy density.

### 5.3.2.2 Transition Metal Dichalcogenide

In recent years, two-dimensional layered sulfide $MS_2$ (M = Ti, Nb, Mo, W, V, etc.) attracts much attention for rechargeable alkali metal batteries (potassium batteries, sodium batteries and lithium batteries), in which $MoS_2$ shows good performance. Based on this background, $MoS_2$ cathode material has attracted great attention in the field of MIBs [81]. $MoS_2$ can be presented in many structures (1T, 2H, etc.), as shown in Figure 5.8, showing different atomic stacking orders and coordination [82]. Among them, the 2H structure is the most stable one at room temperature, and commercial $MoS_2$ is generally the 2H structure.

To improve the reversible capacity of $MoS_2$, Yang et al. [83] synthesized inter-layer expanded 1T/2H-$MoS_2$ nanosheets with high 1T phase content by simple solvothermal method. Its reversible capacity was significantly higher than pure 2H-$MoS_2$ at127 mAh $g^{-1}$, and its cycle stability was likewise excellent. Wu et al. [84] proposed to build a van der Waals heterostructure (vdWHs) to solve the problem of the slow kinetic behavior of $Mg^{2+}$ during the migration of traditional two-dimensional layered materials. In this structure, $MoS_2$ monolayer and graphene alternately overlap to form a vdWHs with a diffusion barrier of 0.4 eV for $Mg^{2+}$, and the diffusion rate is 11 orders of magnitude faster than that of the original $MoS_2$. In addition, it shows

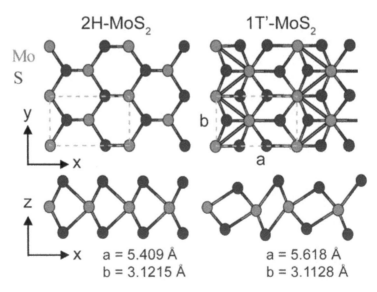

**FIGURE 5.8** Top (top panel) and side view (bottom panel) atomic structures of 2H and 1T' phase $MoS_2$ with lattice constants indicated. The dotted rectangle is the surface unit cell. Reprinted with permission from [82]. Copyright 2018, American Chemical Society.

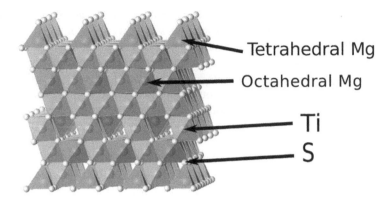

**FIGURE 5.9** Mg insertion into spinel $TiS_2$, accounting for occupancy both octahedrally and tetrahedrally. Reprinted with permission from [85]. Copyright 2018, American Chemical Society.

a specific capacity of 210 mAh $g^{-1}$ at the current density of 20 mA $g^{-1}$, and also provides excellent rate performance with a specific capacity of 90 mAh $g^{-1}$ at the current density of 500 mA $g^{-1}$.

$TiS_2$ spinel structure is the most common crystal structure of intercalation reaction, which can usually realize the rapid diffusion of cations. The volume of the spinel structure will increase with the insertion of Mg at tetrahedrons and octahedrons in Figure 5.9 [85]

Amir et al. [86] pointed out that when $TiS_2$ is used as the cathode material for MIBs, reversible de-intercalation can only be carried out at a high temperature (60°C) due to the high migration barrier and slow diffusion kinetics of $Mg^{2+}$ ions. Liu et al. [87] then reported that the calculated diffusion activation energy of $Mg^{2+}$ in $Ti_2S_4$ is 600 mV, which is smaller than that of $Mg^{2+}$ in $MnO_2$ with spinel structure (>650 mV). At the same time, the diffusion activation energies of $Cr_2S_4$ and $Mn_2S_4$ are calculated, which are 540 and 520 mV, respectively. It showed that spinel sulfides have better $Mg^{2+}$ mobility than spinel oxides, but this high mobility also comes at the cost of lower working voltage and lower theoretical specific energy.

Sun et al. [88] have investigated the electrochemistry of C-$Ti_2S_4$ coin cells with APC electrolyte and Mg negative electrode at 60°C and the obtained specific capacity is about 150 mAh $g^{-1}$. Besides adopting nanometer technology, increasing working temperature and introducing a small amount of water, replacing S in $TiS_2$ with Se or partially replacing S is also an effective way to improve the mobility of $Mg^{2+}$ in $TiS_2$. Gu et al. [89] used $TiSe_2$ as the cathode material of MIBs and obtained a specific capacity of 108 mAh $g^{-1}$ at room temperature.

### 5.3.3 POLYANIONIC COMPOUNDS

Compared with other cathode materials for secondary ion batteries, polyanion compounds have many advantages, such as rich types, diverse structures, adjustable working voltage, good cycle stability, etc., and are ideal cathode materials for developing low-cost environmental protection secondary ion batteries. Its general structural formula is $AM_x[(XO)_y]_z$, where A is an alkali metal element or alkaline earth metal element (such as Li, Na, K, Mg, Ca, etc.), M is a transition metal element (such as Fe, V, Ti, Mn, etc.), and X is a nonmetallic element (such as P, S, etc.) [90]. According to the types of polyanion groups and the current research situation, polyanion cathode materials can be roughly divided into five categories: phosphate series, sulfate series, silicate series, single polyanion series and mixed polyanion series.

Polyanionic silicates and phosphates are widely studied in $Mg^{2+}$ secondary batteries. Because of their large three-dimensional space and stable frame structure, they form gaps that can be occupied by other high-coordination metal ions. They have different crystal structures, composed of transition metal sulfides and metal oxides, showing high magnesium storage capacity and good reversible stability. Feng and Nuli et al. [91–93] prepared silicate cathode materials with olivine structure by sol-gel method and solid-state reaction technology and studied their electrochemical properties in MIBs. The results showed that the discharge voltage plateau of the cathode material prepared by the sol-gel method is 1.6 V, and the discharge specific capacity is 244 mAh $g^{-1}$.

There is a class of compounds with fast ion conduction structure, that is, (NASICON) structure, which has a large enough gap vacancy to accept guest ions and a stable three-dimensional system structure. Makino et al. [94] synthesized NASICON $Mg_{0.5}Ti_2(PO_4)_3$ by sol-gel method for the first time and used it as cathode material of MIBs. The results showed that $Mg^{2+}$ could be reversibly de-intercalated, but its performance was still limited by the diffusion kinetics of $Mg^{2+}$.

The main disadvantages of silicate and phosphate cathode materials are the slow kinetics and electrode diffusion of $Mg^{2+}$, which are mainly caused by the inversion mixing of metal ions and the slow diffusion of cations. The development of polyanionic secondary battery cathode materials with excellent electrochemical performance is still the focus and difficulty of research and development of low-cost energy storage technology. At the same time, it is necessary to speed up the research of wide electrochemical window electrolytes and high-stability electrolytes to match the new high-efficiency and low-cost polyanion cathode materials.

## 5.3.4  PRUSSIAN BLUE ANALOGS

Prussian blue analogs (PBAs) belong to transition metal cyanide, and the general chemical formula is $A_xM[(TCN)_6]_y \cdot mH_2O$ (where A is an alkali metal element, M is a transition metal element, T is generally Fe, $0 \leq x \leq 2$) [95].

Kim et al. [96] prepared a nano Prussian blue compound $Na_{0.69}Fe_2(CN)_6$ as the cathode material of MIBs. When 0.3 mol $L^{-1}$ of an acetonitrile solution of $Mg(TFSI)_2$ was used as the electrolyte, its voltage platform was 3.0 V (vs. $Mg^{2+}/Mg$), and its reversible discharge specific capacity was 70 mAh $g^{-1}$. It still had high capacity retention after 35 cycles. The Nazar [97] research group of the University of Waterloo in Canada first built a magnesium–lithium hybrid battery with Prussian blue analog as the positive electrode material versus Mg anode. Its battery system can achieve

**TABLE 5.1**
**The Performance of Partial Cathode Materials for MIBs**

| Cathode | Category | Voltage (V) | Capacity (mAh g$^{-1}$) |
|---------|----------|-------------|-------------------------|
| $VO_2$ | Layered | 2.17 | 391 |
| $V_2O_5$ | Layered | 2.25 | 150 |
| $\alpha$-$MnO_2$ | Hollandite | 2.00 | 280 |
| $\delta$-$MnO_2$ | Birnessite | 2.00 | 150 |
| $\lambda$-$MnO_2$ | Spinel | 0.6 | 80 |
| $Mo_6S_8$ | Chevrel phase | 1.1 | 135 |
| $MoS_2$ | Layered | 1.0 | 120 |
| $TiS_2$ | Spinel | 1.1 | 150 |
| PBAs | Open framework | 3.0 | 70 |

an average voltage of 2.3 V and an energy density of 290 Wh kg$^{-1}$. Compared with the traditional $Mo_6S_8$ (average voltage of 1.4 V and energy density of 170 Wh kg$^{-1}$) cathode, it has achieved a breakthrough in the cathode field for MIBs.

PBAs are regarded as one of the optimal cathode materials for secondary ion batteries because of their high energy density, abundant resources, low cost and environmental friendliness. In recent years, many efforts have been made to improve the electrochemical performance of PBAs. However, the poor conductivity and low diffusion rate of $Mg^{2+}$ of PBAs cathode materials limit their rate performance.

The performance of some typical cathode materials for MIBs is listed in Table 5.1. Currently, the investigation of cathode materials for MIBs is still limited, facing a major problem of low voltage and low capacity of cathode materials. Thus, it is necessary to optimize the materials and further develop new cathode materials to improve the electrochemical performance. Moreover, further research is required to understand the mechanism of cathode–electrolyte interface interaction, which will aid in the development of advanced magnesium batteries. The improvement of $Mg^{2+}$ ions kinetics in cathode material for magnesium batteries will promote the practical application.

## 5.4 ANODE

The volumetric capacity of Mg anode is almost twice that of Li, and the electrochemical deposition/dissolution process is dendrite free at most experimental conditions. Pure Mg sheets can be used both as an electrode and a collector, which seems to be the simplest approach. However, the passivation film formed at Mg surface during Mg deposition/dissolution process hinders the conduction of $Mg^{2+}$ ions and results in poor cycle stability. The modification and substitution of magnesium anodes are also investigated, while vigorously developing the electrolytes. Chen et al. [98] prepared the various Mg with micro-nano structures using a vapor-transport approach to apply in Mg/air batteries. The reduced polarization and improved cycle performance were observed, which can be attributed to the decreased thickness of the passivation due to the high specific area of Mg nanostructures. However, the fine Mg particles may strongly react with oxygen in the air resulting in the formation of a passivation film when handing the material in an uncontrolled atmosphere. Thus, in this section, the anode research focuses on the progress of Mg anode in constructing interfacial protective layer and alternative anodes including alloy anode and insertion anode.

### 5.4.1 COATED MG ANODES

To bypass passivation, an effective approach is to control the formation of a conductive $Mg^{2+}$ surface film of the Mg electrode. Some studies reported the construction of the protective film on Mg electrode surface before assembling the cell, showing the improved cycle stability and reduced interface resistance between Mg anode and electrolytes. In 2018, Son et al. [99] prepared an $Mg^{2+}$ conducting polymeric interphase and succeeded to enable reversible Mg stripping/deposition in Mg//$V_2O_5$ full cell employing carbonate-based electrolytes containing water. The polymeric coating combining thermally cyclized polyacrylonitrile and magnesium trifluoromethanesulfonate exhibited an ionic conductivity of 1.19 $\mu S^{-1}$. Inspired by the successful application of fluoride in lithium-ion batteries, Li et al. [100] pretreated the Mg electrode with hydrofluoric acid to obtain the $MgF_2$ protective layer, which can conduct $Mg^{2+}$ ions and reduce the side reactions with electrolytes. In addition, the surface modification of Mg is foiled by simply reacting with $SnCl_2$-DME solution to obtain an Sn-based artificial protective layer, which includes halide components combing the insulting nature of halides and the high diffusion coefficient of Sn-based compounds. The Mg modified by Sn-based artificial film can stably cycle over 4,000 cycles at 6 mA cm$^{-2}$ using conventional Mg(TFSI)$_2$/DME as the electrolyte [101]. Zhao et al. [102] reported the Bi-based protective layer employing a similar method and obtained enhanced stability between Mg anode and noncorrosive electrolytes, showing the Mg uniform electrodeposition and a decrease in side reactions (Figure 5.10). Moreover, Wei et al. exhibited the interface protection of Mg metal anode to inhibit dendritic growth through the spontaneous alloy reaction of liquid with Mg metal to form the $Ga_5Mg_2$ alloy layer.

The solid–liquid interphase (SEI) with conductive $Mg^{2+}$ can also be prepared in situ by adding appropriate solvents, salts and/or additives during cycling. In 2018, Li et al. attempted to regulate the composition of SEI by the addition of iodine to the simple salt electrolyte (Mg(TFSI)$_2$-DME). Such a protective layer containing $MgI_2$ is conductive to $Mg^{2+}$ and shows a small overpotential in the Mg//S full cell discharge/ charge tests [103]. $GeCl_4$ was also added to the electrolyte Mg(TFSI)$_2$/DME as an additive to form the Ge-rich artificial protective layer in vivo on Mg metal electrode [104]. The Ge-rich artificial protective layer could conduct $Mg^{2+}$ and hinder the formation of the passivation film. A surplus of $GeCl_4$ in the electrolytes could also aid in the self-healing of the damaged artificial interfacial layer during cycling. In recent years, much attention is paid to magnesium–lithium dual-ion batteries. Here, it is focused on the effect of lithium salt on the interface of the Mg metal anode. Cui and his group [105] synthesized the Mg[B(hfip)$_4$]$_2$ and Li[B(hfip)$_4$] salts and prepared the hybrid $Mg^{2+}$/Li$^+$ electrolytes using Mg[B(hfip)$_4$]$_2$ and Li[B(hfip)$_4$] dissolving in DME solvents in proportion. Their research shows the partial decomposition of Li[B(hfip)$_4$]

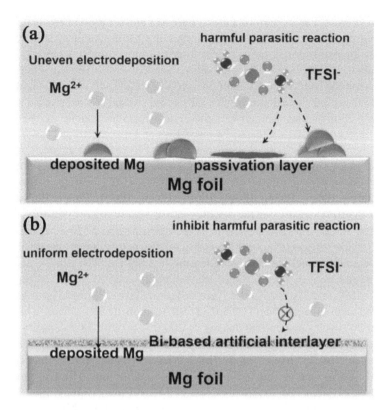

**FIGURE 5.10** Schematic illustration of the Bi-based protective layer. Reprinted with permission from [102]. Copyright 2021, American Chemical Society.

salts to obtain the Li-containing SEI on Mg anode during cycling. This stable Li-containing artificial interfacial layer conducts $Mg^{2+}$ smoothly, hinders the further decomposition of electrolytes on Mg anode and enhances the cycle stability. Thus, we can conclude that preparing an effective and protective layer by pretreating the Mg electrode or the addition of an additive can improve cycle life and reduce electrolyte decomposition and other side reactions.

### 5.4.2 ALLOY ANODES

#### 5.4.2.1 Magnesium Alloy Anodes

In contrast to pure Mg metal, magnesium alloys as negative electrodes may solve the problem of passivation in conventional electrolytes. One of the biggest advantages of MIBs is the high natural abundance of Mg and its low cost. Notably, the higher the magnesium content magnesium alloy anode materials possess, the larger advantage of the abundant of Mg sources can show. In addition, Mg alloys sheet can improve the workability of the metal anode to make fabrication easier and safer. Maddegalla et al. [106] reported that AZ31 Mg alloy films with higher ductility were used as anodes in

MIBs and showed similar performance to pure Mg anode when tested in Mg dissolution/deposition process with full cells comprising $Mo_6S_8$ cathode. However, compatibility with simple magnesium salt electrolytes is not covered in this report. Some research about $Mg_xM$ (M = Bi, Sn, etc.) alloys as anodes for MIBs show the compatibility between the $Mg_xM$ anodes and noncorrosive simple salt electrolytes. Based on the simple two-phase reaction ($3Mg^{2+} + 2Bi + 6e^- \rightarrow Mg_3Bi_2$) at ~0.25 V vs. $Mg^{2+}$/ Mg, Bi–Mg alloy anodes can deliver a theoretical specific capacity of 385 mAh g$^{-1}$. In 2013, Shao et al. [107] prepared Bi nanotubes and assembled an Mg//Mg(BH$_4$)$_2$- LiBH$_4$/diglyme//Bi cell to obtain a pre-magnesiated Bi (Mg$_3$Bi$_2$) anode. The full cell showed about 90 mAh g$^{-1}$ at a rate of 0.1 C using the Mg$_3$Bi$_2$ anode coupling with Mo$_6$S$_8$ cathode in a conventional electrolyte 0.4 M Mg(TFSI)$_2$/diglyme, suggesting the compatibility between the Mg$_3$Bi$_2$ anode and conventional electrolyte. Subsequently, Murgia et al. [108] investigated the discharge/charge properties of ball-milled Mg$_3$Bi$_2$ anode using 0.5 M Mg(TFSI)$_2$/diglyme as the electrolyte and Mo$_6$S$_8$ as the cathode. They proved the reversible process by ex situ XRD study, although the results of discharge/charge show incomplete demagnesiation of Mg$_3$Bi$_2$. The Mg$_3$Bi$_2$ nanocluster has been synthesized by one-step solid-state alloying at 650°C for MIBs reported by Tan et al. [109] The Mg$_3$Bi$_2$ anodes were tested in cells coupling with high-voltage cathode Prussian blue and Mg(TFSI)$_2$-LiTFSI/Acetonitrile electrolyte, which shows more than 200 cycles at a current density of 200 mA g$^{-1}$ with 88% capacity retention and 58 mAh g$^{-1}$ at 2 A g$^{-1}$ (Figure 5.11). In addition, Matsui et al. [110] have studied the performance of the Mg$_3$Bi$_2$ thin films fabricated by magnetron sputtering operating in Mg(TFSI)$_2$/acetonitrile (AN) or glyme-based solutions. The Mg$_3$Bi$_2$ electrode shows no passivation and excellent reversibility with low overpotential. The XPS results reveal the reduced decomposition of TFSI$^-$ anion on the Mg$_3$Bi$_2$ film surface compared with Mg metal. Meng et al. [111] innovatively proposed to construct a Mg$_3$Bi$_2$//S cell by replacing metal Mg with Mg$_3$Bi$_2$ employing in simple Mg(TFSI)$_2$/ DME electrolyte. As shown in Figure 5.12, the cell shows the initial discharge capacity is above 700 mAh g$^{-1}$ (calculated by sulfur cathode) at a rate of C/2 and a capacity of 400 mAh g$^{-1}$ after 30 cycles.

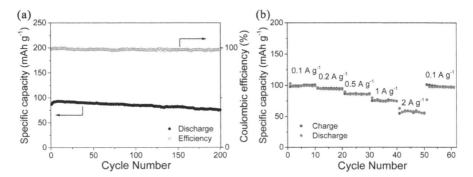

**FIGURE 5.11** Cycling stability and rate performance of PB-MB-650 full cell using Mg(TFSI)$_2$-LiTFSI/AN electrolyte. Reprinted with permission from [109]. Copyright 2018, American Chemical Society.

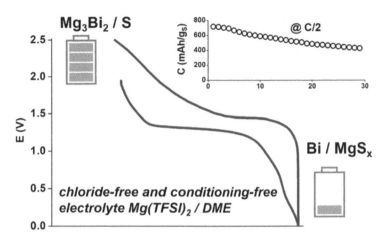

**FIGURE 5.12** Discharge–charge profiles and cycling stability of Mg$_3$Bi$_2$//S full cell. Reprinted with permission from [111]. Copyright 2019, American Chemical Society.

To develop high energy-density Mg battery systems, the anode needs low voltage to ensure the appropriate cell voltage. Singh et al. first reported the Sn electrode displayed a lower Mg$^{2+}$ insertion/extraction voltage (0.15/0.2 V vs. Mg$^{2+}$/Mg) and higher theoretical specific capacity (903 mAh g$^{-1}$) than that of Bi. They assembled the full cell to investigate the compatibility and performance of a pre-magnesiated Sn (Mg$_2$Sn) anode in Mg(TFSI)$_2$/DME coupling with typical Mo$_6$S$_8$ cathode, showing the initial capacity of ~80 mAh g$^{-1}$ similar to that of in EtMgCl-Et$_2$AlCl/THF electrolyte. Nguyen and Song [112] have studied the electrochemical behavior of ball-milled Mg$_2$Sn versus Mg in PhMgCl/THF electrolyte, which demonstrated that Mg$_2$Sn electrode can deliver a capacity of 270 mAh g$^{-1}$. It is reported that the Mg$_2$Sn//V$_2$O$_5$ full cell employing the Mg(TFSI)$_2$/diglyme electrolyte can be reversible. The bulk Mg$_2$Sn as anode for MIBs can form the nanostructured Sn by dealloying in situ. Asl et al. [113] revealed the demagnesiation of Mg$_2$Sn brings about Sn nanoparticles and nanoporous Sn structures. It is noted that the formed nano-Sn by pulverization of active materials can improve the magnesium kinetics and shorten the diffusion length of ions but can also cause irreversible capacity loss as out of contact with the matrix. Ikhe et al. [114] prepared a 3Mg/Mg$_2$Sn alloy anode consisting of three phases, including crystalline Mg-rich phase, amorphous Mg-rich phase and intermetallic Mg$_2$Sn phase. The crystalline Mg-rich phase was irreversibly dissolved during the first discharge, which generated voids and increased the specific surface area of the electrode to support its excellent rate performance and cycling performance. The Mg battery prototype, 3Mg/Mg$_2$Sn anode, coupled with the Mo$_6$S$_8$ cathode in Mg(HMDS)$_2$-MgCl$_2$/THF electrolyte, showed an energy density of ~60 Wh kg$^{-1}$ (based on Mo$_6$S$_8$ mass) for 50 cycles. It is worth mentioning that the spontaneous degradation behavior of Mg$_2$Sn anode in APC electrolytes was revealed experimentally and computationally. The anions present in the APC (e.g. Ph$_4$Al$^-$, Ph$_2$AlCl$^-$, PhAlCl$_3^-$ and AlCl$_4^-$) can extract Mg from Mg$_2$Sn electrodes through galvanic replacement reactions [16].

Murgia et al. [115] first reported that In electrode delivered a specific capacity of 425 mAh g$^{-1}$ at a low rate (C/100) versus the Mg counterelectrode with the EtMgCl-Et$_2$AlCl/THF electrolyte, showing the alloying potential as low as 0.09 V vs. Mg$^{2+}$/Mg. However, the In electrode shows rapid capacity fading with the cycle rate increase. Their studies revealed the two-phase reaction mechanism (Mg$^{2+}$ + In + 2e$^-$ ⇌ MgIn) during the alloy/dealloy process. They also synthesized the MgIn alloy using a mechanical alloying method and observed similar electrochemical behavior. The other $p$-block elements were also investigated such as low melting point (29.8°C) Ga. Mg–Ga alloys can form Mg$_2$Ga$_5$, MgGa$_2$, MgGa, Mg$_2$Ga, and Mg$_5$Ga$_2$ alloys. Wang et al. [116] prepared Mg$_2$Ga$_5$ alloy materials by melting Ga and Mg chips in a stoichiometric ratio under Ar environment at 700°C. The Mg$_2$Ga$_5$ electrodes showed no reactivity at 20°C while delivering a capacity of about 213 mAh g$^{-1}$ after 1,000 cycles at a high rate of 3 C tested at 40 °C. The excellent long-cycle life and outstanding rate capability can be ascribed to the solid–liquid phase transformation process of the Ga/Mg$_2$Ga$_5$ system. The self-healing property of the Mg–Ga alloys can alleviate the capacity loss due to huge volume change, which are typical of Mg alloy anodes. The Mg$_2$Ga$_5$ electrodes were examined in APC electrolyte with/without the addition of LiCl salt. Notably, the important system is to study the performance based on conventional electrolyte in which high-voltage and high-capacity cathode can work.

### 5.4.2.2   Other Alloy Anodes

Alloy anodes have received wide attention due to their particular advantages. Some $p$-block elements (e.g., Bi, Sn, Sb, In and Ga) have been reported to alloy with Mg to form Mg$_x$M delivering high specific capacity at low alloying potentials. The alloy anodes may reduce or avoid surface passivation against conventional electrolytes due to their higher potentials versus Mg$^{2+}$/Mg. Their derivative alloys may improve the Mg storage properties due to the synergistic effect. However, the alloy anodes still face some challenges such as severe volume change, limited cycle life and low Mg$^{2+}$ kinetics. In addition, the alloy materials directly used as anode must require the Mg-containing cathode for the practical application of MIBs.

The fast Mg$^{2+}$ kinetics of Bi alloy anodes is the unique property among the alloy anode materials for MIBs because the diffusion of bivalent Mg$^{2+}$ ions is much more sluggish than that of monovalent cations. The diffusion behavior of Mg$^{2+}$ ions in Bi has been investigated through DFT calculation, which shows a similar magnitude of diffusion coefficient as that of Li$^+$ ions in Bi [117]. Another attractive feature is that the theoretical volumetric capacity of Bi–Mg alloy (3783 mAh cm$^{-3}$) is comparable to that of Mg metal (3,833 mAh cm$^{-3}$) [118]. The reaction process of two-phase Bi/Mg$_3$Bi$_2$ is consistent with the Bi–Mg phase diagram, which shows no other intermetallic compounds [119]. The electrodeposited Bi electrode was first proposed to be an anode material for MIBs by Arthur et al. [120]. The Bi electrodes were tested in a half cell (Bi//Mg) operating EtMgCl-Et$_2$AlCl/THF as the electrolyte. The Bi electrode showed a specific capacity of 222 mAh g$^{-1}$ after 100 cycles at a current density corresponding to 1 C (385 mA g$^{-1}$). Moreover, they also demonstrated the Bi–Mg anodes have good compatibility with the Mg(TFSI)$_2$/CH$_3$CN electrolyte. The Bi nanotubes (Bi-NTs) synthesized by Shao et al. [107] had uniform diameters of ~8 nm

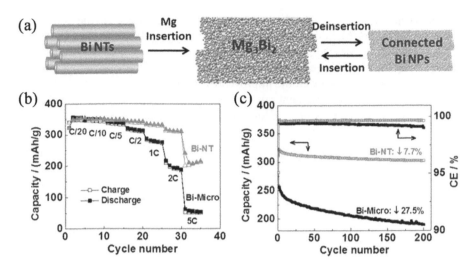

**FIGURE 5.13** Structural transformation of Bi-NTs during discharge–charge process and the rate capability and cycling stability of Bi nanotube and micro-Bi electrodes. Reprinted with permission from [107]. Copyright 2014, American Chemical Society.

through the redox reaction between $BiCl_3$ and Zn powders. The Bi-NTs electrodes were tested versus Mg metal employing the $Mg(BH_4)_2$–$LiBH_4$/diglyme electrolyte in a coin cell. The Bi-NTs electrodes showed more adaptable to large volume changes with less loss of electrical connection of active materials (Figure 5.13). The results in Figure 5.13 showed excellent rate capability (216 mAh g$^{-1}$ at 5 C) and cycling stability (303 mAh g$^{-1}$ after 200 cycles with >90% capacity retention). In contrast, the commercial micro-Bi electrode showed a worse cycle performance with 27% of capacity fading under the same testing conditions. The superior performance of Bi-NTs was attributed to the tubular structure, which can withstand the volume expansion (100%) during the magnesiated process and reduce the capacity loss due to the active material out of the electronic connection. Wang et al. [121] first investigated the Mg ion storage properties of the bismuth oxyfluoride (BiOF) nanosheets synthesized by a solvothermal method. The magnesiated production of the BiOF electrode was proved to be MgO, $MgF_2$ and $Mg_3Bi_2$ after first discharge. The reversible electrochemical magnesium storage of BiOF is mainly due to $Bi/Mg_3Bi_2$. The cycle stability (capacity retention >96% after 100 cycles at 300 mA g$^{-1}$) of the BiOF electrode may be attributed to the space confinement related to inactive MgO and $MgF_2$. In addition to the studies about the nanocrystallization and modification of Bi-based anodes (e.g., Bi nanocrystals [118], mesoporous Bi [122] and Bi nanowires [123]), bismuth-based composites are another design strategy for further improving its electrochemical performance. Penki et al. [124] prepared the Bi and reduced graphene oxide (Bi/RGO) nanocomposites with different component ratios by a solvothermal method. The introduction of RGO can improve the electronic conductivity and reduce the volume changes during the dismagesiated/magnesiated reaction of active material Bi. Among these electrodes, the nanocomposite of Bi60 (60% Bi: 40% RGO) delivered a discharge capacity of 372 mAh g$^{-1}$ at a current density of 39 mA g$^{-1}$ in the 50th cycle. Cen et al. [125] fabricated the Bi nanorods with an N-doped carbon shell (Bi@NC) as the alternative anodes for MIBs. The Bi@NC electrodes were tested in coin-type cells with Mg foil both as reference electrode and counterelectrode operating the APC electrolytes. The results showed the improved cycle stability (320 mAh g$^{-1}$ after 100 cycles at 100 mA g$^{-1}$) and excellent rate capacity (275 mAh g$^{-1}$ at 1 A g$^{-1}$), which are related to the increased electronic conductive and reduced mechanical strain of Bi@NC electrodes due to the carbon-coated shell.

Another promising alloy element for MIBs is Sn, which possessed a lower alloying voltage, higher theoretical specific capacity and more earth abundancy than Bi. Moreover, each Sn atom allows an exchange of four electrons to form $Mg_2Sn$ (Sn + $2Mg^{2+}$ + $4e^-$ → $Mg_2Sn$) more than what occurs for each Bi atom ($2Bi + 3Mg^{2+} + 6e^-$ → $Mg_3Bi_2$). Singh et al. [126] first proposed an Sn anode for MIBs along with good compatibility with conventional electrolytes. However, the discharge/charge results of the Sn electrode in a half cell (Sn// EtMgCl-Et$_2$AlCl/THF//Mg) showed a CE lower than 40% in the first cycle. To design the Sn electrode with better electrochemical performance, the reaction mechanism of the Sn electrode needs to be investigated. The possible mechanism for the capacity fading could be ascribed to the large volume change (214%) during phase transformation between Sn and $Mg_2Sn$ and partial Sn active mass out of matrix contact. Nguyen et al. [127] studied the magnesium storage performance and surface film formation behavior of the bulk and film Sn electrodes for MIBs as anodes. The bulk Sn electrodes were tested in an Mg half-cell in combination with Mg metal in a 0.5 M PhMgCl/THF electrolyte and delivered the reversible capacities of 321–289 mAh g$^{-1}$ at a current density of 52 mA g$^{-1}$. The results of XPS analysis revealed that the surface of cycled Sn thin film electrodes is composed of various inorganic salts of Sn and Mg ($Mg_2Sn$, Mg, $MgCO_3$, SnO, etc.) and organic functionality decomposed by THF. Nacimiento et al. [128] prepared commercial micro-sized and nano-sized Sn electrodes and investigated their electrochemical performance, which was tested with 0.5 M PhMgCl/THF or 0.5 M EtMgCl/THF as the electrolyte versus Mg metal in a half-cell. The micro-sized Sn electrode showed low electrochemical reactivity and delivered a specific capacity of below 10 mAh g$^{-1}$, while the nano-sized Sn electrode exhibited an initial capacity of 225 mAh g$^{-1}$ at a current density of 10 mA g$^{-1}$. They further investigated the nano-sized Sn as a potential anode material employing the Sn//Mg(ClO$_4$)$_2$/AN//MgMn$_2$O$_4$ full cell, showing the reversible discharge capacity of around 25 mAh g$^{-1}$.

Alloys containing two or more metallic elements can be well used as magnesium storage anode materials to obtain new properties due to the possible synergy effects. The introduced metals can be electrochemically active materials that contribute to the capacity, and inactive materials functioned by mitigating large volume changes and increasing the electrical conductivity. The application of multi-element alloys based on Bi and/or Sn components for magnesium ion batteries has received a lot of attention. Arthur et al. [120] attempted to obtain the high energy density anode by alloying low redox potential Bi with high theoretical capacity Sb. However, the $Bi_{0.88}Sb_{0.12}$ and $Bi_{0.55}Sb_{0.45}$ films they fabricated by electrodeposition failed to possess better electrochemical performance than the pure Bi film prepared in the same way. The results showed the strong Mg–Sb bond impeded the reversible Mg storage of

Sb. Murgia et al. [129] studied the electrochemical behavior of the $Sb_{1-x}Bi_x$ prepared by high-energy ball milling. They revealed the one-step reaction from $Sb_{1-x}Bi_x$ to $Mg_3(Sb_{1-x}Bi_x)_2$ by operando XRD and ex situ NMR during the first alloying process. However, the phase separation between Bi and $Mg_3Sb_2$ results in an irreversible capacity loss upon the subsequent charge process. They also prepared InBi alloys using a mechanical alloying method to combine the two active elements [130]. Nevertheless, the cycling stability and rate capability of InBi electrodes are between pure Bi and In electrodes. Pure Sb as the anode material for MIBs is believed to be electrochemical inactivity. Interestingly, the recent research results reported by Blondeau et al. [131] showed the element Sb in InSb alloy anode still exhibited a certain electrochemical reactivity (a reversible capacity of ~30 mAh g⁻¹) after 40 cycles at a C/50 rate, showing the alloying could unlock the reversibility of Sb. Our team also investigated the magnesium storage properties of various ternary micro-sized Bi-Sb-Sn alloys [132]. The findings that the three-step reaction and multi-phase transformation during the discharge/charge process of the Bi-Sb-Sn anodes favor the reversible magnesium storage and the transport of $Mg^{2+}$ ion in Sb and Sn, showing better cycling stability and rate capability than micrometric Bi. In particular, the $Bi_{10}Sb_{10}Sn_{80}$ anode delivered a reversible capacity of 397 mAh g⁻¹ after 60 cycles at a current density of 50 mA g⁻¹ and showed a high capacity of 417 mAh g⁻¹ at a current density of 500 mA g⁻¹. In addition, as reported in [133], another ternary multi-phased of In-Sn-Bi alloys from our team demonstrated that through the design of multiphase fine structure, the reversible storage of magnesium ions in three elements was realized by activating electrochemical reaction with phase transition. Specially, $In_{10}Sn_{10}Bi_{80}$ anode exhibited a ultrahigh stability with the specific capacity retention rate over 98% and the specific capacity of 260 mAh g⁻¹ after 80 cycles were also shown.

Niu et al. [134] reported a feasible strategy to fabricate various porous Bi–Sn alloys through the design of Al–Bi–Sn precursors and selective dealloying, which showed the magnesium storage performance of the porous Bi–Sn electrodes is dependent on the alloy composition and size. Among these porous Bi–Sn alloy anodes, the porous Bi–Sn anode showed superior rate performance and cycling stability with 93% capacity retention after 200 cycles at 1 A g⁻¹. Song et al. [135] prepared eutectic-like biphasic Bi–Sn films on a Cu foil substrate without any conductive agent or adhesive using a one-step magnetron co-sputtering method. The Bi–Sn films delivered a high specific capacity of 538 mAh g⁻¹ at 50 mA g⁻¹ and showed an excellent specific capacity of 417 mAh g⁻¹ at 1 A g⁻¹, which is attached with the favorable synergy of overlapping Bi/Sn phase design and the increased alloy phase interface. They also revealed, combining experimental investigation and DFT calculations, the enhanced electrochemical reactivity of Sn by introducing the second phase of Bi. [136] The results showed that a trace amount of Bi doping can effectively improve the reactivity of Sn with $Mg^{2+}$ ion.

The SnSb/graphene composites proposed by Parent et al. [137] were used as anodes for MIBs, showing that the formed nano-Sn in situ in Figure 5.14 due to the phase separation exhibited high electrochemical reactivity for reversible Mg storage. Subsequently, they further investigated the phase transformation pathways of the SnSb/graphene electrodes during the Mg insertion/extraction process [138]. Their research also showed the $Mg_2Sn/Mg_3Sb_2$ interface stabilized the electrode materials and exhibited good compatibility with $Mg(TFSI)_2$/diglyme solution.

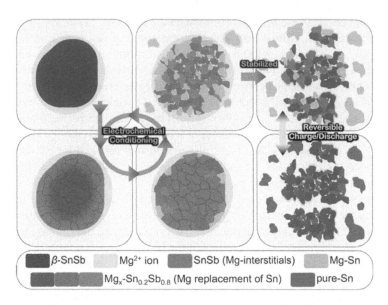

**FIGURE 5.14** Structural and morphological transformations of Sn–Sb alloy during alloying/dealloying processes. Reprinted with permission from [137]. Copyright 2015, American Chemical Society.

### 5.4.3 INSERTION ANODES

#### 5.4.3.1 Metal Oxides

The insertion-type anodes are believed to be applied to a rocking chair system with little volume change and low strain. Some metal oxides have been studied as $Mg^{2+}$ insertion-type anodes for MIBs. Among them, titanium-based compounds have been considered to be promising $Mg^{2+}$ insertion hosts due to their nontoxicity, low cost, low strain and excellent cycling performance.

Inspired by the successful application of spinel $Li_4Ti_5O_{12}$ (LTO) in lithium-ion batteries, Wu et al. [139] reported that the LTO can exhibit good reversible capacity (175 mAh $g^{-1}$) as $Mg^{2+}$ insertion-type anode for MIBs. Notably, they found that the LTO was converted to magnesium titanate, a unique transformation mechanism different from the classical $Li^+/Na^+$ insertion/extraction mechanism in LTO, by the gradual replacement of $Li^+$ by $Mg^{2+}$. In addition, the LTO electrode delivered an outstanding cycling performance of more than 95% capacity retention over 500 cycles with a tiny volume change of about 0.8% after the initial activation. Then, they investigated the effect of the LTO particle size on the electrochemical magnesium storage behavior of MIBs [140]. The results showed experimentally and theoretically that the crystal particle size played a significant role in the Mg storage performance of the LTO electrode and showed a strongly size-dependent.

The layered $Na_2Ti_3O_7$ (NTO) nanoribbons reported by Chen et al. [141] for the first time also showed the potential as a reversible anode for MIBs. The $Mg^{2+}$ ion storage mechanism of NTO was revealed as a two-step process in Figure 5.15. First, NTO reacted with $Mg^{2+}$ to irreversibly form the $MgNaTi_3O_7$ in the first discharge.

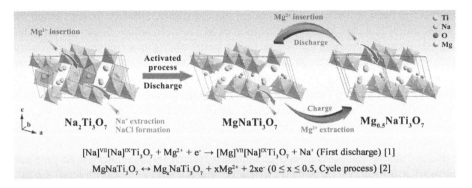

$$[Na]^{VII}[Na]^{IX}Ti_3O_7 + Mg^{2+} + e^- \rightarrow [Mg]^{VII}[Na]^{IX}Ti_3O_7 + Na^+ \text{ (First discharge) [1]}$$

$$MgNaTi_3O_7 \leftrightarrow Mg_xNaTi_3O_7 + xMg^{2+} + 2xe^- \text{ (0} \leq x \leq 0.5, \text{ Cycle process) [2]}$$

**FIGURE 5.15** Schematic diagram of $Mg^{2+}$ insertion-extraction mechanism in NTO structure. Reprinted with permission from [141]. Copyright 2016, American Chemical Society.

In the subsequent cycles, a highly reversible $Mg^{2+}$ ion insertion/extraction process in the activated NTO was shown with a theoretical capacity of 88 mAh g$^{-1}$. Most remarkably, the NTO anode was tested coupling with a $V_2O_5$ cathode operating in $Mg(ClO_4)_2$/diglyme electrolyte, showing good reversible capacity (75 mAh g$^{-1}$) and a stable energy density (53 Wh kg$^{-1}$). The recent research of Luo et al. [142] showed that the $Na_2Ti_6O_{13}$ (NT6) electrode with the three-dimensional microporous structure demonstrated better Mg storage performance than layered $Na_2Ti_3O_7$ (NT3) nanowires. The NT6 electrode reached a high discharge capacity of 165.8 mAh g$^{-1}$ at a current density of 10 mA g$^{-1}$, showing outstanding structural stability during the $Mg^{2+}$ ion insertion/extraction processes due to its favorable structure.

Other metal oxides such as $TiO_2$-B [143] and $Li_3VO_4$ [144] also show the potential for reversible Mg storage but are far away from practical application requirements. In summary, the investigation on the possible use of metal oxides as anodes for MIBs is mainly applying the electrode material suitable for Li$^+$ and Na$^+$ ion insertion. They can provide relatively low reversible capacities as compared to other anode materials. A reasonable structure design would help enhance their performance.

### 5.4.3.2 Two-Dimensional Materials

Since the discovery of graphene in 2004, an increasing interest has been generated in the exploration of two-dimensional (2D) materials. The 2D materials could possess superior physicochemical properties to conduct reversible Mg insertion/exaction than that of bulk materials. Therefore, the possibilities of using various 2D materials as MIB anodes have been investigated by employing first-principles calculations currently.

Er and co-workers studied the adsorption of Mg on defective graphene and graphene allotropes by DFT calculations [145]. They predicted the Mg storage capacity of graphene with a varying degree of divalent and Stone–Wales defects, in which the theoretical capacity is as high as 1042 mAh g$^{-1}$ for graphene with 25% divalent defects. Using the first-principles study, Shomali et al. [146] looked into the absorption of Mg on the surface of graphyne, a novel carbon allotrope. The obtained results from band structure and DOS calculations suggested that the pristine graphyne showed metallic nature after the adsorption of Mg ions. In addition, they investigated the stable adsorption sites and the diffusion paths of Mg in graphyne to predict its electrochemical behavior as an anode material for MIBs. Further study about three graphene-like carbon-nitrogen ($C_2N$, $C_3N$ and $g-C_3N_4$) materials showed that the three types of original structures have semiconductor characteristics [147]. $C_2N$ was considered to be promising anode material for MIBs by the calculated results including the high theoretical capacity (588.4 mAh $g^{-1}$) and low diffusion energy barrier.

Recent theoretical studies about 2D boron-based materials showed their potential application as an anode material for MIBs, of which Cao et al. [148] reported the $BC_3$ as the representative monolayers. The $BC_3$ monolayers exhibited a small open-circuit voltage of 0.05 V and high storage capacity of 796 mAh $g^{-1}$. Moreover, the calculation results showed the migration energy barrier for Mg on the $BC_3$ monolayer as low as 0.096 eV. It is worth mentioning that the antimonene and antimonen/graphene heterostructure also showed the potential application as anode for MIBs according to the calculation study reported by Liang et al. [149]. The calculation results showed the bonding energy of antimonene and antimonen/graphene heterostructure with Mg atoms were respectively –0.72 eV and –0.7 eV, which are thermodynamically acceptable. In particular, the antimonen/graphene heterostructure showed better electrical conductivity and a lower diffusion barrier than that of the monolayer antimonene.

## 5.5　CONCLUSION

MIBs have received much attention in the field of energy storage beyond LIBs in recent years. Much effort has been made to conduct the high energy density Mg batteries. However, the relatively poor compatibility of anode–electrolyte–cathode and slow kinetics of $Mg^{2+}$ ions in cathode materials still need to be improved. For the compatibility issue, optimizing the electrolyte solutions can be promising to match the Mg anode and high voltage cathode for high-performance MIBs. The electrolyte solutions also require noncorrosive, high safety and low cost materials, which remains a challenge. The alternative anodes can be employed to overcome the Mg passivation problem in simple conventional electrolytes from which high voltage/ high capacity cathodes can work effectively. Alloy compounds have certain application prospects, but the increased cost and relatively high anode potential will be far from the expectations of low-cost and high energy density. In addition, alloy anodes without Mg will shrink the optional materials to Mg-containing cathode because the pre-magnesiation can further increase costs. Therefore, modifying the Mg metal surface and seeking appropriate magnesium alloy anodes are more potential directions for the actual applications of MIBs. For the kinetics issue, the focus of research on cathode materials for magnesium batteries is to improve transition metal oxide/sulfide by changing the internal structure to facilitate the rapid insertion and extraction of $Mg^{2+}$ ions at room temperature. The reaction mechanism of cathode materials for MIBs is complex and the influencing factors of $Mg^{2+}$ ion diffusion kinetics still need to be extensively studied. Furthermore, it is necessary to develop new materials adapting to efficient $Mg^{2+}$ insertion/extraction for high capacity and long-life MIBs.

The future breakthrough can be related to the improved compatibility of electrode–electrolyte and rapid kinetics for $Mg^{2+}$ ions, which will accelerate the application of MIBs for large energy storage.

## ACKNOWLEDGMENT

This work is financially supported by National Natural Science Foundation of China (52171100, 51971044), and Fundamental Research Funds for the Central Universities (2021CDJXDJH003).

## REFERENCES

[1] B. Dunn, H. Kamath, J.M. Tarascon, Electrical energy storage for the grid: A battery of choices, *Science*, 334 (2011) 928–935.

[2] J.B. Goodenough, Electrochemical energy storage in a sustainable modern society, *Energy Environ. Sci.*, 7 (2014) 14–18.

[3] A. Manthiram, Electrical energy storage: Materials challenges and prospects, *MRS Bull.*, 41 (2016) 624–631.

[4] S. Su, Z. Huang, Y. NuLi, F. Tuerxun, J. Yang, J. Wang, A novel rechargeable battery with a magnesium anode, a titanium dioxide cathode, and a magnesium borohydride/tetraglyme electrolyte, *Chem. Commun.*, 51 (2015) 2641–2644.

[5] T.J. Carter, R. Mohtadi, T.S. Arthur, F. Mizuno, R. Zhang, S. Shirai, J.W. Kampf, Boron clusters as highly stable magnesium-battery electrolytes, *Angew. Chem. Int. Ed.*, 53 (2014) 3173–3177.

[6] Y.A. Wu, Z. Yin, M. Farmand, Y.-S. Yu, D.A. Shapiro, H.-G. Liao, W.-I. Liang, Y.-H. Chu, H. Zheng, In-situ multimodal imaging and spectroscopy of Mg electrodeposition at electrode–electrolyte interfaces, *Scientific Reports*, 7 (2017) 42527.

[7] O. Tutusaus, R. Mohtadi, N. Singh, T.S. Arthur, F. Mizuno, Study of electrochemical phenomena observed at the Mg metal/electrolyte interface, *ACS Energy Lett.*, 2 (2016) 224–229.

[8] J.S. Lowe, D.J. Siegel, Reaction pathways for solvent decomposition on magnesium anodes, *J. Phys. Chem. C*, 122 (2018) 10714–10724.

[9] H. Kuwata, M. Matsui, N. Imanishi, Passivation layer formation of magnesium metal negative electrodes for rechargeable magnesium batteries, *J. Electrochem. Soc.*, 164 (2017) A3229–A3236.

[10] D. Aurbach, I. Weissman, Y. Gofer, E. Levi, Nonaqueous magnesium electrochemistry and its application in secondary batteries, *Chem. Record*, 3 (2003) 61–73.

[11] D. Aurbach, Y. Gofer, A. Schechter, O. Chusid, H. Gizbar, Y. Cohen, M. Moshkovich, R. Turgeman, A comparison between the electrochemical behavior of reversible magnesium and lithium electrodes, *J. Power Sources*, 97–98 (2001) 269–273.

[12] R. Attias, M. Salama, B. Hirsch, Y. Goffer, D. Aurbach, Anode–electrolyte interfaces in secondary magnesium batteries, *Joule*, 3 (2019) 27–52.

[13] T.D. Gregory, R.J. Hoffman, R.C. Winterton, Nonaqueous electrochemistry of magnesium – applications to energy-storage, *J. Electrochem. Soc.*, 137 (1990) 775–780.

[14] D. Aurbach, Z. Lu, A. Schechter, Y. Gofer, H. Gizbar, R. Turgeman, Y. Cohen, M. Moshkovich, E. Levi, Prototype systems for rechargeable magnesium batteries, *Nature*, 407 (2000) 724–727.

[15] D. Aurbach, G.S. Suresh, E. Levi, A. Mitelman, O. Mizrahi, O. Chusid, M. Brunelli, Progress in rechargeable magnesium battery technology, *Adv. Mater.*, 19 (2007) 4260.

[16] Z. Wang, A. Bandyopadhyay, H. Kumar, M. Li, A. Venkatakrishnan, V.B. Shenoy, E. Detsi, Degradation of magnesium-ion battery anodes by galvanic replacement reaction in all-phenyl complex electrolyte, *J. Energy Storage*, 23 (2019) 195–201.

[17] L.W. Gaddum, H.E. French, The electrolysis of Grignard solution, *J. Am. Chem. Soc.*, 49 (1927) 1295–1299.

[18] J.M. Nelson, W.V. Evans, The electromotive force developed in cells containing nonaqueous liquids – [Preliminary paper], *J. Am. Chem. Soc.*, 39 (1917) A82–A83.

[19] H.S. Kim, T.S. Arthur, G.D. Allred, J. Zajicek, J.G. Newman, A.E. Rodnyansky, A.G. Oliver, W.C. Boggess, J. Muldoon, Structure and compatibility of a magnesium electrolyte with a sulphur cathode, *Nat. Commun.*, 2 (2011) 427.

[20] Z. Zhao-Karger, X.Y. Zhao, O. Fuhr, M. Fichtner, Bisamide based non-nucleophilic electrolytes for rechargeable magnesium batteries, *RSC Adv.*, 3 (2013) 16330–16335.

[21] C. Liao, N. Sa, B. Key, A.K. Burrell, L. Cheng, L.A. Curtiss, J.T. Vaughey, J.J. Woo, L.B. Hu, B.F. Pan, Z.C. Zhang, The unexpected discovery of the Mg(HMDS)(2)/ MgCl2 complex as a magnesium electrolyte for rechargeable magnesium batteries, *J. Mater. Chem. A*, 3 (2015) 6082–6087.

[22] Y. Viestfrid, M.D. Levi, Y. Gofer, D. Aurbach, Microelectrode studies of reversible Mg deposition in THF solutions containing complexes of alkylaluminum chlorides and dialkylmagnesium, *J. Electroanal. Chem.*, 576 (2005) 183–195.

[23] R.E. Doe, R. Han, J. Hwang, A.J. Gmitter, I. Shterenberg, H.D. Yoo, N. Pour, D. Aurbach, Novel, electrolyte solutions comprising fully inorganic salts with high anodic stability for rechargeable magnesium batteries, *Chem. Commun.*, 50 (2014) 243–245.

[24] S.Y. Ha, Y.W. Lee, S.W. Woo, B. Koo, J.S. Kim, J. Cho, K.T. Lee, N.S. Choi, Magnesium(II) bis(trifluoromethane sulfonyl) imide-based electrolytes with wide electrochemical windows for rechargeable magnesium batteries, *ACS Appl. Mater. Interfaces*, 6 (2014) 4063–4073.

[25] R. Jay, A.W. Tomich, J. Zhang, Y.F. Zhao, A. De Gorostiza, V. Lavallo, J.C. Guo, Comparative study of Mg(CB11H12)(2) and Mg(TFSI)(2) at the magnesium/electrolyte interface, *ACS Appl. Mater. Interfaces*, 11 (2019) 11414–11420.

[26] Y.W. Cheng, R.M. Stolley, K.S. Han, Y.Y. Shao, B.W. Arey, N.M. Washton, K.T. Mueller, M.L. Helm, V.L. Sprenkle, J. Liu, G.S. Li, Highly active electrolytes for rechargeable Mg batteries based on a [Mg-2(mu-Cl)(2)](2+) cation complex in dimethoxyethane, *PCCP*, 17 (2015) 13307–13314.

[27] J.G. Connell, B. Genorio, P.P. Lopes, D. Strmcnik, V.R. Stamenkovic, N.M. Markovic, Tuning the reversibility of Mg anodes via controlled surface passivation by H2O/Cl- in organic electrolytes, *Chem. Mater.*, 28 (2016) 8268–8277.

[28] E. Peled, The electrochemical-behavior of alkali and alkaline-earth metals in nonaqueous battery systems – The solid electrolyte interphase model, *J. Electrochem. Soc.*, 126 (1979) 2047–2051.

[29] E. Peled, H. Straze, Kinetics of magnesium electrode in thionyl chloride solutions, *J. Electrochem. Soc.*, 124 (1977) 1030–1035.

[30] Z. Ma, M. Kar, C.L. Xiao, M. Forsyth, D.R. MacFarlane, Electrochemical cycling of Mg in Mg[TFSI](2)/tetraglyme electrolytes, *Electrochem. Commun.*, 78 (2017) 29–32.

[31] J.H. Connor, W.E. Reid, G.B. Wood, Electrodeposition of metals from organic solutions .5. Electrodeposition of magnesium and magnesium alloys, *J. Electrochem. Soc.*, 104 (1957) 38–41.

[32] D. Samuel, C. Steinhauser, J.G. Smith, A. Kaufman, M.D. Radin, J. Naruse, H. Hiramatsu, D.J. Siegel, Ion pairing and diffusion in magnesium electrolytes based on magnesium borohydride, *ACS Appl. Mater. Interfaces*, 9 (2017) 43755–43766.

[33] H.M. Xu, Z.H. Zhang, J.J. Li, L.X. Qiao, C.L. Lu, K. Tang, S.M. Dong, J. Ma, Y.J. Liu, X.H. Zhou, G.L. Cui, Multifunctional additives improve the electrolyte properties of magnesium borohydride toward magnesium-sulfur batteries, *ACS Appl. Mater. Interfaces*, 10 (2018) 23757–23765.

[34] R. Mohtadi, M. Matsui, T.S. Arthur, S.J. Hwang, Magnesium borohydride: From hydrogen storage to magnesium battery, *Angew. Chem. Int. Edit.*, 51 (2012) 9780–9783.

[35] S.J. Su, Y. Nuli, N. Wang, D. Yusipu, J. Yang, J.L. Wang, Magnesium borohydride-based electrolytes containing 1-butyl-1-methylpiperidinium bis(trifluoromethyl sulfonyl)imide ionic liquid for rechargeable magnesium batteries, *J. Electrochem. Soc.*, 163 (2016) D682–D688.

[36] R. Horia, D.T. Nguyen, A.Y.S. Eng, Z.W. Seh, Using a chloride-free magnesium battery electrolyte to form a robust anode-electrolyte nanointerface, *Nano Lett.*, 21 (2021) 8220–8228.

[37] O. Tutusaus, R. Mohtadi, T.S. Arthur, F. Mizuno, E.G. Nelson, Y.V. Sevryugina, An efficient halogen-free electrolyte for use in rechargeable magnesium batteries, *Angew. Chem. Int. Ed.*, 54 (2015) 7900–7904.

[38] Z. Zhao-Karger, M.E.G. Bardaji, O. Fuhr, M. Fichtner, A new class of non-corrosive, highly efficient electrolytes for rechargeable magnesium batteries, *J. Mater. Chem. A*, 5 (2017) 10815–10820.

[39] A.B. Du, Z.H. Zhang, H.T. Qu, Z.L. Cui, L.X. Qiao, L.L. Wang, J.C. Chai, T. Lu, S.M. Dong, T.T. Dong, H.M. Xu, X.H. Zhou, G.L. Cui, An efficient organic magnesium borate-based electrolyte with non-nucleophilic characteristics for magnesium-sulfur battery, *Energy Environ. Sci.*, 10 (2017) 2616–2625.

[40] Z.H. Zhang, Z.L. Cui, L.X. Qiao, J. Guan, H.M. Xu, X.G. Wang, P. Hu, H.P. Du, S.Z. Li, X.H. Zhou, S.M. Dong, Z.H. Liu, G.L. Cui, L.Q. Chen, Novel design concepts of efficient mg-ion electrolytes toward high-performance magnesium-selenium and magnesium-sulfur batteries, *Adv. Energy Mater.*, 7 (2017) 1602055.

[41] S. Ikeda, M. Takahashi, J. Ishikawa, K. Ito, Solid electrolytes with multivalent cation conduction .1. Conducting species in Mg-Zr-PO$_4$ system, *Solid State Ionics*, 23 (1987) 125–129.

[42] J. Kawamura, K. Morota, N. Kuwata, Y. Nakamura, H. Maekawa, T. Hattori, N. Imanaka, Y. Okazaki, G.Y. Adachi, High temperature P-31 NMR study on Mg2+ ion conductors, *Solid State Commun.*, 120 (2001) 295–298.

[43] N. Imanaka, Y. Okazaki, G.Y. Adachi, Bivalent magnesium ionic conduction in the magnesium phosphate based composites, *Chem. Lett.* (1999) 939–940.

[44] N. Imanaka, Y. Okazaki, G. Adachi, Divalent magnesium ion conducting characteristics in phosphate based solid electrolyte composites, *J. Mater. Chem.*, 10 (2000) 1431–1435.

[45] S. Higashi, K. Miwa, M. Aoki, K. Takechi, A novel inorganic solid state ion conductor for rechargeable Mg batteries, *Chem. Commun.*, 50 (2014) 1320–1322.

[46] P. Canepa, S.H. Bo, G.S. Gautam, B. Key, W.D. Richards, T. Shi, Y.S. Tian, Y. Wang, J.C. Li, G. Ceder, High magnesium mobility in ternary spinel chalcogenides, *Nat. Commun.*, 8 (2017) 1759.

[47] L.P. Wang, Z. Zhao-Karger, F. Klein, J. Chable, T. Braun, A.R. Schur, C.R. Wang, Y.G. Guo, M. Fichtner, MgSc(2)Se(4)A Magnesium solid ionic conductor for all-solid-state Mg batteries?, *ChemSusChem*, 12 (2019) 2286–2293.

[48] O. Chusid, Y. Gofer, H. Gizbar, Y. Vestfrid, E. Levi, D. Aurbach, I. Riech, Solid-state rechargeable magnesium batteries, *Adv. Mater.*, 15 (2003) 627–630.

[49] M. Morita, T. Shirai, N. Yoshimoto, M. Ishikawa, Ionic conductance behavior of poly-meric gel electrolyte containing ionic liquid mixed with magnesium salt, *J. Power Sources*, 139 (2005) 351–355.

[50] A.B. Du, H.R. Zhang, Z.H. Zhang, J.W. Zhao, Z.L. Cui, Y.M. Zhao, S.M. Dong, L.L. Wang, X.H. Zhou, G.L. Cui, A crosslinked polytetrahydrofuran-borate-based polymer electrolyte enabling wide-working-temperature-range rechargeable magnesium batteries, *Adv. Mater.*, 31 (2019) 1805930.

[51] T. Luo, Y.P. Liu, H.F. Su, R.C. Xiao, L.T. Huang, Q. Xiang, Y. Zhou, C.G. Chen, Nanostructured-VO2(B): A high-capacity magnesium-ion cathode and its electro-chemical reaction mechanism, *Electrochim. Acta*, 260 (2018) 805–813.

[52] D.M. Gao, J.R. Dong, R.C. Xiao, B. Shang, D.M. Yu, C.G. Chen, Y.P. Liu, K. Zheng, F.S. Pan, Fast kinetics of monoclinic VO2(B) bulk upon magnesiation via DFT plus U calculations, *PCCP*, 24 (2022) 2150–2157.

[53] S.F. Vyboishchikov, J. Sauer, (V2O5)(n) gas-phase clusters (n=1–12) compared to V2O5 crystal: DFT calculations, *J. Phys. Chem. A*, 105 (2001) 8588–8598.

[54] X.L. Li, P. Meduri, X.L. Chen, W. Qi, M.H. Engelhard, W. Xu, F. Ding, J. Xiao, W. Wang, C.M. Wang, J.G. Zhang, J. Liu, Hollow core-shell structured porous Si-C nanocomposites for Li-ion battery anodes, *J. Mater. Chem.*, 22 (2012) 11014–11017.

[55] A. Esmanski, G.A. Ozin, Silicon inverse-opal-based macroporous materials as nega-tive electrodes for lithium ion batteries, *Adv. Funct. Mater.*, 19 (2009) 1999–2010.

[56] Y. Yao, M.T. McDowell, I. Ryu, H. Wu, N.A. Liu, L.B. Hu, W.D. Nix, Y. Cui, Interconnected silicon hollow nanospheres for lithium-ion battery anodes with long cycle life, *Nano Lett.*, 11 (2011) 2949–2954.

[57] D.Y. Chen, X. Mei, G. Ji, M.H. Lu, J.P. Xie, J.M. Lu, J.Y. Lee, Reversible lithium-ion storage in silver-treated nanoscale hollow Porous Silicon Particles, *Angew. Chem. Int. Ed.*, 51 (2012) 2409–2413.

[58] J.K. Yoo, J. Kim, Y.S. Jung, K. Kang, Scalable fabrication of silicon nanotubes and their application to energy storage, *AM*, 24 (2012) 5452–5456.

[59] N. Liu, H. Wu, M.T. McDowell, Y. Yao, C.M. Wang, Y. Cui, A yolk-shell design for stabilized and scalable Li-ion battery alloy anodes, *Nano Lett.*, 12 (2012) 3315–3321.

[60] G. Gershinsky, H.D. Yoo, Y. Gofer, D. Aurbach, Electrochemical and spectroscopic analysis of Mg2+ intercalation into thin film electrodes of layered oxides: V2O5 and MoO3, *Langmuir*, 29 (2013) 10964–10972.

[61] R.G. Zhang, X.Q. Yu, K.W. Nam, C. Ling, T.S. Arthur, W. Song, A.M. Knapp, S.N. Ehrlich, X.Q. Yang, M. Matsui, Alpha-MnO2 as a cathode material for rechargeable Mg batteries, *Electrochem. Commun.*, 23 (2012) 110–113.

[62] N. Kumagai, S. Komaba, H. Sakai, N. Kumagai, Preparation of todorokite-type manganese-based oxide and its application as lithium and magnesium rechargeable battery cathode, *J. Power Sources*, 97–8 (2001) 515–517.

[63] Q. Feng, H. Kanoh, K. Ooi, Manganese oxide porous crystals, *J. Mater. Chem.*, 9 (1999) 319–333.

[64] Y.F. Shen, R.P. Zerger, R.N. Deguzman, S.L. Suib, L. Mccurdy, D.I. Potter, C.L. Oyoung, Manganese oxide octahedral molecular-sieves – preparation, characteriza-tion, and applications, *Science*, 260 (1993) 511–515.

[65] S. Rasul, S. Suzuki, S. Yamaguchi, M. Miyayama, Synthesis and electrochemical behavior of hollandite MnO2/acetylene black composite cathode for secondary Mg-ion batteries, *Solid State Ionics*, 225 (2012) 542–546.

[66] S. Rasul, S. Suzuki, S. Yamaguchi, M. Miyayama, High capacity positive electrodes for secondary Mg-ion batteries, *Electrochim. Acta*, 82 (2012) 243–249.

[67] K.W. Nam, S. Kim, S. Lee, M. Salama, I. Shterenberg, Y. Gofer, J.S. Kim, E. Yang, C.S. Park, J.S. Kim, S.S. Lee, W.S. Chang, S.G. Doo, Y.N. Jo, Y. Jung, D. Aurbach, J.W. Choi, The high performance of crystal water containing manganese birnessite cathodes for magnesium batteries, *Nano Lett.*, 15 (2015) 4071–4079.

[68] S. Kim, K.W. Nam, S. Lee, W. Cho, J.S. Kim, B.G. Kim, Y. Oshima, J.S. Kim, S.G. Doo, H. Chang, D. Aurbach, J.W. Choi, Direct observation of an anomalous spinel-to-layered phase transition mediated by crystal water intercalation, *Angew. Chem. Int. Ed.*, 54 (2015) 15094–15099.

[69] S.Q. Zhao, T.M. Liu, D.W. Hou, W. Zeng, B. Miao, S. Hussain, X.H. Peng, M.S. Javed, Controlled synthesis of hierarchical birnessite-type $MnO2$ nanoflowers for supercapacitor applications, *Appl. Surf. Sci.*, 356 (2015) 259–265.

[70] J.L. Zhou, L. Yu, M. Sun, S.Y. Yang, F. Ye, J. He, Z.F. Hao, Novel synthesis of birnessite-type $MnO2$ nanostructure for water treatment and electrochemical capacitor, *Ind. Eng. Chem. Res.*, 52 (2013) 9586–9593.

[71] M.Q. Wang, S.S.K. Yagi, Layered birnessite $MnO2$ with enlarged interlayer spacing for fast Mg-ion storage, *J. Alloys Compd.*, 820 (2020) 153135.

[72] D.A. Kitchaev, S.T. Dacek, W.H. Sun, G. Ceder, Thermodynamics of phase selection in $MnO2$ framework structures through alkali intercalation and hydration, *J. Am. Chem. Soc.*, 139 (2017) 2672–2681.

[73] H. Kurihara, T. Yajima, S. Suzuki, Preparation of cathode active material for rechargeable magnesium battery by atmospheric pressure microwave discharge using carbon felt pieces, *Chem. Lett.*, 37 (2008) 376–377.

[74] C. Fong, B.J. Kennedy, M.M. Elcombe, A powder neutron-diffraction study of lambda-manganese-dioxide and gamma-manganese-dioxide and of $LiMn_2O_4$, *Z. Kristallogr.*, 209 (1994) 941–945.

[75] H.G. Wang, S. Yuan, D.L. Ma, X.B. Zhang, J.M. Yan, Electrospun materials for lithium and sodium rechargeable batteries: from structure evolution to electrochemical performance, *Energy Environ. Sci.*, 8 (2015) 1660–1681.

[76] Z.F. Wang, Z.H. Ruan, W.S. Ng, H.F. Li, Z.J. Tang, Z.X. Liu, Y.K. Wang, H. Hu, C.Y. Zhi, , *Small Methods*, 2 (2018) 1800150.

[77] P. Saha, P.H. Jampani, M.K. Datta, C.U. Okoli, A. Manivannan, P.N. Kumta, A convenient approach to Mo6S8 chevrel phase cathode for rechargeable magnesium battery, *J. Electrochem. Soc.*, 161 (2014) A593–A598.

[78] E. Lancry, E. Levi, Y. Gofer, M. Levi, G. Salitra, D. Aurbach, Leaching chemistry and the performance of the Mo6S8 cathodes in rechargeable Mg batteries, *Chem. Mater.*, 16 (2004) 2832–2838.

[79] J. Richard, A. Benayad, J.F. Colin, S. Martinet, Charge transfer mechanism into the Chevrel phase Mo6S8 during Mg intercalation, *J. Phys. Chem. C*, 121 (2017) 17096–17103.

[80] H.D. Yoo, I. Shterenberg, Y. Gofer, G. Gershinsky, N. Pour, D. Aurbach, Mg rechargeable batteries: an on-going challenge, *Energy Environ. Sci.*, 6 (2013) 2265–2279.

[81] X.J. He, R.C. Wang, H.M. Yin, Y.F. Zhang, W.K. Chen, S.P. Huang, 1T-MoS2 monolayer as a promising anode material for (Li/Na/Mg)-ion batteries, *Appl. Surf. Sci.*, 584 (2022) 152537.

[82] S.J.R. Tan, S. Sarkar, X.X. Zhao, X. Luo, Y.Z. Luo, S.M. Poh, I. Abdelwahab, W. Zhou, T. Venkatesan, W. Chen, S.Y. Quek, K.P. Loh, Temperature- and phase-dependent phonon renormalization in 1T '-MoS2, *Acs Nano*, 12 (2018) 5051–5058.

[83] J.D. Yang, J.X. Wang, L. Zhu, X. Wang, X.Y. Dong, W. Zeng, J.F. Wang, F.S. Pan, Boosting magnesium storage in MoS(2)via a 1T phase introduction and interlayer expansion strategy: theoretical prediction and experimental verification, *Sustainable Energy Fuels*, 5 (2021) 5471–5480.

[84] C.L. Wu, G.Y. Zhao, X.B. Yu, C. Liu, P.B. Lyu, G. Maurin, S.R. Le, K.N. Sun, N.Q. Zhang, MoS2/graphene heterostructure with facilitated Mg-diffusion kinetics for high-performance rechargeable magnesium batteries, *Chem. Eng. J.*, 412 (2021) 128736.

[85] S.K. Kolli, A. Van der Ven, First-principles study of spinel MgTiS2 as a cathode material, *Chem. Mater.*, 30 (2018) 2436–2442.

[86] N. Amir, Y. Vestfrid, O. Chusid, Y. Gofer, D. Aurbach, Progress in nonaqueous magnesium electrochemistry, *J. Power Sources*, 174 (2007) 1234–1240.

[87] M. Liu, A. Jain, Z.Q. Rong, X.H. Qu, P. Canepa, R. Malik, G. Ceder, K.A. Persson, Evaluation of sulfur spinel compounds for multivalent battery cathode applications, *Energy Environ. Sci.*, 9 (2016) 3201–3209.

[88] X.Q. Sun, P. Bonnick, V. Duffort, M. Liu, Z.Q. Rong, K.A. Persson, G. Ceder, L.F. Nazar, A high capacity thiospinel cathode for Mg batteries, *Energy Environ. Sci.*, 9 (2016) 2273–2277.

[89] Y.P. Gu, Y. Katsura, T. Yoshino, H. Takagi, K. Taniguchi, Rechargeable magnesium-ion battery based on a TiSe2-cathode with d-p orbital hybridized electronic structure, *Sci. Rep.*, 5 (2015) 12486.

[90] T. Jin, H.X. Li, K.J. Zhu, P.F. Wang, P. Liu, L.F. Jiao, Polyanion-type cathode materials for sodium-ion batteries, *Chem. Soc. Rev.*, 49 (2020) 2342–2377.

[91] Z.Z. Feng, J. Yang, Y. Nuli, J.L. Wang, X.J. Wang, Z.X. Wang, Preparation and electrochemical study of a new magnesium intercalation material Mg1.03Mn0.97SiO4, *Electrochem. Commun.*, 10 (2008) 1291–1294.

[92] Z.Z. Feng, J. Yang, Y.N. NuLi, J.L. Wang, Sol-gel synthesis of Mg1.03Mn0.97SiO4 and its electrochemical intercalation behavior, *J. Power Sources*, 184 (2008) 604–609.

[93] Y.N. Nuli, J. Yang, J.L. Wang, Y. Li, Electrochemical intercalation of $Mg^{2+}$ in magnesium manganese silicate and its application as high-energy rechargeable magnesium battery cathode, *J. Phys. Chem. C*, 113 (2009) 12594–12597.

[94] K. Makino, Y. Katayama, T. Miura, T. Kishi, Magnesium insertion into Mg0.5+y(FeyTi1-y)2(PO4)(3), *J. Power Sources*, 97–8 (2001) 512–514.

[95] Y. Mizuno, M. Okubo, E. Hosono, T. Kudo, K. Oh-ishi, A. Okazawa, N. Kojima, R. Kurono, S. Nishimura, A. Yamada, Electrochemical $Mg^{2+}$ intercalation into a bimetallic CuFe Prussian blue analog in aqueous electrolytes, *J. Mater. Chem. A*, 1 (2013) 13055–13059.

[96] D.M. Kim, Y. Kim, D. Arumugam, S.W. Woo, Y.N. Jo, M.S. Park, Y.J. Kim, N.S. Choi, K.T. Lee, Co-intercalation of Mg2+ and Na+ in Na0.69Fe2(CN)(6) as a high-voltage cathode for magnesium batteries, *ACS Appl. Mater. Interfaces*, 8 (2016) 8554–8560.

[97] X.Q. Sun, V. Duffort, L.F. Nazar, Prussian blue Mg-Li hybrid batteries, *Adv. Sci.*, 3 (2016) 1600044.

[98] W.Y. Li, C.S. Li, C.Y. Zhou, H. Ma, J. Chen, Metallic magnesium nano/mesoscale structures: Their shape-controlled preparation and Mg/air battery applications, *Angew. Chem. Int. Ed.*, 45 (2006) 6009–6012.

[99] S.-B. Son, T. Gao, S.P. Harvey, K.X. Steirer, A. Stokes, A. Norman, C. Wang, A. Cresce, K. Xu, C. Ban, An artificial interphase enables reversible magnesium chemistry in carbonate electrolytes, *Nat. Chem.*, 10 (2018) 532–539.

[100] B. Li, R. Masse, C. Liu, Y. Hu, W. Li, G. Zhang, G. Cao, Kinetic surface control for improved magnesium-electrolyte interfaces for magnesium ion batteries, *Energy Storage Mater.*, 22 (2019) 96–104.

[101] J. Luo, Y. Xia, J. Zhang, X. Guan, R. Lv, Enabling Mg metal anodes rechargeable in conventional electrolytes by fast ionic transport interphase, *Natl. Sci. Rev.*, 7 (2020) 333–341.

[102] Y.M. Zhao, A.B. Du, S.M. Dong, F. Jiang, Z.Y. Guo, X.S. Ge, X.L. Qu, X.H. Zhou, G.L. Cui, A bismuth-based protective layer for magnesium metal anode in noncorrosive electrolytes, *ACS Energy Lett.*, 6 (2021) 2594–2601.

[103] X. Li, T. Gao, F. Han, Z. Ma, X. Fan, S. Hou, N. Eidson, W. Li, C. Wang, Reducing Mg anode overpotential via ion conductive surface layer formation by iodine additive, *Adv. Energy Mater.*, 8 (2018) 1701728.

[104] J. Zhang, X. Guan, R. Lv, D. Wang, P. Liu, J. Luo, Rechargeable Mg metal batteries enabled by a protection layer formed in vivo, *Energy Storage Mater.*, 26 (2020) 408–413.

[105] K. Tang, A. Du, S. Dong, Z. Cui, X. Liu, C. Lu, J. Zhao, X. Zhou, G. Cui, A stable solid electrolyte interphase for magnesium metal anode evolved from a bulky anion lithium salt, *Adv. Mater.*, 32 (2019) 1904987.

[106] A. Maddegalla, A. Mukherjee, J.A. Blazquez, E. Azaceta, O. Leonet, A.R. Mainar, A. Kovalevsky, D. Sharon, J.F. Martin, D. Sotta, Y. Ein-Eli, D. Aurbach, M. Noked, AZ31 Magnesium alloy foils as thin anodes for rechargeable magnesium batteries, *ChemSusChem*, 14 (2021) 4690–4696.

[107] Y. Shao, M. Gu, X. Li, Z. Nie, P. Zuo, G. Li, T. Liu, J. Xiao, Y. Cheng, C. Wang, J.-G. Zhang, J. Liu, Highly reversible Mg insertion in nanostructured Bi for Mg ion batteries, *Nano Lett.*, 14 (2013) 255–260.

[108] F. Murgia, L. Stievano, L. Monconduit, R. Berthelot, Insight into the electrochemical behavior of micrometric Bi and Mg3Bi2 as high performance negative electrodes for Mg batteries, *J. Mater. Chem. A*, 3 (2015) 16478–16485.

[109] Y.-H. Tan, W.-T. Yao, T. Zhang, T. Ma, L.-L. Lu, F. Zhou, H.-B. Yao, S.-H. Yu, High voltage magnesium-ion battery enabled by nanocluster Mg3Bi2 alloy anode in noncorrosive electrolyte, *ACS Nano*, 12 (2018) 5856–5865.

[110] M. Matsui, H. Kuwata, D. Mori, N. Imanishi, M. Mizuhata, Destabilized passivation layer on magnesium-based intermetallics as potential anode active materials for magnesium ion batteries, *Front. Chem.*, 7 (2019) 00007.

[111] Z. Meng, D. Foix, N. Brun, R. Dedryvère, L. Stievano, M. Morcrette, R. Berthelot, Alloys to replace Mg anodes in efficient and practical Mg-ion/sulfur batteries, *ACS Energy Lett.*, 4 (2019) 2040–2044.

[112] D.-T. Nguyen, S.-W. Song, Magnesium stannide as a high-capacity anode for magnesium-ion batteries, *J. Power Sources*, 368 (2017) 11–17.

[113] H. Yaghoobnejad Asl, J. Fu, H. Kumar, S.S. Welborn, V.B. Shenoy, E. Detsi, In situ dealloying of bulk Mg2Sn in Mg-ion half cell as an effective route to nanostructured Sn for high performance Mg-ion battery anodes, *Chem. Mater.*, 30 (2018) 1815–1824.

[114] A.B. Ikhe, S.C. Han, S.J.R. Prabakar, W.B. Park, K.-S. Sohn, M. Pyo, 3Mg/Mg2Sn anodes with unprecedented electrochemical performance towards viable magnesium-ion batteries, *J. Mater. Chem. A*, 8 (2020) 14277–14286.

[115] F. Murgia, E.T. Weldekidan, L. Stievano, L. Monconduit, R. Berthelot, First investigation of indium-based electrode in Mg battery, *Electrochem. Commun.*, 60 (2015) 56–59.

[116] L. Wang, S.S. Welborn, H. Kumar, M. Li, Z. Wang, V.B. Shenoy, E. Detsi, High-rate and long cycle-life alloy-type magnesium-ion battery anode enabled through (de) magnesiation-induced near-room-temperature solid–liquid phase transformation, *Adv. Energy Mater.*, 9 (2019) 1902086.

[117] S.C. Jung, Y.-K. Han, Fast magnesium ion transport in the bi/Mg3Bi2 two-phase electrode, *J. Phys. Chem. C*, 122 (2018) 17643–17649.

[118] K.V. Kravchyk, L. Piveteau, R. Caputo, M. He, N.P. Stadie, M.I. Bodnarchuk, R.T. Lechner, M.V. Kovalenko, Colloidal bismuth nanocrystals as a model anode material for rechargeable Mg-ion batteries: Atomistic and mesoscale insights, *ACS Nano*, 12 (2018) 8297–8307.

[119] A.A. Nay.eb-Hashemi, J.B. Clark, The Bi-Mg (bismuth-magnesium) system, *Bull. Alloy Phase Diagrams* 6(1985) 528–533.

[120] T.S. Arthur, N. Singh, M. Matsui, Electrodeposited Bi, Sb and Bi1-xSbx alloys as anodes for Mg-ion batteries, *Electrochem. Commun.*, 16 (2012) 103–106.

[121] W. Wang, L. Liu, P.F. Wang, T.T. Zuo, Y.X. Yin, N. Wu, J.M. Zhou, Y. Wei, Y.G. Guo, A novel bismuth-based anode material with a stable alloying process by the space confinement of an in situ conversion reaction for a rechargeable magnesium ion battery, *Chem. Commun.*, 54 (2018) 1714–1717.

[122] X. Xu, D.L. Chao, B. Chen, P. Liang, H. Li, F.X. Xie, K. Davey, S.Z. Qiao, Revealing the magnesium-storage mechanism in mesoporous bismuth via spectroscopy and ab-initio simulations, *Angew. Chem. Int. Ed.*, 59 (2020) 21728–21735.

[123] Z.G. Liu, J. Lee, G.L. Xiang, H.F.J. Glass, E.N. Keyzer, S.E. Dutton, C.P. Grey, Insights into the electrochemical performances of Bi anodes for Mg ion batteries using Mg-25 solid state NMR spectroscopy, *Chem. Commun.*, 53 (2017) 743–746.

[124] T.R. Penki, G. Valurouthu, S. Shivakumara, V.A. Sethuraman, N. Munichandraiah, In situ synthesis of bismuth (Bi)/reduced graphene oxide (RGO) nanocomposites as high-capacity anode materials for a Mg-ion battery, *New J. Chem.*, 42 (2018) 5996–6004.

[125] Y. Cen, J.R. Dong, T.T. Zhu, X. Cai, X. Wang, B.B. Hu, C.A.L. Xu, D.M. Yu, Y.P. Liu, C.G. Chen, Bi nanorods anchored in N-doped carbon shell as anode for high-performance magnesium ion batteries, *Electrochim. Acta*, 397 (2021).

[126] N. Singh, T.S. Arthur, C. Ling, M. Matsui, F. Mizuno, A high energy-density tin anode for rechargeable magnesium-ion batteries, *Chem. Commun.*, 49 (2013) 149–151.

[127] D.-T. Nguyen, X.M. Tran, J. Kang, S.-W. Song, Magnesium storage performance and surface film formation behavior of tin anode material, *ChemElectroChem*, 3 (2016) 1813–1819.

[128] F. Nacimiento, M. Cabello, C. Pérez-Vicente, R. Alcántara, P. Lavela, G. Ortiz, J. Tirado, On the mechanism of magnesium storage in micro- and nano-particulate tin battery electrodes, *Nanomaterials*, 8 (2018) 501.

[129] F. Murgia, D. Laurencin, E.T. Weldekidan, L. Stievano, L. Monconduit, M.-L. Doublet, R. Berthelot, Electrochemical Mg alloying properties along the Sb1-xBix solid solution, *Electrochim. Acta*, 259 (2018) 276–283.

[130] F. Murgia, L. Monconduit, L. Stievano, R. Berthelot, Electrochemical magnesiation of the intermetallic InBi through conversion-alloying mechanism, *Electrochim. Acta*, 209 (2016) 730–736.

[131] L. Blondeau, E. Foy, H. Khodja, M. Gauthier, Unexpected behavior of the InSb alloy in Mg-ion batteries: Unlocking the reversibility of Sb, *J. Phys. Chem. C*, 123 (2019) 1120–1126.

[132] D.C. Gu, Y. Yuan, J.W. Liu, D.J. Li, W.B. Zhang, L. Wu, F.Y. Cao, J.F. Wang, G.S. Huang, F.S. Pan, The electrochemical properties of bismuth-antimony-tin alloy anodes for magnesium ion batteries, *J. Power Sources*, 548 (2022) 232076.

[133] X.W. Zheng, C. Song, Y. Yuan, D.J. Li, D.C. Gu, L. Wu, G.S. Huang, J.F. Wang, F.S. Pan, High stability In–Sn–Bi multi-element alloy anode for Mg ion batteries, *J. Power Sources*, 575 (2023) 233141.

[134] J. Niu, K. Yin, H. Gao, M. Song, W. Ma, Z. Peng, Z. Zhang, Composition- and size-modulated porous bismuth–tin biphase alloys as anodes for advanced magnesium ion batteries, *Nanoscale*, 11 (2019) 15279–15288.

[135] M. Song, J. Niu, K. Yin, H. Gao, C. Zhang, W. Ma, F. Luo, Z. Peng, Z. Zhang, Self-supporting, eutectic-like, nanoporous biphase bismuth-tin film for high-performance magnesium storage, *Nano Res.*, 12 (2019) 801–808.

[136] M. Song, T. Zhang, J. Niu, H. Gao, Y. Shi, Y. Zhang, W. Ma, Z. Zhang, Boosting electrochemical reactivity of tin as an anode for Mg ion batteries through introduction of second phase, J. Power Sources, 451 (2020) 227735.

[137] L.R. Parent, Y. Cheng, P.V. Sushko, Y. Shao, J. Liu, C.-M. Wang, N.D. Browning, Realizing the full potential of insertion anodes for Mg-ion batteries through the nanostructuring of Sn, *Nano Lett.*, 15 (2015) 1177–1182.

[138] Y. Cheng, Y. Shao, L.R. Parent, M.L. Sushko, G. Li, P.V. Sushko, N.D. Browning, C. Wang, J. Liu, Interface promoted reversible mg insertion in nanostructured tin-antimony alloys, *Adv. Mater.*, 27 (2015) 6598–6605.

[139] N. Wu, Y.C. Lyu, R.J. Xiao, X.Q. Yu, Y.X. Yin, X.Q. Yang, H. Li, L. Gu, Y.G. Guo, A highly reversible, low-strain Mg-ion insertion anode material for rechargeable Mg-ion batteries, *NPG Asia Mater.*, 6 (2014) e120.

[140] N. Wu, Y.X. Yin, Y.G. Guo, Size-dependent electrochemical magnesium storage performance of spinel lithium titanate, *Chem-Asian J.*, 9 (2014) 2099–2102.

[141] C.C. Chen, J.B. wang, Q. Zhao, Y.J. Wang, J. Chen, Layered Na2Ti3O7/MgNaTi3O7/Mg0.5NaTi3O7 nanoribbons as high-performance anode of rechargeable Mg-ion batteries, *ACS Energy Lett.*, 1 (2016) 1165–1172.

[142] L. Luo, Y.C. Zhen, Y.Z. Lu, K.Q. Zhou, J.X. Huang, Z.G. Huang, S. Mathur, Z.S. Hong, Structural evolution from layered Na2Ti3O7 to Na2Ti6O13 nanowires enabling a highly reversible anode for Mg-ion batteries, *Nanoscale*, 12 (2020) 230–238.

[143] Y. Meng, D.S. Wang, Y.Y. Zhao, R.Q. Lian, Y.J. Wei, X.F. Bian, Y. Gao, F. Du, B.B. Liu, G. Chen, Ultrathin TiO2-B nanowires as an anode material for Mg-ion batteries based on a surface Mg storage mechanism, *Nanoscale*, 9 (2017) 12934–12940.

[144] J. Zeng, Y. Yang, C. Li, J.Q. Li, J.X. Huang, J. Wang, J.B. Zhao, Li3VO4: an insertion anode material for magnesium ion batteries with high specific capacity, *Electrochim. Acta*, 247 (2017) 265–270.

[145] D. Er, E. Detsi, H. Kumar, V.B. Shenoy, Defective graphene and graphene allotropes as high-capacity anode materials for Mg ion batteries, *ACS Energy Lett.*, 1 (2016) 638–645.

[146] E. Shomali, I.A. Sarsari, F. Tabatabaei, M. Mosaferi, N. Seriani, Graphyne as the anode material of magnesium-ion batteries: Ab initio study, *Comput. Mater. Sci.*, 163 (2019) 315–319.

[147] J.H. Zhang, G. Liu, H.C. Hu, L.Y. Wu, Q. Wang, X.J. Xin, S.J. Li, P.F. Lu, Graphene-like carbon-nitrogen materials as anode materials for Li-ion and mg-ion batteries, *Appl. Surf. Sci.*, 487 (2019) 1026–1032.

[148] Y. Cao, S. Ahmai, A.G. Ebadi, N.Y. Xu, A. Issakhov, M. Derakhshandeh, Boron carbide hexagonal monolayer as promising anode material for magnesium-ion batteries, *Inorg. Chem. Commun.*, 133 (2021) 108888.

[149] Y.B. Liang, K. Liu, Z. Liu, J. Wang, C.S. Liu, Y. Liu, A First-Principles study of monolayer and heterostructure antimonene as potential anode materials for Magnesium-ion batteries, *Appl. Surf. Sci.*, 577 (2022) 151880.

# 6 Aluminum-Ion Batteries
## New Attractive Emerging Energy Storage Devices

*Hongsen Li, Huaizhi Wang, Hao Zhang, Zhengqiang Hu and Yongshuai Liu*

## 6.1 INTRODUCTION

In modern society, the development of energy storage technologies is indispensable for the efficient integration and application of intermittent renewable energies. Among various energy storage systems, electrochemical energy storage employing rechargeable batteries has been one of the most efficient systems for electrical energy storage [1]. During the past few decades, LIBs have become the most widely used devices for electrochemical energy storage ranging from portable devices to large energy grids due to their high energy density [2, 3]. Unfortunately, further applications of LIBs are encumbered severely on account of the cost and somewhat limited resources of lithium in the earth [4]. Additionally, the current LIBs still couldn't meet the increasing demand of safety concerns from thermal runaway. Therefore, low-cost batteries employing other metal anode materials including sodium [5, 6], potassium [7, 8], magnesium [9, 10], calcium [11, 12] and aluminum [13, 14] have attracted current research interest and been regarded as suitable candidates for electrochemical storage devices [15]. Among these nonlithium metals, Al could deliver an optimal specific capacity per mass unit (2980 mAh g$^{-1}$) and highest volumetric capacity (8,046 mAh cm$^{-3}$) due to three-electron redox properties (Equation (1)) [16].

$$Al^{3+} + 3e^- \leftrightarrow Al \tag{1}$$

Moreover, as AI is the most abundant metal element on the earth's crust, it is inexpensive for constructing anode materials and safe to be handled in the ambient environment, which is significant for large-scale applications. Hence, AIBs have been perceived as reliable alternative energy storage devices in recent years [17, 18]. Electrolytes reported for AIBs can be divided into two types: nonaqueous systems and aqueous systems. According to previous reports, the nonaqueous electrolyte dominated the application of electrolyte for AIBs because its low vapor pressure and wide electrochemical window are beneficial to achieve highly reversible plating/stripping efficiencies of Al [16, 19]. As the development of nonaqueous system becoming cumulatively mature, the demand for a more friendly and safer electrolyte has caused the aqueous electrolyte to emerge. The aqueous electrolyte with insensitivity to air and water simplifies the assembly process and explores primary aqueous

138    DOI: 10.1201/9781003208108-6

AIBs such as Al–air system. Another vital influencing factor of AIBs is cathode materials. They are the pivotal constituent in constructing battery systems with great electrochemical performance for widespread applications. To date, various types of cathode materials have been investigated to fabricate practical aluminum storage systems. The cathode materials for AIBs are categorized into intercalation-type electrode materials and conversion-type electrode materials based on the different energy storage mechanisms. However, the development of cathode materials for AIBs was still impeded by numerous challenges including low discharge voltage, capacitive behavior and rapid capacity decay. So far to solve these issues, many researchers have devoted themselves to designing cathode materials based on the principles including possessing appropriate and stable tunnel construction that is susceptible to intercalation, a high redox potential and desirable electrochemical performance.

In this chapter, we introduce the full scope of the state-of-the-art of AIBs, with consideration from fundamental mechanism to practical technologies, with the aim of fully understanding the electrochemistry in the AIBs. Initially, electrochemical charge storage mechanisms from the aqueous AIBs and nonaqueous AIBs are provided. Later, we would mainly focus on the development of electrodes and electrolyte for AIBs in the past years. Some inherent properties of many cathodes (carbonaceous materials, transition metal compounds and organic compounds) and electrolytes (ionic liquid (IL) electrolyte, molten salt electrolyte, solid electrolyte and hybrid electrolyte) for AIBs are also analyzed in this chapter. Through the in-depth understanding of the above topics, the causes behind the drawbacks of the existing AIBs, and how these issues can be addressed are revealed and explained. In the last section, challenges and future prospects are incorporated to provide insight and guide the readers to design and develop more reliable AIBs with great performance.

## 6.2 ELECTROCHEMICAL CHARGE STORAGE MECHANISMS

In the case of nonaqueous AIBs, the electrochemical charge storge mechanisms can be divided into three main types in the IL electrolyte of $AlCl_3$/[EMIM]Cl, including reversible intercalation/extraction reactions, adsorption/desorption reactions, and conversion reactions [20]. As for aqueous aluminum-ion batteries (AAIBs), the energy storage in cathodes is mainly based on the intercalation reactions of $Al^{3+}$ [21].

### 6.2.1 Intercalation Mechanism

This type of reaction mechanism in nonaqueous AIBs refers to the reversible intercalation of mobile guest ions ($AlCl_4^-$ anions, $Al^{3+}$ or other cations containing Al) into the lattice structure of host materials. For AAIBs, $Al^{3+}$ is found as the main guest ion in intercalation.

#### 6.2.1.1 Intercalation of $AlCl_4^-$ Anions

The process of reversible intercalation/extraction of $AlCl_4^-$ anions mostly occurs in the carbon-based materials involving graphite, graphene, metal–organic framework

(MOF) or other amorphous carbon [22–29]. The typical reaction equations can be listed as follows:

$$C_n + AlCl_4^- \leftrightarrow C_n\left[AlCl_4\right] + e^- \tag{2}$$

where $n$ is the number of carbon atoms per intercalated anion.

Figure 6.1a exhibits that $AlCl_4^-$ anions are reversibly embedded into the cathode material during the charge–discharge process, while the metal aluminum can be deposited on/stripped from the anode accompanying by the reaction with $Al_2Cl_7^-$. Moreover, the conducting polymers [30–33] are also confirmed that they can reversibly accommodate $AlCl_4^-$ anions, much like carbon-based materials, which may have potential application value in $AlCl_3$/[EMIM]Cl electrolyte [34].

### 6.2.1.2   Intercalation of Al³⁺ or Other Cations Containing Al

$Al^{3+}$ intercalation/de-intercalation reaction mechanism, existing in both nonaqueous AIBs and AAIBs, which involves the reaction between $Al^{3+}$ and certain cathode materials, such as $VO_2$ [35], $V_2O_5$ [36], $TiO_2$ [37, 38], $FeS_2$ [39], $Mo_6S_8$ [40], $CoSe_2$ [41], $MoSe_2$ [42] etc, and the cathode reaction can be expressed as follows:

$$M + 4nAl_2Cl_7^- + 3ne^- \leftrightarrow Al_nM + 7nAlCl_4^- \tag{3}$$

where M represents the host material in nonaqueous/aqueous system.

As shown in Figure 6.1b, $Al^{3+}$ cations participate in the intercalation process in cathode crystal during cycling, and the resulting $Al^{3+}$ stripped from the anode side of AIBs can combine with $AlCl_4^-$ to form $Al_2Cl_7^-$.

In addition, many other cations containing Al can also be intercalated into the host materials in nonaqueous AIBs (Figure 6.1c), including $AlCl_2^+$ [43], $AlCl^{2+}$ [44], etc. The insertion/extraction of this type of cations mainly occurs in the organic positive electrode, the charge storage mechanism of which is similar to the "rocking-chair" reactions as elaborated above.

### 6.2.2   Adsorption Mechanism

The adsorption/desorption mechanism involving $AlCl_4^-$ anions always takes place in the micro- and mesopores of the electrode materials that are mostly porous carbon materials such as carbon nanotubes (CNTs), N-doped microporous carbon and ordered mesoporous carbon (CMK-3) [20, 45]. As exhibited in Figure 6.1d, the electrochemical charge/discharge proceeds by adsorption reactions involving two different charge carriers, $AlCl_4^-$ and $Al^{3+}$, among which the former was absorbed onto the electrode surface compactly from the IL electrolyte during the charging process while the latter was absorbed onto the surface of Al anode simultaneously [46].

### 6.2.3   Conversion Mechanism

The traditional conversion reaction usually refers to the conversion of species containing multivalent elements, which can be summarized as follows:

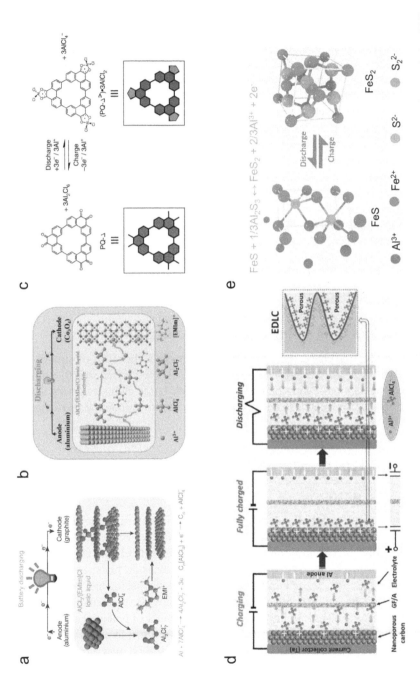

**FIGURE 6.1** Typical schematic diagrams of different reaction mechanisms. (a) Intercalation of $AlCl_4^-$ anions. Adapted with permission.[66] Copyright 2015, Springer Nature. (b) Intercalation of $Al^{3+}$ cations. Adapted with permission of $AlCl_2^+$ cations. Adapted with permission.[43] Copyright 2018, Springer Nature. (c) Intercalation of $AlCl_4^-$ anions. Adapted with permission.[46] Copyright 2017, IOP Publishing. (e) Conversion reaction for $FeS_2$ cathode in nonaqueous AIBs. Adapted with permission. [55] Copyright 2019, Elsevier.

$$mn\text{Al}^{3+} + \text{M}_n\text{X}_m + 3mne^- \leftrightarrow m\text{A}_n\text{X} + n\text{M}^0 \tag{4}$$

where M stands for transition metal/higher-valence cations ($\text{Fe}^{3+}$, $\text{Co}^{2+}$, $\text{Ni}^{2+}$, $\text{Mn}^{2+}$, $\text{V}^{3+}$, $\text{Cu}^{2+}$, etc.) [47–51] and X denotes certain anions ($\text{S}^{2-}$, $\text{P}^{3-}$, $\text{Cl}^-$, etc.) [52–54].

The charge storage mechanism in these conversion-type materials means a solid-state redox process, inevitably accompanied by the phase transition. As shown in Figure 6.1e, the $\text{FeS}_2$ cathode goes to two different phases (FeS and $\text{Al}_2\text{S}_3$) by intercalating with $\text{Al}^{3+}$ during the discharge process, which would be restored to $\text{FeS}_2$ in the subsequent charge process [55]. The irreversible phase-transfer process with structure collapse and electrode material pulverization always gives rise to the rapid capacity fading and unstable cycle performance.

In addition, sulfur-based materials (S, Se, and Te) can also act as the conversion-type cathodes in nonaqueous AIBs, which display good operating potentials and excellent theoretical capacities [47, 56]. The nature of electrochemical charge storage mechanism in these materials is still a conversion reaction that involves a transition process of elemental chalcogens to various chain polysulfides and followed by conversion into $\text{Al}_2\text{S}_3$, $\text{Se}_2\text{Cl}_2$, and $\text{TeCl}_3\text{AlCl}_4$ [57–59]. Notably, the polysulfides have been confirmed to dissolve in the ILs, generating a "shuttle effect" that is best to be prevented in metal-ion battery (MIBs) system.

### 6.2.4   Multiple Ions Involved Intercalation-Conversion Mechanism

In these AIBs cathodes, the charge storage mechanism involves the redox of multiple cationic and anionic ($\text{Cl}^-$, $\text{AlCl}_4^-$, and $\text{Al}^{3+}$). As confirmed by a recent report [60], both $\text{Cl}^-$ and $\text{AlCl}_4^-$ can act as charge carriers in a metal selenide based cathodes during the charge process, and the $\text{Al}^{3+}$ can also be embedded into the host upon the discharge process.

## 6.3   POTENTIAL ELECTRODES AND ELECTROLYTES

### 6.3.1   Potential Electrodes

On the basis of the electrochemical charge storage mechanisms, the potential electrodes in AIBs can be divided into different types, such as acidic $\text{AlCl}_3$-based and aqueous electrolyte according to the guest ions. The electrochemical behavior of guest ions is dependent on the crystal structure of electrode materials, which determines whether the cathode can provide enough interlayer spacing and withstand the structural collapse caused by volume expansion to achieve the reversible cycle of mobile guest ions [18, 61, 62]. Until now, a fairly large number of potential electrode materials have being extensively investigated globally in an effort to establish state-of-the art and commercially viable AIBs [63, 64], including carbon materials, [25, 65, 66] transition metal oxides [51, 67], metal sulfides/selenides [41, 50, 68], elemental chalcogens [57, 69, 70], and other cathodes [71, 72]. In the following subsection, we introduce and summarize the development of various types of cathode materials for nonaqueous AIBs/AAIBs.

### 6.3.1.1 Carbon Materials

As a typical anion-type cathode in nonaqueous AIBs, carbon materials with extremely high conductivity and stable structure can reversibly intercalate and de-intercalate the $AlCl_4^-$ anions, which demonstrate excellent cycling stability and high rate durability [19]. The Al/graphite AIBs with high voltage and long cycle life using a room temperature molten salt electrolyte of 1.5:1 $AlCl_3$:1,2-dimethyl-3-propylimidazolium chloride was first investigated by Gifford and Palmisano in 1988; however, the generation of gaseous $Cl_2$ in the positive electrode limits the further development of this system [73]. Fortunately, since Dai et al. [66] reported a three-dimensional graphitic-foam cathode to fabricate an ultrafast rechargeable AIBs in 2015, this significant breakthrough in AIBs has revived worldwide attention. As shown in Figure 6.2, the AIB displays an innovative performance in terms of rate capability, cycling stability, and operating voltage plateaus (~2 V). No evidence for capacity fading was observed up to 7,500 cycles at an ultrahigh current density of 4 A $g^{-1}$. Furthermore, the AIB maintained its specific capacity (~60 mAh $g^{-1}$) and cycling stability over a range of charge and discharge rates (1–6 A $g^{-1}$), which is infinitely close to the performance of certain supercapacitors.

In the next year, Dai et al. [74] further explored various graphite materials for positive electrodes in AIBs, among which natural graphite is proved to have better electrochemical performance involving high discharge capacity and well-defined voltage plateaus. In addition to graphite, graphene-based materials are also widely developed as AIB cathodes, the most representative research is a "trihigh tricontinuous" (3H3C) graphene film material reported by Gao et al. [75]. As displayed in Figure 6.3a, the cathode was synthesized with the "trihigh tricontinuous (3H3C) design". Such a cathode with targeted design demonstrates a high discharge capacity of 120 mAh $g^{-1}$ with 91.7% retention after 250,000 cycles (Figure 6.3b). Moreover, the aluminum–graphene battery operates normally within a wide temperature range, showing the splendid high and low temperature performance benefits not only from the high ionic conductivity of IL electrolyte but also from the special design (3H3C) of the cathode (Figure 6.3c). Some other graphene materials with outstanding electrochemical properties have been subsequently developed as cathodes for AIBs, such as graphene nanoribbons on highly porous 3D graphene (GNHPG) foam [25], few-layer graphene [76], and graphene nanosheet (GN) paper [77], etc.

It should be noted that the graphite-based materials are also widely employed in the AAIB system. Wu et al. [78] prepared graphite nanosheets with extensive interlayer spacing by a facile expanded method. Cyclic voltammograms (CVs) of the graphite nanosheet was tested in the electrolyte without $Al^{3+}$ (0.750 mol/L $Na_2SO_4$, 0.005 mol/L $H_2SO_4$, 0.500 mol/L $Zn(CH_3COO)_2$) and no obvious redox peak was found (Figure 6.4a), indicating that the redox reaction in the $Al_2(SO_4)_3$/$Zn(CH_3COO)_2$ aqueous electrolyte corresponds to the intercalation of $Al^{3+}$. Besides, the graphite nanosheet also exhibits good cycling behavior (94% capacity retention after 200 cycles at 0.5 A $g^{-1}$) and remarkable rate performance (60 mAh $g^{-1}$ even at 2 A $g^{-1}$). Pan et al. [79] reported a Al/graphite AAIB using "water-in-salt" high-concentration aqueous $AlCl_3$ solution as the electrolyte, which can reduce the hydrogen-evolving reaction (HER) potential to less than −2.3 V (vs. Ag/AgCl) and then achieve a uniform

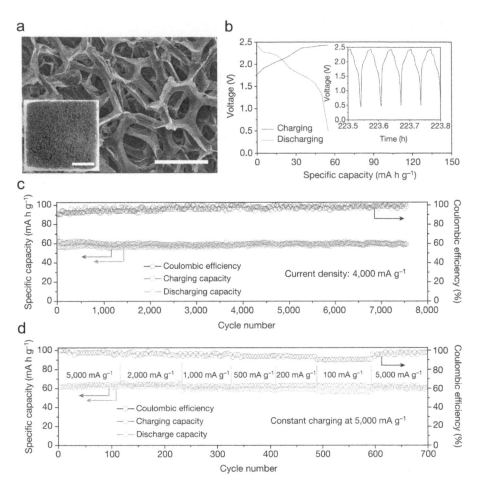

**FIGURE 6.2** An ultrafast and stable rechargeable Al/graphite cell. (a) A scanning electron microscopy image showing a graphitic foam with an open frame structure; inset, photograph of graphitic foam; (b) galvanostatic charge and discharge curves of an Al/graphitic-foam pouch cell at a current density of 4,000 mA g$^{-1}$; (c) long-term stability test over 7,500 charging and discharging cycles at a current density of 4,000 mA g$^{-1}$; and (d) an Al/graphitic-foam pouch cell charging at 5,000 mA g$^{-1}$ and discharging at current densities ranging from 100 to 5,000 mA g$^{-1}$. Adapted with permission.[66] Copyright 2015, Springer Nature.

deposition/stripping of Al (Figure 6.4b). Moreover, Al/graphite AAIBs demonstrate a high discharge capacity (165 mAh g$^{-1}$ at a current density of 0.5 A g$^{-1}$) and stable cycling life (more than 1,000 cycles with high capacity retention of 99%).

### 6.3.1.2 Transition Metal Oxides

Transition metal oxides can be used as cathode materials for both nonaqueous AIBs and AAIBs. For nonaqueous systems, vanadium oxides have attracted adequate attention in the research community due to their layered structure with anisotropic

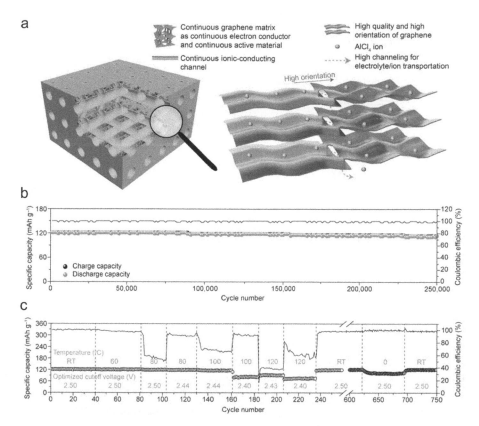

**FIGURE 6.3** AIBs fabricated with 3H3C graphene film material. (a) Illustration of 3H3C design for a desired graphene cathode; (b) stable cycling performance of the GF-HC cathode at current density of 100 A g$^{-1}$ within 250,000 cycles; and (c) specific capacities and Coulombic efficiencies of GF-HC cathodes at different temperatures from 0 to 120°C with a cutoff voltage optimization strategy. RT, room temperature. Adapted with permission.[75] Copyright 2015, American Association for the Advancement of Science.

properties [80]. Vanadium pentoxide ($V_2O_5$) can used as an enhanced cathode for AIBs, which consists of alternating layers of edge-shared and corner-shared [$VO_5$] pyramids [80]. Of note, the interlayer of such materials can host monovalent/multi-valent ions but needs to be handled with discreetness for its toxicity [81]. In 2011, Archer et al. [82] fabricated a novel AIB with $V_2O_5$ nanowire as cathode against an Al anode, delivering high discharge-specific capacity of 305 mAh g$^{-1}$ in the first discharge and charge state and 273 mAh g$^{-1}$ after 20 cycles. More importantly, the Al/ILs/$V_2O_5$ paradigm outlined a promising rechargeable battery system, and the expected high specific capacity of 442 mAh g$^{-1}$ (normalized by the mass of $V_2O_5$, corresponding to a stoichiometry of Al$V_2O_5$) spurs the rapid growth of AIB research [56]. Nevertheless, the accuracy of the above results was challenged by Menke et al. [83]. They indicate that the capacity is ascribed to corrosion reactions occurring at the stainless steel current collector, and the $V_2O_5$ electrode is electrochemically inactive

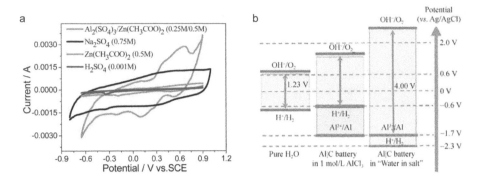

**FIGURE 6.4** Characteristics of different cathodes in AAIBs system. (a) CV curves of graphite nanosheet in different electrolytes at 1 mV s$^{-1}$. Adapted with permission.[78] Copyright 2016, American Chemical Society. (b) Illustration of an expanded electrochemical stability window for the "water-in-salt" electrolytes. Adapted with permission.[79] Copyright 2019, Royal Society of Chemistry.

to Al. When replacing the current collector with other metals (Ni foam or Mo), the electrochemical activity of $V_2O_5$ cathode in AIBs is confirmed to be reproduced [84]. Furthermore, Amine et al. [36] used Ni foam as the current collector to load the binder-free $V_2O_5$ positive electrode synthesized by direct deposition process, which demonstrates the optimized charge exchange between the active material and the current collector with good electrical contact and an enhanced $Al^{3+}$ intercalation/de-intercalation into the host lattice (Figure 6.5a). Moreover, the binder-free treatment accelerates the migration and diffusion of the electrolyte ions in the cathode with large-scale three-dimensional network structure, and finally reduces the electrochemical polarization. As shown in Figure 6.5b, the rechargeable AIBs with binder-free $V_2O_5$ deposited on Ni foam exhibit improved electrochemical performance when compared with the cell using the common binders (PVDF and PTFE), involving relatively high operating potential (0.6 V) and high initial discharge-specific capacity (239 mAh g$^{-1}$). Another vanadium oxide ($VO_2$) with metastable monoclinic structure is also reported as a positive electrode material for AIBs [35]. The charge storage mechanism inside the AIB assembled with $VO_2$ as the cathode and Al foil as the anode is based on the insertion/extraction of $Al^{3+}$, as presented in Figure 6.5c. The charge/discharge curves tested at 50 mA g$^{-1}$ (Figure 6.5d) are similar to that of $V_2O_5$, where the voltage plateaus occur at 0.5 V and 0.2 V, apart from the different cutoff voltage set in the operation protocols.

For aqueous systems, some metal oxides ($V_2O_5$ [85], $TiO_2$ [38], $MnO_2$ [86], $MoO_3$ [87], etc.) have also been developed as potential electrodes. Wu and his coworkers [88] designed an in-situ electrochemical transformation strategy involving the spinel-to-layered transition, which makes the spinel $Mn_3O_4$ transform into layered and amorphous mixed phase of $Al_xMnO_2 \cdot nH_2O$ (Figure 6.6a). The AAIB assembled with $Al_xMnO_2 \cdot nH_2O$ cathode and Al anode displays a short plateau at 1.3 V and a long plateau at 1.65 V (Figure 6.6b), corresponding to $Al^{3+}$ de-intercalation from the host lattices accompanied by the oxidation of manganese. As demonstrated in Figure 6.6c,

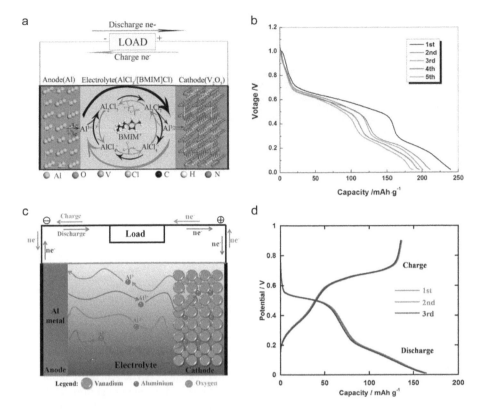

**FIGURE 6.5** The electrochemical behavior of $V_2O_5$ and $VO_2$ cathode. (a) Schematic diagram of reaction sequence in aluminum battery during charge/discharge; (b) galvanostatic discharge profiles of cell using cathode with binder-free Ni-$V_2O_5$. Adapted with permission.[36] Copyright 2015, American Chemical Society. (c) Schematic representation of the super-valent battery during charge/discharge process; and (d) initial three charge/discharge curves at a current density of 50 mA $g^{-1}$. Adapted with permission.[35] Copyright 2013, Springer Nature.

the discharge capacity in the first cycle is as high as 467 mAh $g^{-1}$ and it can still be maintained at 272 mAh $g^{-1}$ after 60 cycles. Moreover, the high discharge plateaus with an average potential of 1.1 V make the energy density of the positive electrode material up to 481 Wh $kg^{-1}$.

### 6.3.1.3 Metal Sulfides and Selenides

Metal sulfide and selenide materials are mainly exploited as promising cathodes for nonaqueous systems due to their low electronegativity and high specific capacity [89–91]. Hu et al. [68] reported a novel 3D reduced graphene oxide (rGO)-supported $SnS_2$ electrode (G-$SnS_2$) for superior nonaqueous AIBs (Figure 6.7a). The electrochemical behavior of G-$SnS_2$ and bare $SnS_2$, as shown in Figure 6.7b, demonstrate that the discharge plateau of the former (0.68 V vs Al) is much higher than for the latter (~ 0.45 V vs Al). More importantly, the G-$SnS_2$ composite displays both high specific

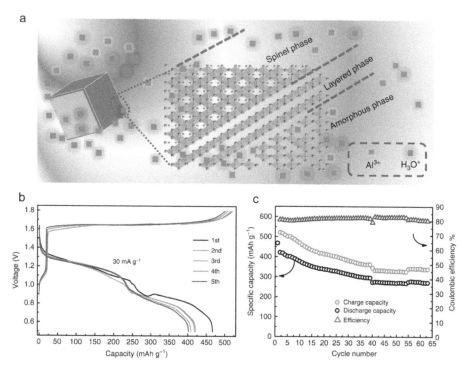

**FIGURE 6.6** AAIBs assembled with $Al_xMnO_2 \cdot nH_2O$ cathode. (a) The schematic profile of the structure of $Al_xMnO_2 \cdot nH_2O$; (b) galvanostatic charge and discharge profile; (c) efficiency and cycling ability of the rechargeable AAIB. Adapted with permission.[88] Copyright 2019, Springer Nature.

capacity (392 mAh $g^{-1}$) at the current density of 100 mA $g^{-1}$ and excellent cycling performance (maintained 70 mAh $g^{-1}$ after 100 cycles at 200 mA $g^{-1}$.) (Figure 6.7c). The splendid electrochemical performance can be assigned to two aspects. First, owing to the layered structure of $SnS_2$ materials, the chloroaluminate anion intercalation/de-intercalation during charging and discharging process is highly reversible. Second, the introduction of the rGO matrix increases the specific surface area and offers more active reaction sites at which chloroaluminate anions can interact [47]. Hence, ultrahigh performance is achieved based on a simple modification strategy.

Another novel composite electrode ($Ni_3S_2$/graphene) was designed as the cathode for a soft package AIB [92]. When assembled with Al anode in the IL electrolyte, the $Ni_3S_2$/graphene composite exhibited a high voltage plateau (~ 1.0 V) and high initial discharge-specific capacity (~ 350 mAh $g^{-1}$) as well as an outstanding cycle stability (60 mAh $g^{-1}$ after 100 cycles). In addition to composite cathodes, some positive materials with their own unique lattice structure were also exploited as the potential electrode for AIBs. Geng et al. [40] investigated a Chevrel phase material ($Mo_6S_8$) with unique 3D channels for ion diffusion, acquainted for the ability to afford multi-electron transfer due to its inherent electron-deficient characteristics [93]. Figure 6.8a exhibits two distinct discharge voltage plateaus (0.55 and 0.37 V), which can be

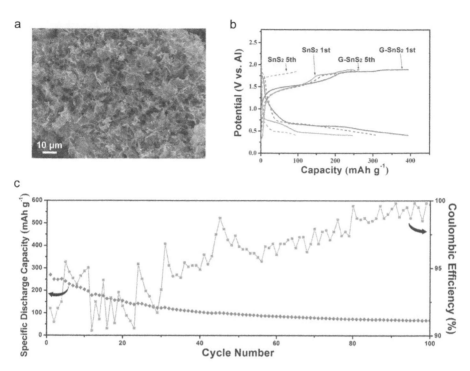

**FIGURE 6.7** Optimization of $SnS_2$-based cathode. (a) SEM images of G-$SnS_2$; (b) first/fifth charge–discharge curves of AIBs assembled with the G-$SnS_2$ and $SnS_2$ electrodes at a specific current of 100 mA $g^{-1}$; and (c) specific discharge capacity and coulombic efficiency versus cycle number with G-$SnS_2$ electrode for AIBs at a specific current of 200 mA $g^{-1}$. Adapted with permission.[68] Copyright 2017, John Wiley and Sons.

attributed to different intercalation sites similar to $Mg^{2+}$ intercalation (Figure 6.8b) [40]. The Al storage capacity of $Mo_6S_8$ cathode is up to 148 mAh $g^{-1}$ at a current density of 12 mA $g^{-1}$ (Figure 6.8a). Furthermore, the discharge capacity can be maintained at 70 mAh $g^{-1}$ after 50 cycles. Afterwards, the group further prepared and tested the layered $TiS_2$ and thiospinel $Cu_{0.31}Ti_2S_4$ cathode for AIBs system [94]. The result revealed that the layered phase ($TiS_2$) can deliver faster ionic diffusion and more stable discharge-specific capacity than the cubic phase ($Cu_{0.31}Ti_2S_4$). However, the research on layered cathodes goes far beyond the electrode material reported above. In 2020, Zhao et al. [95] designed and synthesized a star-shaped two-dimensional (2D) $WS_2$ layered cathode for high-performance AIBs, which can enable the rapid and stable storage of the large $AlCl_4^-$ anions (Figure 6.8c). As the cathode with a stable crystal structure, the star-shaped 2D $WS_2$ electrode exhibits an outstanding electrochemical performance in terms of remarkable rate capability (254 mAh $g^{-1}$ at 0.1 A $g^{-1}$, 86 mAh $g^{-1}$ at 5 A $g^{-1}$) (Figure 6.8d) and durable cycling performance (119 mAh $g^{-1}$ remained after 500 cycles), which exceeds most of the AIBs counterparts reported recently.

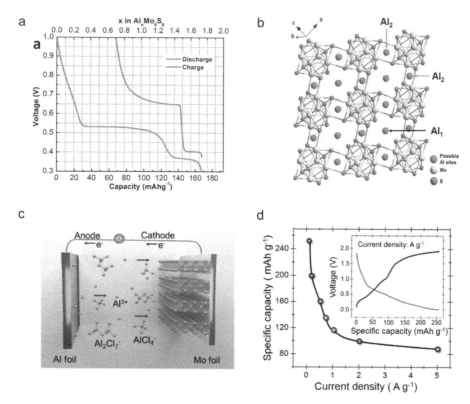

**FIGURE 6.8** Characterizations of $Mo_6S_8$ and $WS_2$ cathode materials. (a) Composition of Al intercalated $Mo_6S_8$ vs capacity; (b) crystal structure of Al intercalated $Mo_6S_8$. Adapted with permission.[40] Copyright 2015, American Chemical Society. (c) Schematic of the charging process in $Al//WS_2$ battery; and (d) rate capabilities with various current densities from 0.1 to 5 A $g^{-1}$ and the inset of (d) is the corresponding charging/discharging profiles for the star-shaped 2D $WS_2$ microsheet assemblies at 0.1 A $g^{-1}$. Adapted with permission.[95] Copyright 2020, Elsevier.

Such excellent electrochemical behavior also exists in other cathode materials with optimized structure for rechargeable AIBs, such as $VS_2$ nanosheet [96], $MoS_2$ microsphere [97], SnS porous film [98], hexagonal NiS nanobelt [49], $Co_3S_4$ microsphere [99], 2D porous $Co_9S_8$ nanosheet [100], etc. Apart from metal sulfide cathodes, metal selenides are also confirmed to be electrochemically active in AIBs. Wang et al. [41] fabricated an AIB with $CoSe_2$/C-ND@rGO as cathode materials and revealed the energy-storage mechanism involving the incorporation of $Al^{3+}$ into $CoSe_2$ lattice in the formation of $Al_mCo_nSe_2$ and cobalt species, whereas the dissolution and pulverization of active materials resulted in the capacity deterioration. Figure 6.9a illustrates the synthesis process of the $CoSe_2$/C-ND@rGO cathode, where the rGO wrapping layer enhances the structural stability and conductivity of electrode materials upon cycling. As a result, the cathode demonstrated a desirable performance (143 mAh $g^{-1}$ at 1 A $g^{-1}$ after 500 cycles) with a discharge platform at 1.8 V (Figure 6.9b).

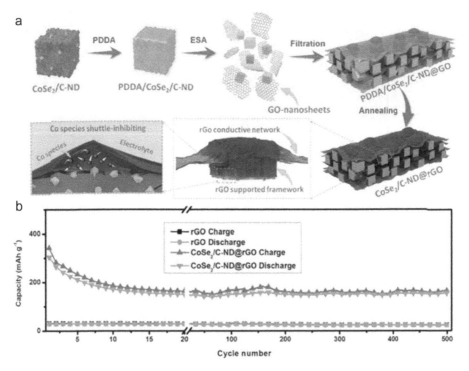

**FIGURE 6.9** Optimization of CoSe₂/C-ND electrode. (a) Schematic illustration of the synthetic strategy used to prepare CoSe₂/C-ND@rGO and the cycling performance enhancement mechanism proposed; and (b) cycling performance of CoSe₂/C-ND@rGO and pure rGO at 1000 mA g⁻¹. Adapted with permission.[41] Copyright 2018, Royal Society of Chemistry.

MoSe₂ with distinct sandwiched structure is an engaging material among the metal selenides, where the Mo layer is situated between two layers of Se and is stacked together via weak van der Waals interaction [101, 102]. In this case, Li et al. [42] synthesized a nanosized MoSe₂@carbon matrix (N-MoSe₂@C) through ion complexation and selenization strategy. Owing to the small and uniform MoSe₂ nanoparticles produced by this unique method, the reaction kinetics is greatly accelerated and the Al storage performance is further improved. Moreover, the test on the high–low temperature performance for the AIB shows that the capacity can be maintained at various temperatures. Some one-dimensional (1D) electrode materials have also been developed as cathodes for AIBs. In 2018, Zhao et al. [103] prepared a 1D $Cu_{2-x}Se$ nanorod as the cathode for AIBs with acidic ILs as the electrolyte, which revealed excellent electrochemical properties and prominent kinetics due to the high conductivity and unique 1D structure. To be specific, the discharge-specific capacity in the first cycle reaches 260 mAh g⁻¹ at the current density of 50 mA g⁻¹ with a coulombic efficiency (CE) of 93.9%. This research verifies both high electrochemical performance and in-depth charge/discharge storage mechanism, which provides a direction

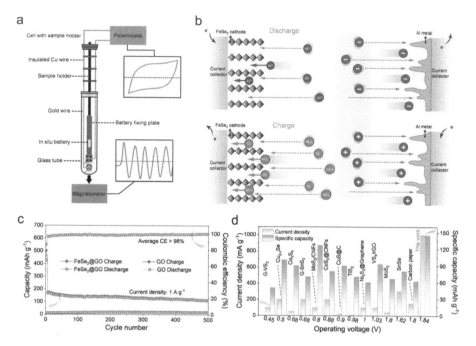

**FIGURE 6.10** Schematics of in situ magnetometry technology, reaction mechanisms and electrochemical behavior in an AIB. (a) Schematic of advanced in situ magnetic testing setup; (b) the electrochemical process for the $FeSe_2$ cathode in AIBs; (c) cycling stability of the $FeSe_2$-based cathode material over 500 cycles at 1 A g$^{-1}$; and (d) the comparison of operating voltage, capacity versus current densities of the $FeSe_2$@GO electrode with various reported electrodes for AIBs. Adapted with permission.[60] Copyright 2022, Royal Society of Chemistry.

for the choice of future AIBs cathodes. More critically, Li et al. [60] synthesized a $FeSe_2$-based cathode with 1D microrod structure, in 2021, which is adopted as the model electrode material to explore the energy storage mechanism in AIBs from an integrated chemical and physical point of view. Based on the previous studies by Li et al. on the origin of extra capacity in transition metal oxides for LIBs by in situ magnetometry [104, 105] in this work, they first extended this methodology to the field of AIBs (the setup is shown in Figure 6.10a). Surprisingly, the spin-polarized surface capacitance in $FeSe_2$-based AIBs is observed by in situ magnetometry for the first time, showing that $Al^{3+}$ can also be used as a charge compensator in the formation of space charge zones with electrons. More importantly, the comprehensive experimental characterization and theoretical calculations reveal that the electrochemical charge–discharge processes of $FeSe_2$ involve three different charge carriers including the anions $Cl^-$ and $AlCl_4^-$ and the cation $Al^{3+}$. The former two work during the charge process, while the last one participates in the reaction in the discharge process (Figure 6.10b). Furthermore, the cell assembled with $FeSe_2$ delivers an excellent electrochemical performance. As shown in Figure 6.10c, the $FeSe_2$@GO cathode displays a nearly constant specific capacity at a current density of 1 A g$^{-1}$ over 500 cycles with merely 0.06% capacity decay per cycle. In addition, the $FeSe_2$@GO electrode is able to achieve a CE of over 98% after 500 cycles. In contrast to the low operating voltage and low specific capacity at moderate current densities observed in typical transition-metal composite-based AIBs, $FeSe_2$@GO exhibits the most outstanding electrochemical performances, which is possibly due to the advantage of 1D structure of the $FeSe_2$@GO composite and the enhanced conductivity resulting from the introduction of GO. In general, this innovative work opens the door for rethinking the key rules of the design and preparation of electrode materials with high electrochemical performance to obtain significant improvement.

### 6.3.1.4  Elemental Chalcogens

Elemental chalcogen-based electrode materials with reasonable discharge plateau have attracted considerable attention due to their high cell capacity and the low cost. This class of rechargeable nonaqueous AIBs mainly includes Al–S, Al–Se, and Al–Te batteries [57, 58, 106]. Nevertheless, the nonaqueous AIB assembled with elemental chalcogen cathode and IL electrolyte usually exhibits poor reversibility and low lifespan, thus limiting its practical application. In 2018, Wan et al. [107] designed a novel host material for Al–S batteries composed of S anchored on a carbonized HKUST-1 matrix (S@HKUST-1-C). The results involving the Cu in HKUST-1 can form S–Cu ionic clusters with polysulfide confirmed by XRD and Auger spectrum, indicating that Cu enhances reaction kinetics and improves the reversibility of S in the charge–discharge process. Furthermore, the S@HKUST-1-C composite indicates high reversible capacity and outstanding capacity retention (600 mAh g$^{-1}$ at the 75th cycle and 460 mAh g$^{-1}$ at 500th cycle) as well as high CE over 95%. Compared with elemental sulfur, selenium not only possesses comparable chemical properties but also has higher conductivity and lower ionization potential ($1 \times 10^{-10}$ S m$^{-1}$, 9.7 eV). In 2019, Yu and his coworkers [108] introduced a mesoporous carbon fiber (MCF) with controllable pore structures as the cathode for Al–Se batteries. The experimental results revealed that the diffusivity of chloroaluminate ions depends on the size of mesoporous in composite Se cathodes, while the key to obtaining high electrical conductivity of the composite cathode lies in the intimate connectivity between mesoporous carbon shell and CNT core. Specifically, the composite Se cathode exhibited both high discharge-specific capacity (366 mAh g$^{-1}$ at 2 A g$^{-1}$) and stable cycle performance (retained 152 mAh g$^{-1}$ after 500 cycles) after the size was adjusted to 7.1 nm. Apart from S and Se, Te was also proposed as cathode material for high-performance AIBs due to its high electrical conductivity ($2 \times 10^{-4}$ S m$^{-1}$) and ultrahigh theoretical specific capacity (1,260.27 mAh g$^{-1}$ and voltage plateau at 1.5 V). Jiao et al. [57] investigated Te nanowire (TeNW) cathodes for a novel Al–Te battery, as denoted in Figure 6.11a. In order to solve the capacity fading and promote the rechargeable capabilities in Te cathodes, the chemical dissolution of TeNWs and their compound conversion during cycling must be optimized. Consequently, rGO nanosheets and functionalized single-walled carbon nanotubes (SWCNTs) were employed to modify the Te-based compounds and separators, respectively. As presented in Figure 6.11b, Type III Al–Te battery displays a more obvious voltage plateau with capacity of 1,150 mAh g$^{-1}$, which can ascribe to the high electrical conductivity of the rGO interlayer.

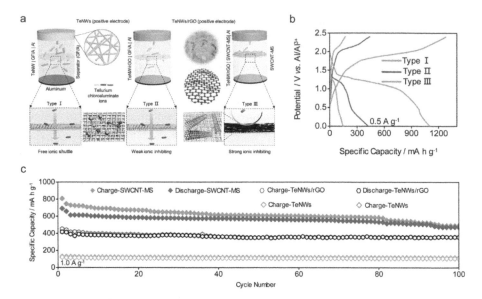

**FIGURE 6.11** The schemes for a stable Al-Te battery. (a) Schematic illustrations of various cell configurations for Al-Te battery; (b) the charge/discharge curves at the current density of 0.5 A $g^{-1}$; and (c) cycling performance of the three types of TABs at a specific current of 1 A $g^{-1}$. Adapted with permission.[57] Copyright 2019, Royal Society of Chemistry.

The considerable electrochemical performance (550 mAh $g^{-1}$ after 100 cycles at 1 A $g^{-1}$) is further manifested in Figure 6.11c, reflecting that the strategy of utilizing the rGO-modified cathodes coupled with SWCNT-modified separators can enhance the reversible energy storage ability of TeNWs.

### 6.3.1.5 Other Cathodes

In addition to carbon, transition metal oxides, metal sulfides/selenides and elemental chalcogen cathode materials, and other types of potential electrodes are also investigated for use in nonaqueous AIBs/AAIBs, including Prussian blue analogs (PBAs) [109, 110] chloride-based materials [48, 111] conducting polymers [31–33], etc. In 2015, Gao et al. [109] found that the copper hexacyanoferrate (CuHCF) could electrochemically insert Al ions reversibly in an aqueous system. This system exhibits a discharge capacity of 62.9 mAh $g^{-1}$ at 50 mA $g^{-1}$ and maintains 46.9 mAh $g^{-1}$ at 400 mA $g^{-1}$, making it a promising cathode for AAIBs. Also, the experimental results demonstrate that the electrochemical capacity is highly relied on the $Fe^{3+}/Fe^{2+}$ redox couple in the CuHCF framework rather than the guest species. Donahue and his coworkers [112] prepared an iron chloride ($FeCl_3$) with low-temperature molten salt IL electrolyte composed of the mixture of $AlCl_3$ and [EMIM]Cl. Nevertheless, the discharge-specific capacity of electrode materials is limited by the low utilization of the active material in the IL electrolyte. Moreover, the phenomenon of self-discharge also occurs in the $Al/AlCl_3$ system, which can be attributed to the dissolution of $FeCl_3$ into the electrolyte and further reaction with the Al anode. Different from the above materials, conductive polymer as an anion intercalation cathode material occupies an important position in various positive electrodes for rechargeable AIBs. Hudak et al. [30] designed polypyrrole and polythiophene cathodes and further investigated their Al storage performance, which displays the electrode capacities at near-theoretical levels (in the range of 30–100 mAh g$^{-1}$). They also proposed that the mechanism involves $AlCl_4^-$ and $Al_2Cl_7^-$ intercalation, which is confirmed by some characterization methods, such as Fourier transform infrared (FTIR) spectroscopy and elemental analysis. In addition, some nonaqueous AIB cathode materials ($LiFePO_4$ [113], $V_2CT_x$ MXene [114], $Li_3VO_4$ [115], etc.) and AAIB-positive electrode materials ($Na_3V_2(PO_4)_3$ [116], $FeVO_4$ [72], phenazine (PZ) [117], etc.) were also developed for high-performance energy storage devices. These important findings constitute a major advancement in the area of practical AIBs cathode design and facilitate the solution to fabricate affordable large-scale energy storage systems.

## 6.3.2 Typical Electrolytes

### 6.3.2.1 Ionic Liquids (IL)

The most studied systems are the nonaqueous IL electrolytes due to their thermal stability and moderate potential window. ILs used in aluminum batteries are generally Lewis acidic melts of aluminum chloride and an organic salt such as imidazolium, pyridinium or pyrrolidinium, among which electrolytes containing alkyl imidazolium cations such as 1-ethyl-3-methylimidazolium ([EMIm]$^+$) and 1-butyl-3-methylimidazolium ([BMIm]$^+$) stand out due to the high coulombic efficiency close to 100% and stable deposition of aluminum. Generally, the soluble halides of Al species in IL electrolyte would generate $Al^{3+}$, Cl$^-$ and $AlCl_4^-$ ions in solution through an asymmetric cleavage process [43]. When the molar ratio of $AlCl_3$:[EMIC]$^+$ is greater than 1, the predominant anion in the liquid is $Al_2Cl_7^-$. When the molar ratio of $AlCl_3$:[EMIC] is equal to 1, $AlCl_4^-$ anions are the only category present in the electrolyte. When the molar ratio is less than 1, $AlCl_4^-$ and Cl$^-$ can coexist [56]. The electrodeposition and stripping of aluminum atoms can only occur under acidic conditions, and the reaction mechanism is listed as follows:

$$4Al_2Cl_7^- + 3e^- \rightleftharpoons Al + 7AlCl_4^- \tag{5}$$

Thus, the Lewis acidity of chloroaluminate ILs refers to the molar ratio of $AlCl_3$ to the organic salt, which determines the corresponding speciation and its physical and electrochemical properties. The Lewis acidity of ionic liquids can be adjusted by the $AlCl_3$:[EMIC]$^+$ ratio. Dai et al. [66] compared $AlCl_3$/[EMIM]Cl with different molar ratios and found that the ionic liquid with molar ratio of 1:1.3 showed the most stable cycling performance as the electrolyte for AIB, and no capacity decay were observed after 7,500 cycles (Figure 6.12). The battery prepared with this kind of electrolyte can work normally in 0–2.45 V (versus $Al^{3+}$/Al) range. The IL electrolyte will decompose due to side reactions occurring at high voltage, which will affect the overall performance of the battery. In addition, IL $AlCl_3$/[BMIM]Cl electrolyte is also a common electrolyte with similar physicochemical properties to $AlCl_3$/[EMIM]Cl. Jiao and

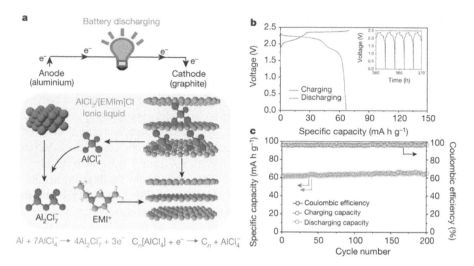

**FIGURE 6.12** Schematic drawing of Al/graphite cell during discharge, using the optimal composition of $AlCl_3$/[EMIm]Cl ionic liquid electrolyte. Adapted with permission.[66] Copyright 2015, Springer Nature.

coworkers [118] assembled an Al/graphite battery using $AlCl_3$/[BMIM]Cl electrolyte. After 100 cycles at 50 mA $g^{-1}$, the specific capacity was maintained at 86.39 mAh $g^{-1}$ and the energy density reached 148.2 Wh $kg^{-1}$. A capacity retention after 300 cycles was 90%, indicating the [BMIM]-based IL electrolyte processed good cycle stability. Besides, the [BMIM]-based IL electrolyte is widely used in transition metal chalcogenide cathode materials, such as $V_2O_5$ nanowires [119], $VO_2$ nanorods [35], $Mo_{2.5+y}VO_{9+z}$ [120] and $Cu_{0.31}Ti_2S_4$ [94].

It is noted that IL electrolytes are not limited to imidazolium-based ionic liquids. The high cost and hygroscopicity of these mixtures have prompted the investigation of ILs formed from alternative organic salts. Gao etal. [121] prepared $AlCl_3$/$Et_3NHCl$ electrolyte by using cheap commercial triethylamine hydrochloride ($Et_3NHCl$), A detailed study proved that the binding energy of $[Et_3NH]^+AlCl_4^-$ was 4.4 kcal/mol, higher than that of $[EMIM]^+AlCl_4^-$ by DFT calculation and NMR test (Figure 6.13a), which implies that $AlCl_3$/$Et_3NHCl$ electrolyte exhibits a more stable electrochemical performance than the imidazolium-based IL electrolyte. Meanwhile, shown in Figure 6.13b, DFT calculations demonstrate that $[Et_3NH]^+AlCl_4^-$ possesses a higher highest occupied molecular orbital (HOMO) and a lower lowest unoccupied molecular orbital (LUMO) than $[EMIM]^+AlCl_4^-$, indicating a wider electrochemical window, which has been confirmed by He and his colleagues. Other attempts on cations in IL electrolytes such as 1-methyl-1-propylpyrrolidinium chloride (Py13Cl) [122], imidazole hydrochloride (ImidazoleHCl) [123], benzyltriethylammonium chloride (TEBAC) [124], 1-ethyl-3methylimidazolium chloride ([C2mim]Cl) [125] and many others [126, 127] also achieved some satisfactory progress.

Undoubtedly, IL electrolyte provides new possibilities for nonaqueous AIBs, but the decomposition and dissolution of cathode materials in chloroaluminate IL

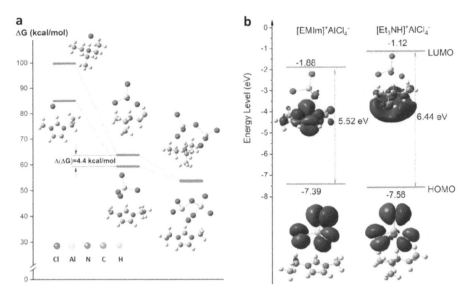

**FIGURE 6.13** DFT calculated bonding energies of $Et_3NH^+$ and $[EMIm]^+$ with $Cl^-$, $AlCl_4^-$, $Al_2Cl_7^-$ and HOMO and LUMO plots of $[EMIm]^+AlCl_4^-$ and $[Et_3NH]^+AlCl_4^-$. Adapted with permission.[121] Copyright 2019, Elsevier.

as well as some inevitable side reactions lead to relatively low Coulomb efficiency. Another strategy for designing new IL electrolyte is the replacement of $Cl^-$, which not only reduces the corrosiveness and toxicity of electrolyte but also broadens the electrochemical window of the whole system. Wu and coworkers [128] first explored the use of $OTF^-$ ion to replace $Cl^-$ in IL, and widened the voltage window to 3.25 V. However, the oxide film on the Al surface cannot be removed by $Al(OTF)_3/[BMIM]OTF$ ILs; thus the metal Al anode must be specially treated. As shown in Figure 6.14, a channel suitable for $Al^{3+}$ transport can be fabricated on the anode surface by corrosive $AlCl_3$-based electrolyte pretreatment and then a stable Al/electrolyte interface in non-corrosive $Al(OTF)_3$-based electrolyte would be built. This strategy has a very high reference significance in designing an appropriate electrolyte.

### 6.3.2.2 Deep Eutectic Solvents

The high cost and strong corrosivity of the most commonly used imidazolium-based IL electrolytes have stimulated the research of alternative electrolytes. Deep eutectic solvents have, thus, attracted much attention due to similar physical properties with ILS, including wide stable range and non-flammability but with much affordable costs, making them suitable for battery applications.

In 2017, by mixing Lewis acidic metal salt $AlCl_3$ with urea at a certain ratio of 1.5:1, Jiao et al. [129] prepared a deep eutectic solvent electrolyte ($AlCl_3$/urea). The cell with carbon paper as the positive electrode and $AlCl_3$/urea as the electrolyte demonstrated a high coulombic efficiency (>99%) and discharge plateau (1.9 V),

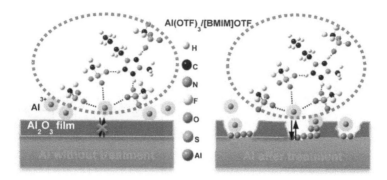

**FIGURE 6.14** Schematic diagram of Al deposition/dissolution on surface of untreated and treated Al anode. Adapted with permission.[128] Copyright 2016, American Chemical Society.

indicating that this electrolyte has a relatively ideal application potential. However, due to poor electrical conductivity, the operating temperature of 120°C is still a restricted condition for both commercial and industrial applications.

In order to improve the application performance of the urea-based deep eutectic solvents at low temperature, Dai et al. proposed the use of dichloroethyl aluminum as an additive for optimization [130] and further expanded the halogen-based urea derivative electrolyte (UR-ILA) in the follow-up work [131]. The UR-ILA electrolyte contains fewer types of ions ($AlCl_4^-$, $Al_2Cl_7^-$ and $[AlCl_2 \cdot (urea)_n]^+$), combining the light concentration of ions, resulting in less interaction between the cation and anion, leading to the lower viscosity and higher ionic conductivity of UR–ILA electrolyte. Additionally, acetamide- [132] and urea-based [133] deep eutectic solvents have also been shown to be promising for improving the cycling stability and cycle life in Al-S batteries, maintaining specific capacities of 500 mAh g$^{-1}$ for 60 and 100 cycles, respectively.

### 6.3.2.3   Inorganic Molten Salts

Inorganic molten salts electrolytes have been shown to accumulate rapid aluminum storage at elevated temperatures. Jiao and coworkers [134] mixed $AlCl_3$ with NaCl at a specific molar ratio of 1.63, resulting in an electrolyte with the lowest melting point (close to the eutectic temperature of 108°C). When the operating temperature is set at 120°C, $AlCl_4^-$ and $Al_2Cl_7^-$ coexist in the electrolyte, thus showing high ionic conductivity. During the discharge process, the two types of polyanions can be co-embedded in the graphite positive electrode, thus the specific capacity could reach two to three times that of the IL system at low current density. An advanced ternary $AlCl_3$/LiCl/KCl inorganic salts were exploited by Yu et al. [135] with low molten point (95°C) and relatively low operation temperature 99°C (Figure 6.15). By forming a ternary inorganic salt, the eutectic point can be lowered below the boiling point of water, allowing the use of water-based heating systems. Considering the required operating conditions, these electrolytes will be more suitable for large-scale energy storage applications at the present stage.

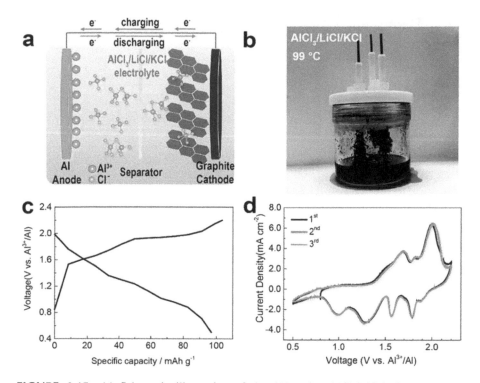

**FIGURE 6.15** (a) Schematic illustration of the AIB using AlCl₃/LiCl/KCl molten salt electrolyte. (b) Optical photo of the battery showing the liquid electrolyte recorded at 99°C. (c) The typical galvanostatic charge/discharge curves of the battery at 500 mA g⁻¹ and 99°C. (d) CV curves of the battery scanned with a rate of 1.0 mV s⁻¹ at 99°C. Adapted with permission. [135] Copyright 2019, Royal Society of Chemistry.

### 6.3.2.4 Polymer and Gel Electrolytes

It is urgent to solve problems like the traditional solid–liquid interface, which is beneficial to enhance the safety and potential application ability of AIBs. In addition to composition optimization, researchers gradually focus on the construction of quasi-solid polymer electrolyte and gel electrolyte. For example, a potential Al-conductive poly(ethylene oxide) (PEO)-based hybrid solid electrolyte was reported, in which nanometer-sized $SiO_2$ and IL of 1-ethyl-3-methylimidazolium bis(fluorosulfonyl) imide ([EMI]FSI) was used to plasticize the PEO chains and to improve the ionic conductivities of the hybrid solid electrolytes [136]. A high ionic conductivity of $9.6×10^{-4}$ S cm⁻¹ can be achieved at room temperature, and an electrochemical stability window of 3 V is observed. However, aluminum deposition and dissolution have not been found in such PEO-based electrolyte, possibly because the coordination of ether group in the PEO with the $Al_2Cl_7^-$ species reduces the electrochemical activity.

Gel polymer electrolytes formed using polymer frames and plasticizers generally have higher ionic conductivity than solid polymer electrolytes and are characterized

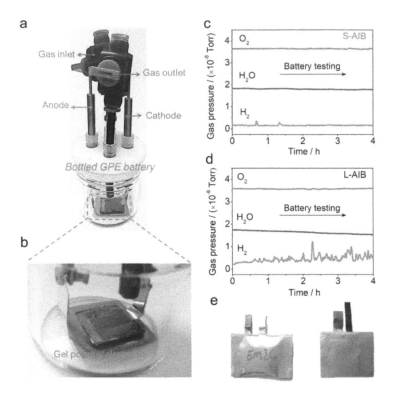

**FIGURE 6.16** (a) Schemes of the preparation process of the GPEs. (b) The bottled battery with GPE electrolytes; (c) the GPE electrolyte system. (d) liquid electrolyte system; (e) the photos of the pouch cells after 300 cycles: (left) the liquid electrolyte system and (right) GPE electrolyte system. Adapted with permission.[138] Copyright 2017, John Wiley and Sons.

by the ease of machining into a variety of desired shapes and sizes. In 2016, Dai et al. [137] synthesized a gel polymer electrolyte by radical polymerization using $AlCl_3$ complexed acrylamide as functional monomer, $AlCl_3$/[EMIM]Cl ionic liquid with molar ratio of 1–1.5 as plasticizer. Reversible aluminum deposition can be achieved at 50°C under the condition of the content of ionic liquid being 80 wt%. The ionic conductivity of the electrolyte increases linearly with the concentration of ionic liquid and can reach $1.66 \times 10^{-3}$ S cm$^{-1}$ at 20°C. Jiao et al. [138] optimized the electrolyte using dichloromethane (DCM) as the solvent and improved the assembly process of AIB (Figure 6.16 a,b). The same group integrated the as-prepared gel polymer electrolyte into the AIB system and multiple tests on the safety of the battery were conducted. The prepared quasi-solid flexible battery reduces the water sensitivity of the battery system, enhances the stability of the electrode–electrolyte interface and enables it to effectively adapt the battery itself to the mechanical stress during the industrial process, which greatly improves the safety and stability of AIBs. Shown in Figure 6.16(c) and (d), the AIB assembled with gel polymer electrolyte hardly produces $H_2$ in the procedure of charging and discharging, where the content of oxygen and water is kept almost the same as that of IL AIBs. Figure 6.16(e) shows the optical photographs of the soft-packaged AIB with different electrolytes after 300 cycles. There is hardly any volume expansion in the AIB with gel polymer electrolyte wherein the initial state of compaction is basically maintained. These results proved that gel polymer electrolyte can effectively inhibit gas production and prevent volume expansion and electrolyte leakage of AIBs.

The ultimate goal of developing solid and quasi-solid electrolytes is to solve the safety problems of AIBs and improve their energy density [47]. Ion migration in (quasi-)solid electrolyte is a major problem in the development of multivalent ion batteries; among them, large ion radius and carrier charge are highlighted. The multivalent ions with larger radius are more likely to interact with the energetic/steric trap, resulting in lower ion mobility. It can be foreseen that in the development of AIB, all solid electrolytes will gradually become the future research trend, which requires a lot of basic research to develop a suitable solid electrolyte, paving the way for the development of AIB.

### 6.3.2.5 Aqueous Electrolytes

The application of aqueous electrolytes in aluminum batteries has been largely hindered by the relatively low standard electrode potential ($-1.66$ V versus SHE) although aqueous liquid electrolytes are highly conductive ($10^1$–$10^2$ mS cm$^{-1}$), non-flammable and inexpensive while the narrow electrochemical stability, decomposition of electrolyte and evolution of hydrogen are deemed to place severe limitations on the operation voltage. Furthermore, rapid and irreversible formation of a high electrical bandgap passivating $Al_2O_3$ oxide film on aluminum have, to date, created frustration in all efforts to create aqueous Al-based electrochemical cells with high reversibility. The compact oxide film grown on the surface of metal aluminum introduces a contact resistance of 35 kilo-ohm, resulting a large overpotential and huge Coulombic efficiency. To avoid this problem, researchers have struggled to find alternative anode materials for metal aluminum, such as anatase $TiO_2$ [38, 139–142], $MoO_3$ [143] and $WO_3$ [144]. These aluminum batteries based on three-electrode system ducked the problem of inert anode; however, the limitation from aqueous liquid electrolyte persists and requires a break through.

## 6.4 BENEFITS AND CHALLENGES

### 6.4.1 Benefits

Compared with other batteries, AIBs with three-electron-transfer electrode reaction exhibit high capacity and acceptable energy density. The volumetric capacity of an Al anode is up to 8,035 mAh cm$^{-3}$ is much higher than Li metal. In addition, aluminum is the most abundant metal on the earth's crust and is far safer than other alkali metals. The nonaqueous AIBs with a wide electrochemical window and highly reversible stripping/plating efficiencies of aluminum are considered very promising. Compared with nonaqueous AIBs, AAIBs also have some advantages in practical applications. First, aqueous electrolytes cost less and require less assembly environment. The battery can be assembled in open air, which is more convenient than assembling nonaqueous batteries. Second, the ionic conductivity of the aqueous electrolyte is higher, which helps to increase the power density of battery [145]. Third, aqueous electrolyte has low flammability and higher safety. Moreover, aqueous electrolyte helps to shield the electrostatic force caused by the high charge density of trivalent aluminum ions [146]. Finally, aqueous electrolyte is non-toxic and environmentally friendly, so it has attracted more studies.

## 6.4.2 CHALLENGES

Although the aluminum battery system has made great progress, there are still several key challenges to be solved. Firstly, lack of cathode materials suitable for the insertion and extraction of trivalent aluminum ions. Although the radius of aluminum-ion is smaller than that of lithium ion, the trivalent aluminum ions can encounter higher charge density and stronger electrostatic interaction with the cathode materials compared to monovalent Li ions, which inevitably hinders ionic diffusion and intercalation [47]. Therefore, many materials suitable for lithium-ion insertion are not suitable for the aluminum ion. The cathode materials that have been used in nonaqueous aluminum battery still have problems such as no stable discharge voltage plateau [147], low discharge voltages [82], low discharge capacity [148] and poor cycle performance [149]. Secondly, in the ambient atmosphere, aluminum is protected by a passivation aluminum oxide layer ≈5 nm thick [150]. The covering layer will limit the exposed active sites. In nonaqueous AIBs, it will affect the performance of the first few charge–discharge cycles [151]. For AAIBs, it can affect the deposition of aluminum ions and cause battery failure. Finally, there are still many problems with the electrolyte. For nonaqueous aluminum batteries, the extremely strong corrosivity of chloroaluminate species greatly limits the choices for current collector and battery packaging materials. Side reactions also occur during charging and discharging, such as the production of chlorine gas, which can lead to a relatively low coulombic efficiency [152]. In addition, chloroaluminate-based ionic liquid also has the characteristics of high moisture absorption, high viscosity and high price [153–155]. These drawbacks impede the commercial application of nonaqueous AIBs. For AAIBs, the narrow electrochemical window of the aqueous electrolyte limits the working voltage of the AAIBs and reduces its energy density [21]. Additionally, because the standard reduction potential of aluminum is lower than the hydrogen evolution potential, the metal aluminum anode in the aqueous electrolyte will inevitably undergo a hydrogen evolution reaction, which consumes water in the aqueous electrolyte, resulting in poor cycle stability of the battery. All of the above-mentioned problems need to be solved in subsequent research.

## 6.5   CURRENT DEVELOPMENTS AND FUTURE PROSPECTS

In recent years, increasing number of studies prove that AIBs have the potential to be a hopeful energy storage system. Researchers have carried out research on AIBs and made some progress. In this chapter, we summarized the development of cathode materials and electrolytes for aqueous nonaqueous AIBs. Meanwhile, prospects and directions are provided for the development in the future.

The general reactions of electrode materials in AIBs are intercalation reaction, conversion reaction and alloy reaction. Among the three reactions, the intercalation process involves the participation of polyanions with large radius or $Al^{3+}$ with high electrostatic repulsion, which results in extremely slow reaction kinetics. For intercalation-type electrodes, materials that have large ion diffusion channels or spacious layered structures stand out, benefiting from the shorter diffusion path and expanding interlayer space. Organic electrode materials with controllable adjustment on structure characteristics could effectively respond to this major challenge; however, thermal stability, poor conductivity and high solubility need to be further improved. The conversion-type electrodes are beneficial to achieve high energy densities. However, consecutive phase transitions are caused by the highly electrostatic repulsive $Al^{3+}$, which will destroy the structure during cycling. This is a common problem in AIBs electrode materials. Researchers are trying to solve the problem in a variety of ways. For instance, improving the morphology of the materials can ameliorate the electrochemical properties of the materials to some extent, such as nanosizing. Besides modifications by heteroatom doping, carbon coating may also help to solve the problem.

Electrolytes are important components of AIBs. They should also be taken seriously since the electrolyte of batteries is one of the critical conditions that will limit the development in practical application. Studies have shown that electrolytes have a great effect on the electrochemical properties of AIBs. Nonaqueous electrolytes that are characterized by high stability but low safety and high costs are the most extensively applied electrolytes in AIBs. The aqueous electrolytes represented by water-in-salt electrolytes have relatively lower costs and significantly improved safety; however, the narrow voltage window and restricted cycle stability make it still a dissatisfactory substitute to IL electrolytes. But it's necessary to emphasize the importance of studies on aqueous electrolytes for the development and commercialization of AIBs. In the future, more suitable electrolytes, binders and other reagents for batteries should be developed. In this process, the solubility of electrode materials in electrolyte and the decomposition reaction of electrolyte at high voltage should be considered. Besides, the selected additives in electrolytes should be favorable to the formation of stable SEI films. The synergistic effect of these reagents and electrode materials can be used to improve the electrochemical performance of batteries.

In summary, both challenges and opportunities exist in the future research of AIBs. The development of nonaqueous AIB is still in its infancy, the commercial availability of this battery requires further scientific and technical development. More research should focus on the overall design of AIBs, including optimization of cathodes structure, modification of aluminum anode and regulation of electrolyte to achieve the goal of market penetration. In addition, simple, sustainable and scalable fabrication approaches for the components in the AIBs are greatly needed because cost-effectiveness is very important for industrial-scale production. It is believed that AIB technology could be applicable in transportation and large-scale grid energy storage in the near future.

## REFERENCES

[1] M. Armand, J. M. Tarascon. Building better batteries. *Nature*. 2008, *451* (7179), 652.

[2] D. Larcher, J. M. Tarascon. Towards greener and more sustainable batteries for electrical energy storage. *Nat. Chem.* 2015, *7* (1), 19.

[3] N. Wang, Z. Bai, Z. Fang, X. Zhang, X. Xu, Y. Du, L. Liu, S. Dou, G. Yu. General synthetic strategy for pomegranate-like transition-metal phosphides@N-doped carbon nanostructures with high lithium storage capacity. *ACS Mater. Lett.* 2019, *1* (2), 265.

[4] P. Albertus, S. Babinec, S. Litzelman, A. Newman. Status and challenges in enabling the lithium metal electrode for high-energy and low-cost rechargeable batteries. *Nat. Energy.* 2017, *3* (1), 16.

[5] H. Li, X. Zhang, Z. Zhao, Z. Hu, X. Liu, G. Yu. Flexible sodium-ion based energy storage devices: Recent progress and challenges. *Energy Storage Mater.* 2020, *26*, 83.

[6] H. Fu, Q. Yin, Y. Huang, H. Sun, Y. Chen, R. Zhang, Q. Yu, L. Gu, J. Duan, W. Luo. Reducing interfacial resistance by na-$SiO_2$ composite anode for NASICON-based solid-state sodium battery. *ACS Mater. Lett.* 2020, *2* (2), 127.

[7] K. Liang, M. Li, Y. Hao, W. Yan, M. Cao, S. Fan, W. Han, J. Su. Reduced graphene oxide with 3D interconnected hollow channel architecture as high-performance anode for Li/Na/K-ion storage. *Chem. Eng. J.* 2020, *394*, 124956.

[8] J. Ge, B. Wang, J. Wang, Q. Zhang, B. Lu. Nature of $FeSe_2$ /N-C anode for high performance potassium ion hybrid capacitor. *Adv. Energy. Mater.* 2019, *10* (4), 1903277.

[9] Z. Guo, S. Zhao, T. Li, D. Su, S. Guo, G. Wang. Recent advances in rechargeable magnesium-based batteries for high-efficiency energy storage. *Adv. Energy. Mater.* 2020, *10* (21), 1903591.

[10] M. Xu, S. Lei, J. Qi, Q. Dou, L. Liu, Y. Lu, Q. Huang, S. Shi, X. Yan. Opening magnesium storage capability of two-dimensional MXene by intercalation of cationic surfactant. *ACS Nano.* 2018, *12* (4), 3733.

[11] J. Park, Z.-L. Xu, G. Yoon, S. K. Park, J. Wang, H. Hyun, H. Park, J. Lim, Y.-J. Ko, Y. S. Yun, K. Kang. Calcium-ion batteries: Stable and high-power calcium-ion batteries enabled by calcium intercalation into graphite. *Adv. Mater.* 2020, *32* (4), 2070029.

[12] R. J. Gummow, G. Vamvounis, M. B. Kannan, Y. He. Calcium-ion batteries: Current state-of-the-art and future perspectives. *Adv. Mater.* 2018, *30* (39), 1801702.

[13] Y. Ru, S. Zheng, H. Xue, H. Pang. Potassium cobalt hexacyanoferrate nanocubic assemblies for high-performance aqueous aluminum ion batteries. *Chem. Eng. J.* 2020, *382*, 122853.

[14] X. Huo, X. Wang, Z. Li, J. Liu, J. Li. Two-dimensional composite of D-$Ti_3C_2T_x$@S@ $TiO_2$ (MXene) as the cathode material for aluminum-ion batteries. *Nanoscale.* 2020, *12* (5), 3387.

[15] S.-W. Kim, D.-H. Seo, X. Ma, G. Ceder, K. Kang. Electrode materials for rechargeable sodium-ion batteries: Potential alternatives to current lithium-ion batteries. *Adv. Energy. Mater.* 2012, *2* (7), 710.

[16] S. Li, P. Ge, F. Jiang, H. Shuai, W. Xu, Y. Jiang, Y. Zhang, J. Hu, H. Hou, X. Ji. The advance of nickel–cobalt–sulfide as ultra-fast/high sodium storage materials: The influences of morphology structure, phase evolution and interface property. *Energy Storage Mater.* 2019, *16*, 267.

[17] G. A. Elia, K. Marquardt, K. Hoeppner, S. Fantini, R. Lin, E. Knipping, W. Peters, J.-F. Drillet, S. Passerini, R. Hahn. An overview and future perspectives of aluminum batteries. *Adv. Mater.* 2016, *28* (35), 7564.

[18] H. Yang, H. Li, J. Li, Z. Sun, K. He, H. M. Cheng, F. Li. The rechargeable aluminum battery: Opportunities and challenges. *Angew. Chem. Int. Edit.* 2019, *58* (35), 11978.

[19] F. Ambroz, T. J. Macdonald, T. Nann. Trends in aluminium-based intercalation batteries. *Adv. Energy. Mater.* 2017, *7* (15), 1602093.

[20] J. Tu, W. L. Song, H. Lei, Z. Yu, L. L. Chen, M. Wang, S. Jiao. Nonaqueous rechargeable aluminum batteries: Progresses, challenges, and perspectives. *Chem. Rev.* 2021, *121* (8), 4903.

[21] P. Xu, X. Guo, Y. Bai, C. Wu. Research progress and challenges of aqueous aluminum ion batteries. *J. Chin. Ceram. Soc.* 2020, *48* (7), 1034.

[22] C. Zhang, R. He, J. Zhang, Y. Hu, Z. Wang, X. Jin. Amorphous carbon-derived nanosheet-bricked porous graphite as high-performance cathode for aluminum-ion batteries. *ACS Appl. Mater. Inter.* 2018, *10* (31), 26510.

[23] M. L. Agiorgousis, Y.-Y. Sun, S. Zhang. The role of ionic liquid electrolyte in an aluminum–graphite electrochemical cell. *ACS Energy. Lett.* 2017, *2* (3), 689.

[24] Z. Liu, J. Wang, X. Jia, W. Lo, Q. Zhang, L. Fan, H. Ding, H. Yang, X. Yu, X. Li, B. Lu. Graphene armored with a crystal carbon shell for ultrahigh-performance potassium ion batteries and aluminum batteries. *ACS Nano.* 2019, *13* (9), 10631.

[25] X. Yu, B. Wang, D. Gong, Z. Xu, B. Lu. Graphene nanoribbons on highly porous 3D graphene for high-capacity and ultrastable al-ion batteries. *Adv. Mater.* 2017, *29* (4), 1604118.

[26] C. Li, S. Dong, R. Tang, X. Ge, Z. Zhang, C. Wang, Y. Lu, L. Yin. Heteroatomic interface engineering in MOF-derived carbon heterostructures with built-in electric-field effects for high performance Al-ion batteries. *Energ. Environ. Sci.* 2018, *11* (11), 3201.

[27] J. Qiao, H. Zhou, Z. Liu, H. Wen, J. Yang. Defect-free soft carbon as cathode material for Al-ion batteries. *Ionics.* 2019, *25* (3), 1235.

[28] J. Wei, W. Chen, D. Chen, K. Yang. An amorphous carbon-graphite composite cathode for long cycle life rechargeable aluminum ion batteries. *J. Mater. Sci. Technol.* 2018, *34* (6), 983.

[29] J. Tu, J. Wang, S. Li, W.-L. Song, M. Wang, H. Zhu, S. Jiao. High-efficiency transformation of amorphous carbon into graphite nanoflakes for stable aluminum-ion battery cathodes. *Nanoscale.* 2019, *11* (26), 12537.

[30] N. S. Hudak. Chloroaluminate-doped conducting polymers as positive electrodes in rechargeable aluminum batteries. *J. Phys. Chem. C.* 2014, *118* (10), 5203.

[31] M. Walter, K. V. Kravchyk, C. Böfer, R. Widmer, M. V. Kovalenko. Polypyrenes as high-performance cathode materials for aluminum batteries. *Adv. Mater.* 2018, *30* (15), 1705644.

[32] T. Schoetz, M. Ueda, A. Bund, C. P. de Leon. Preparation and characterization of a rechargeable battery based on poly-(3,4-ethylenedioxythiophene) and aluminum in ionic liquids. *J. Solid. State. Electr.* 2017, *21* (11), 3237.

[33] S. Zhao, H. Chen, J. Li, J. Zhang. Synthesis of polythiophene/graphite composites and their enhanced electrochemical performance for aluminum ion batteries. *New. J. Chem.* 2019, *43* (37), 15014.

[34] L. Zhang, L. Chen, H. Luo, X. Zhou, Z. Liu. Large-sized few-layer graphene enables an ultrafast and long-life aluminum-ion battery. *Adv. Energy. Mater.* 2017, *7* (15), 1700034.

[35] W. Wang, B. Jiang, W. Xiong, H. Sun, Z. Lin, L. Hu, J. Tu, J. Hou, H. Zhu, S. Jiao. A new cathode material for super-valent battery based on aluminium ion intercalation and deintercalation. *Sci. Rep-UK.* 2013, *3* (1), 3383.

[36] H. Wang, Y. Bai, S. Chen, X. Luo, C. Wu, F. Wu, J. Lu, K. Amine. Binder-free $V_2O_5$ cathode for greener rechargeable aluminum battery. *ACS Appl. Mater. Inter.* 2015, *7* (1), 80.

[37] V. Verma, S. Kumar, W. Manalastas Jr, R. Satish, M. Srinivasan. Progress in rechargeable aqueous zinc-and aluminum-ion battery electrodes: Challenges and outlook. *Adv. Sustain. Syst.* 2019, *3* (1), 1800111.

[38] S. Liu, J. J. Hu, N. F. Yan, G. L. Pan, G. R. Li, X. P. Gao. Aluminum storage behavior of anatase $TiO_2$ nanotube arrays in aqueous solution for aluminum ion batteries. *Energ. Environ. Sci.* 2012, *5* (12), 9743.

[39] T. Mori, Y. Orikasa, K. Nakanishi, K. Chen, M. Hattori, T. Ohta, Y. Uchimoto. Discharge/charge reaction mechanisms of $FeS_2$ cathode material for aluminum rechargeable batteries at 55°C. *J. Power Sources.* 2016, *313*, 9.

[40] L. Geng, G. Lv, X. Xing, J. Guo. Reversible electrochemical intercalation of aluminum in $Mo_6S_8$. *Chem. Mater.* 2015, *27* (14), 4926.

[41] T. Cai, L. Zhao, H. Hu, T. Li, X. Li, S. Guo, Y. Li, Q. Xue, W. Xing, Z. Yan, L. Wang. Stable $CoSe_2$/carbon nanodice@reduced graphene oxide composites for high-performance rechargeable aluminum-ion batteries. *Energ. Environ. Sci.* 2018, *11* (9), 2341.

[42] Z. Zhao, Z. Hu, H. Liang, S. Li, H. Wang, F. Gao, X. Sang, H. Li. Nanosized $MoSe_2$@carbon matrix: A stable host material for the highly reversible storage of potassium and aluminum ions. *ACS Appl. Mater. Inter.* 2019, *11* (47), 44333.

[43] D. J. Kim, D.-J. Yoo, M. T. Otley, A. Prokofjevs, C. Pezzato, M. Owczarek, S. J. Lee, J. W. Choi, J. F. Stoddart. Rechargeable aluminium organic batteries. *Nat. Energy.* 2019, *4* (1), 51.

[44] D. J. Yoo, M. Heeney, F. Glocklhofer, J. W. Choi. Tetradiketone macrocycle for divalent aluminium ion batteries. *Nat. Commun.* 2021, *12* (1), 2386.

[45] X. Yu, M. J. Boyer, G. S. Hwang, A. Manthiram. Room-temperature aluminum-sulfur batteries with a lithium-ion-mediated ionic liquid electrolyte. *Chem.* 2018, *4* (3), 586.

[46] S. Wang, Z. Kang, S. Li, J. Tu, J. Zhu, S. Jiao. High specific capacitance based on N-doped microporous carbon in $[EMIm]Al_xCl_y$ ionic liquid electrolyte. *J. Electrochem. Soc.* 2017, *164* (13), A3319.

[47] Y. Zhang, S. Liu, Y. Ji, J. Ma, H. Yu. Emerging nonaqueous aluminum-ion batteries: Challenges, status, and perspectives. *Adv. Mater.* 2018, *30* (38), 1706310.

[48] Q. Yan, Y. Shen, Y. Miao, M. Wang, M. Yang, X. Zhao. Vanadium oxychloride as cathode for rechargeable aluminum batteries. *J. Alloys Compd.* 2019, *806*, 1109.

[49] Z. Yu, Z. Kang, Z. Hu, J. Lu, Z. Zhou, S. Jiao. Hexagonal NiS nanobelts as advanced cathode materials for rechargeable Al-ion batteries. *Chem. Commun.* 2016, *52* (68), 10427.

[50] H. Hong, J. Liu, H. Huang, C. Atangana Etogo, X. Yang, B. Guan, L. Zhang. Ordered macro-microporous metal-organic framework single crystals and their derivatives for rechargeable aluminum-ion batteries. *J. Am. Chem. Soc.* 2019, *141* (37), 14764.

[51] X. Zhang, G. Zhang, S. Wang, S. Li, S. Jiao. Porous CuO microsphere architectures as high-performance cathode materials for aluminum-ion batteries. *J. Mater. Chem. A.* 2018, *6* (7), 3084.

[52] G. Li, J. Tu, M. Wang, S. Jiao. $Cu_3P$ as a novel cathode material for rechargeable aluminum-ion batteries. *J. Mater. Chem. A.* 2019, *7* (14), 8368.

[53] Z. Li, C. Gao, J. Zhang, A. Meng, H. Zhang, L. Yang. Mountain-like nanostructured 3D $Ni_3S_2$ on Ni foam for rechargeable aluminum battery and its theoretical analysis on charge/discharge mechanism. *J. Alloys Compd.* 2019, *798*, 500.

[54] J. Tu, M. Wang, X. Xiao, H. Lei, S. Jiao. Nickel phosphide nanosheets supported on reduced graphene oxide for enhanced aluminum-ion batteries. *ACS Sustain. Chem. Eng.* 2019, *7* (6), 6004.

[55] Z. Zhao, Z. Hu, R. Jiao, Z. Tang, P. Dong, Y. Li, S. Li, H. Li. Tailoring multi-layer architectured $FeS_2$@C hybrids for superior sodium-, potassium-and aluminum-ion storage. *Energy Storage Mater.* 2019, *22*, 228.

[56] F. Wu, H. Yang, Y. Bai, C. Wu. Paving the path toward reliable cathode materials for aluminum-ion batteries. *Adv. Mater.* 2019, *31* (16), 1806510.

[57] X. Zhang, S. Jiao, J. Tu, W.-L. Song, X. Xiao, S. Li, M. Wang, H. Lei, D. Tian, H. Chen, D. Fang. Rechargeable ultrahigh-capacity tellurium-aluminum batteries. *Energ. Environ. Sci.* 2019, *12* (6), 1918.

[58] X. Yu, A. Manthiram. Electrochemical energy storage with a reversible nonaqueous room-temperature aluminum-sulfur chemistry. *Adv. Energy. Mater.* 2017, *7* (18), 1700561.

[59] Y. Kong, A. K. Nanjundan, Y. Liu, H. Song, X. Huang, C. Yu. Modulating ion diffusivity and electrode conductivity of carbon nanotube@mesoporous carbon fibers for high performance aluminum-selenium batteries. *Small.* 2019, *15* (51) e1904310.

[60] H. Wang, L. Zhao, H. Zhang, Y. Liu, L. Yang, F. Li, W. Liu, X. Dong, X. Li, Z. Li, X. Qi, L. Wu, Y. Xu, Y. Wang, K. Wang, H. Yang, Q. Li, S. Yan, X. Zhang, F. Li, H. Li. Revealing the multiple cathodic and anodic involved charge storage mechanism in an FeSe$_2$ cathode for aluminium-ion batteries by in situ magnetometry. *Energ. Environ. Sci.* 2022, *15* (1), 311.

[61] Z. Hu, H. Zhang, H. Wang, F. Zhang, Q. Li, H. Li. Nonaqueous aluminum ion batteries: Recent progress and prospects. *ACS Mater. Lett.* 2020, *2* (8), 887.

[62] J. Xu, Y. Dou, Z. Wei, J. Ma, Y. Deng, Y. Li, H. Liu, S. Dou. Recent progress in graphite intercalation compounds for rechargeable metal (Li, Na, K, Al)-ion batteries. *Adv. Sci.* 2017, *4* (10), 1700146.

[63] A. Eftekhari, P. Corrochano. Electrochemical energy storage by aluminum as a lightweight and cheap anode/charge carrier. Sustain. *Energy Fuels* 2017, *1* (6), 1246.

[64] S. K. Das. Graphene: A cathode material of choice for aluminum-ion batteries. *Angew. Chem. Int. Edit.* 2018, *57* (51), 16606.

[65] H. Chen, F. Guo, Y. Liu, T. Huang, B. Zheng, N. Ananth, Z. Xu, W. Gao, C. Gao. A defect-free principle for advanced graphene cathode of aluminum-ion battery. *Adv. Mater.* 2017, *29* (12), 1605958.

[66] M.-C. Lin, M. Gong, B. Lu, Y. Wu, D.-Y. Wang, M. Guan, M. Angell, C. Chen, J. Yang, B.-J. Hwang, H. Dai. An ultrafast rechargeable aluminium-ion battery. *Nature.* 2015, *520* (7547), 325.

[67] J. Liu, Z. Li, X. Huo, J. Li. Nanosphere-rod-like Co$_3$O$_4$ as high performance cathode material for aluminium ion batteries. *J. Power Sources.* 2019, *422*, 49.

[68] Y. Hu, B. Luo, D. Ye, X. Zhu, M. Lyu, L. Wang. An innovative freeze-dried reduced graphene oxide supported SnS$_2$ cathode active material for aluminum-ion batteries. *Adv. Mater.* 2017, *29* (48), 1606132.

[69] H. Chen, C. Chen, Y. Liu, X. Zhao, N. Ananth, B. Zheng, L. Peng, T. Huang, W. Gao, C. Gao. High-quality graphene microflower design for high-performance Li–S and Al-ion batteries. *Adv. Energy. Mater.* 2017, *7* (17), 1700051.

[70] T. Gao, X. Li, X. Wang, J. Hu, F. Han, X. Fan, L. Suo, A. J. Pearse, S. B. Lee, G. W. Rubloff, K. J. Gaskell, M. Noked, C. Wang. A rechargeable Al/S battery with an ionic-liquid electrolyte. *Angew. Chem. Int. Edit.* 2016, *55* (34), 9898.

[71] X. Wen, Y. Liu, A. Jadhav, J. Zhang, D. Borchardt, J. Shi, B. M. Wong, B. Sanyal, R. J. Messinger, J. Guo. Materials compatibility in rechargeable aluminum batteries: Chemical and electrochemical properties between vanadium pentoxide and chloroaluminate ionic liquids. *Chem. Mater.* 2019, *31* (18), 7238.

[72] S. Kumar, R. Satish, V. Verma, H. Ren, P. Kidkhunthod, W. Manalastas, Jr, M. Srinivasan. Investigating FeVO$_4$ as a cathode material for aqueous aluminum-ion battery. *J. Power Sources.* 2019, *426*, 151.

[73] P. R. Gifford, J. B. Palmisano. An aluminum/chlorine rechargeable cell employing a room temperature molten salt electrolyte. *J. Electrochem. Soc.* 1988, *135* (3), 650.

[74] D.-Y. Wang, C.-Y. Wei, M.-C. Lin, C.-J. Pan, H.-L. Chou, H.-A. Chen, M. Gong, Y. Wu, C. Yuan, M. Angell, Y.-J. Hsieh, Y.-H. Chen, C.-Y. Wen, C.-W. Chen, B.-J. Hwang, C.-C. Chen, H. Dai. Advanced rechargeable aluminium ion battery with a high-quality natural graphite cathode. *Nat. Commun.* 2017, *8* (1), 14283.

[75] H. Chen, H. Xu, S. Wang, T. Huang, J. Xi, S. Cai, F. Guo, Z. Xu, W. Gao, C. Gao. Ultrafast all-climate aluminum-graphene battery with quarter-million cycle life. *Sci. Adv.* 2017, *3* (12), eaao7233.

[76] A. Childress, P. Parajuli, S. Eyley, W. Thielemans, R. Podila, A. M. Rao. Effect of nitrogen doping in the few layer graphene cathode of an aluminum ion battery. *Chem. Phys. Lett.* 2019, *733*, 136669.

[77] P. Wang, H. Chen, N. Li, X. Zhang, S. Jiao, W.-L. Song, D. Fang. Dense graphene papers: Toward stable and recoverable Al-ion battery cathodes with high volumetric and areal energy and power density. *Energy Storage Mater.* 2018, *13*, 103.

[78] F. Wang, F. Yu, X. Wang, Z. Chang, L. Fu, Y. Zhu, Z. Wen, Y. Wu, W. Huang. Aqueous rechargeable zinc/aluminum ion battery with good cycling performance. *ACS Appl. Mater. Inter.* 2016, *8* (14), 9022.

[79] W. Pan, Y. Wang, Y. Zhang, H. Y. H. Kwok, M. Wu, X. Zhao, D. Y. C. Leung. A low-cost and dendrite-free rechargeable aluminium-ion battery with superior performance. *J. Mater. Chem. A.* 2019, *7* (29), 17420.

[80] J. Livage. Vanadium pentoxide gels. *Chem. Mater.* 1991, *3* (4), 578.

[81] N. A. Chernova, M. Roppolo, A. C. Dillon, M. S. Whittingham. Layered vanadium and molybdenum oxides: batteries and electrochromics. *J. Mater. Chem.* 2009, *19* (17), 2526.

[82] N. Jayaprakash, S. K. Das, L. A. Archer. The rechargeable aluminum-ion battery. *Chem. Commun.* 2011, *47* (47), 12610.

[83] L. D. Reed, E. Menke. The roles of $V_2O_5$ and stainless steel in rechargeable Al–Ion batteries. *J. Electrochem. Soc.* 2013, *160* (6), A915.

[84] M. Chiku, H. Takeda, S. Matsumura, E. Higuchi, H. Inoue. Amorphous vanadium oxide/carbon composite positive electrode for rechargeable aluminum battery. *ACS App. Mater. Inter.* 2015, *7* (44), 24385.

[85] Q. Zhao, L. Liu, J. Yin, J. Zheng, D. Zhang, J. Chen, L. A. Archer. Proton intercalation/de-intercalation dynamics in vanadium oxides for aqueous aluminum electrochemical cells. *Angew. Chem. Int. Edit.* 2020, *59* (8), 3048.

[86] S. He, J. Wang, X. Zhang, J. Chen, Z. Wang, T. Yang, Z. Liu, Y. Liang, B. Wang, S. Liu, L. Zhang, J. Huang, J. Huang, L. A. O'Dell, H. Yu. A high-energy aqueous aluminum-manganese battery. *Adv. Funct. Mater.* 2019, *29* (45), 1905228.

[87] J. Joseph, A. P. O'Mullane, K. Ostrikov. Hexagonal molybdenum trioxide (h-$MoO_3$) as an electrode material for rechargeable aqueous aluminum-ion batteries. *ChemElectroChem.* 2019, *6* (24), 6002.

[88] C. Wu, S. Gu, Q. Zhang, Y. Bai, M. Li, Y. Yuan, H. Wang, X. Liu, Y. Yuan, N. Zhu, F. Wu, H. Li, L. Gu, J. Lu. Electrochemically activated spinel manganese oxide for rechargeable aqueous aluminum battery. *Nat. Commun.* 2019, *10* (1), 73.

[89] M. Liu, A. Jain, Z. Rong, X. Qu, P. Canepa, R. Malik, G. Ceder, K. A. Persson. Evaluation of sulfur spinel compounds for multivalent battery cathode applications. *Energ. Environ. Sci.* 2016, *9* (10), 3201.

[90] T. Mori, Y. Orikasa, K. Nakanishi, C. Kezheng, M. Hattori, T. Ohta, Y. Uchimoto. Discharge/charge reaction mechanisms of $FeS_2$ cathode material for aluminum rechargeable batteries at 55°C. *J. Power Sources.* 2016, *313*, 9.

[91]  W. Guan, L. Wang, H. Lei, J. Tu, S. Jiao. $Sb_2Se_3$ nanorods with N-doped reduced graphene oxide hybrids as high-capacity positive electrode materials for rechargeable aluminum batteries. *Nanoscale*. 2019, *11* (35), 16437.

[92]  S. Wang, Z. Yu, J. Tu, J. Wang, D. Tian, Y. Liu, S. Jiao. A novel aluminum-ion battery: $Al/AlCl_3$-[EMIm]Cl/$Ni_3S_2$@graphene. *Adv. Energy. Mater.* 2016, *6* (13), 1600137.

[93]  F. Thöle, L. F. Wan, D. Prendergast. Re-examining the Chevrel phase $Mo_6S_8$ cathode for Mg intercalation from an electronic structure perspective. *Phys. Chem. Chem. Phys.* 2015, *17* (35), 22548.

[94]  L. Geng, J. P. Scheifers, C. Fu, J. Zhang, B. P. T. Fokwa, J. Guo. Titanium sulfides as intercalation-type cathode materials for rechargeable aluminum batteries. *ACS Appl. Mater. Inter.* 2017, *9* (25), 21251.

[95]  Z. Zhao, Z. Hu, Q. Li, H. Li, X. Zhang, Y. Zhuang, F. Wang, G. Yu. Designing two-dimensional $WS_2$ layered cathode for high-performance aluminum-ion batteries: From micro-assemblies to insertion mechanism. *Nano Today*. 2020, *32*, 100870.

[96]  L. Wu, R. Sun, F. Xiong, C. Pei, K. Han, C. Peng, Y. Fan, W. Yang, Q. An, L. Mai. A rechargeable aluminum- ion battery based on a $VS_2$ nanosheet cathode. *Phys. Chem. Chem. Phys.* 2018, *20* (35), 22563.

[97]  Z. Li, B. Niu, J. Liu, J. Li, F. Kang. Rechargeable aluminum-ion battery based on $MoS_2$ microsphere cathode. *ACS Appl. Mater. Inter.* 2018, *10* (11), 9451.

[98]  K. Liang, L. Ju, S. Koul, A. Kushima, Y. Yang. Self-supported tin sulfide porous films for flexible aluminum-ion batteries. *Adv. Energy. Mater.* 2019, *9* (2), 1802543.

[99]  H. Li, H. Yang, Z. Sun, Y. Shi, H.-M. Cheng, F. Li. A highly reversible $Co_3S_4$ microsphere cathode material for aluminum-ion batteries. *Nano Energy*. 2019, *56*, 100.

[100]  Z. Hu, K. Zhi, Q. Li, Z. Zhao, H. Liang, X. Liu, J. Huang, C. Zhang, H. Li, X. Guo. Two-dimensionally porous cobalt sulfide nanosheets as a high-performance cathode for aluminum-ion batteries. *J. Power Sources*. 2019, *440*, 227147.

[101]  S. Tongay, J. Zhou, C. Ataca, K. Lo, T. S. Matthews, J. Li, J. C. Grossman, J. Wu. Thermally driven crossover from indirect toward direct bandgap in 2D semiconductors: $MoSe_2$ versus $MoS_2$. *Nano Lett.* 2012, *12* (11), 5576.

[102]  H. Schmidt, F. Giustiniano, G. Eda. Electronic transport properties of transition metal dichalcogenide field-effect devices: surface and interface effects. *Chem. Soc. Rev.* 2015, *44* (21), 7715.

[103]  J. Jiang, H. Li, T. Fu, B. J. Hwang, X. Li, J. Zhao. One-dimensional $cu_{2-x}Se$ nanorods as the cathode material for high-performance aluminum-ion battery. *ACS Appl. Mater. Inter.* 2018, *10* (21), 17942.

[104]  H. Li, Z. Hu, Q. Xia, H. Zhang, Z. Li, H. Wang, X. Li, F. Zuo, F. Zhang, X. Wang, W. Ye, Q. Li, Y. Long, Q. Li, S. Yan, X. Liu, X. Zhang, G. Yu, G. X. Miao. Operando magnetometry probing the charge storage mechanism of CoO lithium-ion batteries. *Adv. Mater.* 2021, *33* (12), e2006629.

[105]  Q. Li, H. Li, Q. Xia, Z. Hu, Y. Zhu, S. Yan, C. Ge, Q. Zhang, X. Wang, X. Shang, S. Fan, Y. Long, L. Gu, G. X. Miao, G. Yu, J. S. Moodera. Extra storage capacity in transition metal oxide lithium-ion batteries revealed by in situ magnetometry. *Nat Mater*. 2020, *20*, 76.

[106]  S. Liu, X. Zhang, S. He, Y. Tang, J. Wang, B. Wang, S. Zhao, H. Su, Y. Ren, L. Zhang, J. Huang, H. Yu, K. Amine. An advanced high energy-efficiency rechargeable aluminum-selenium battery. *Nano Energy*. 2019, *66*, 104159.

[107]  Y. Guo, H. Jin, Z. Qi, Z. Hu, H. Ji, L.-J. Wan. Carbonized-MOF as a sulfur host for aluminum-sulfur batteries with enhanced capacity and cycling life. *Adv. Funct. Mater.* 2019, *29* (7), 1807676.

[108] Y. Kong, A. K. Nanjundan, Y. Liu, H. Song, X. Huang, C. Yu. Modulating ion diffusivity and electrode conductivity of carbon nanotube@mesoporous carbon fibers for high performance aluminum–selenium batteries. *Small.* 2019, *15* (51), 1904310.

[109] S. Liu, G. L. Pan, G. R. Li, X. P. Gao. Copper hexacyanoferrate nanoparticles as cathode material for aqueous Al-ion batteries. *J. Mater. Chem. A.* 2015, *3* (3), 959.

[110] K. Zhang, T. H. Lee, B. Bubach, H. W. Jang, M. Ostadhassan, J.-W. Choi, M. Shokouhimehr. Graphite carbon-encapsulated metal nanoparticles derived from Prussian blue analogs growing on natural loofa as cathode materials for rechargeable aluminum-ion batteries. *Sci. Rep-UK.* 2019, *9* (1), 13665.

[111] K. Suto, A. Nakata, H. Murayama, T. Hirai, J.-i. Yamaki, Z. Ogumi. Electrochemical properties of al/vanadium chloride batteries with $AlCl_3$-1-ethyl-3-methylimidazolium chloride electrolyte. *J. Electrochem. Soc.* 2016, *163* (5), A742.

[112] F. M. Donahue, S. E. Mancini, L. Simonsen. Secondary aluminium-iron (III) chloride batteries with a low temperature molten salt electrolyte. *J. Appl. Electrochem.* 1992, *22* (3), 230.

[113] X.-G. Sun, Z. Bi, H. Liu, Y. Fang, C. A. Bridges, M. P. Paranthaman, S. Dai, G. M. Brown. A high performance hybrid battery based on aluminum anode and $LiFePO_4$ cathode. *Chem. Commun.* 2016, *52* (8), 1713.

[114] A. VahidMohammadi, A. Hadjikhani, S. Shahbazmohamadi, M. Beidaghi. Two-dimensional vanadium carbide (MXene) as a high-capacity cathode material for rechargeable aluminum batteries. *ACS Nano.* 2017, *11* (11), 11135.

[115] J. Jiang, H. Li, J. Huang, K. Li, J. Zeng, Y. Yang, J. Li, Y. Wang, J. Wang, J. Zhao. Investigation of the reversible intercalation/deintercalation of al into the novel $Li_3VO_4$@C microsphere composite cathode material for aluminum-ion batteries. *ACS Appl. Mater. Inter.* 2017, *9* (34), 28486.

[116] F. Nacimiento, M. Cabello, R. Alcántara, P. Lavela, J. L. Tirado. NASICON-type $Na_3V_2(PO4)_3$ as a new positive electrode material for rechargeable aluminium battery. *Electrochim. Acta.* 2018, *260*, 798.

[117] J. Chen, Q. Zhu, L. Jiang, R. Liu, Y. Yang, M. Tang, J. Wang, H. Wang, L. Guo. Rechargeable aqueous aluminum organic batteries. *Angew. Chem. Int. Edit.* 2021, *60* (11), 5794.

[118] Z. Yu, J. Tu, C. Wang, S. Jiao. A rechargeable al/graphite battery based on $AlCl_3$/1-butyl-3-methylimidazolium chloride ionic liquid electrolyte. *Chemistryselect.* 2019, *4* (11), 3018.

[119] S. Gu, H. Wang, C. Wu, Y. Bai, H. Li, F. Wu. Confirming reversible $Al^{3+}$ storage mechanism through intercalation of $Al^{3+}$ into $V_2O_5$ nanowires in a rechargeable aluminum battery. *Energy Storage Mater.* 2017, *6*, 9.

[120] W. Kaveevivitchai, A. Huq, S. Wang, M. J. Park, A. Manthiram. Rechargeable aluminum-ion batteries based on an open-tunnel framework. *Small* 2017, *13* (34), 1701296.

[121] H. Xu, T. Bai, H. Chen, F. Guo, J. Xi, T. Huang, S. Cai, X. Chu, J. Ling, W. Gao, Z. Xu, C. Gao. Low-cost $AlCl_3$/$Et_3NHCl$ electrolyte for high-performance aluminum-ion battery. *Energy Storage Mater.* 2019, *17*, 38.

[122] G. Zhu, M. Angell, C.-J. Pan, M.-C. Lin, H. Chen, C.-J. Huang, J. Lin, A. J. Achazi, P. Kaghazchi, B.-J. Hwang, H. Dai. Rechargeable aluminum batteries: effects of cations in ionic liquid electrolytes. *RSC Adv.* 2019, *9* (20), 11322.

[123] C. Xu, S. Zhao, Y. Du, W. Zhang, P. Li, H. Jin, Y. Zhang, Z. Wang, J. Zhang. A high capacity aluminum-ion battery based on imidazole hydrochloride electrolyte. *Chemelectrochem.* 2019, *6* (13), 3350.

[124] C. Xu, J. Li, H. Chen, J. Zhang. Benzyltriethylammonium chloride electrolyte for high-performance al-ion batteries. *ChemNanoMat*. 2019, *5* (11), 1367.

[125] T. Tsuda, Y. Uemura, C.-Y. Chen, Y. Hashimoto, I. Kokubo, K. Sutani, K. Muramatsu, S. Kuwabata. Graphene-coated activated carbon fiber cloth positive electrodes for aluminum rechargeable batteries with a chloroaluminate room-temperature ionic liquid. *J. Electrochem. Soc.* 2017, *164* (12), A2468.

[126] N. Canever, N. Bertrand, T. Nann. Acetamide: A low-cost alternative to alkyl imidazolium chlorides for aluminium-ion batteries. *Chem. Commun.* 2018, *54* (83), 11725.

[127] C. Yang, S. Wang, X. Zhang, Q. Zhang, W. Ma, S. Yu, G. Sun. Substituent effect of imidazolium ionic liquid: A potential strategy for high coulombic efficiency al battery. *J. Phys. Chem. C.* 2019, *123* (18), 11522.

[128] H. Wang, S. Gu, Y. Bai, S. Chen, F. Wu, C. Wu. High-voltage and noncorrosive ionic liquid electrolyte used in rechargeable aluminum battery. *ACS Appl. Mater. Inter.* 2016, *8* (41), 27444.

[129] H. Jiao, C. Wang, J. Tu, D. Tian, S. Jiao. A rechargeable Al-ion battery: Al/molten AlCl$_3$-urea/graphite. *Chem. Commun.* 2017, *53* (15), 2331.

[130] M. Angell, C.-J. Pan, Y. Rong, C. Yuan, M.-C. Lin, B.-J. Hwang, H. Dai. High Coulombic efficiency aluminum-ion battery using an AlCl$_3$-urea ionic liquid analog electrolyte. *P. Natl. Acad. Sci. USA.* 2017, *114* (5), 834.

[131] M. Angell, G. Zhu, M.-C. Lin, Y. Rong, H. Dai. Ionic liquid analogs of AlCl$_3$ with urea derivatives as electrolytes for aluminum batteries. *Adv. Funct. Mater.* 2020, *30* (4), 1901928.

[132] W. Chu, X. Zhang, J. Wang, S. Zhao, S. Liu, H. Yu. A low-cost deep eutectic solvent electrolyte for rechargeable aluminum-sulfur battery. *Energy Storage Mater.* 2019, *22*, 418.

[133] Y. Bian, Y. Li, Z. Yu, H. Chen, K. Du, C. Qiu, G. Zhang, Z. Lv, M.-C. Lin. Using an AlCl$_3$/urea ionic liquid analog electrolyte for improving the lifetime of aluminum-sulfur batteries. *ChemElectroChem*. 2018, *5* (23), 3607.

[134] J. Tu, S. Wang, S. Li, C. Wang, D. Sun, S. Jiao. The effects of anions behaviors on electrochemical properties of Al/graphite rechargeable aluminum-ion battery via molten AlCl$_3$-NaCl liquid electrolyte. *J. Electrochem. Soc.* 2017, *164* (13), A3292.

[135] J. Wang, X. Zhang, W. Chu, S. Liu, H. Yu. A sub-100 °C aluminum ion battery based on a ternary inorganic molten salt. *Chem. Commun.* 2019, *55* (15), 2138.

[136] S. Song, M. Kotobuki, F. Zheng, Q. Li, C. Xu, Y. Wang, W. D. Z. Li, N. Hu, L. Lu. Al conductive hybrid solid polymer electrolyte. *Solid State Ion.* 2017, *300*, 165.

[137] X. G. Sun, Y. Fang, X. Jiang, K. Yoshii, T. Tsuda, S. Dai. Polymer gel electrolytes for application in aluminum deposition and rechargeable aluminum ion batteries. *Chem Commun.* 2016, *52* (2), 292.

[138] Z. Yu, S. Jiao, S. Li, X. Chen, W.-L. Song, T. Teng, J. Tu, H.-S. Chen, G. Zhang, D.-N. Fang. Flexible stable solid-state Al-ion batteries. *Adv. Funct. Mater.* 2019, *29* (1), 1806799.

[139] Y. Liu, S. Sang, Q. Wu, Z. Lu, K. Liu, H. Liu. The electrochemical behavior of Cl$^-$ assisted Al$^{3+}$ insertion into titanium dioxide nanotube arrays in aqueous solution for aluminum ion batteries. *Electrochim. Acta.* 2014, *143*, 340.

[140] Y. J. He, J. F. Peng, W. Chu, Y. Z. Li, D. G. Tong. Retracted article: Black mesoporous anatase TiO$_2$ nanoleaves: a high capacity and high rate anode for aqueous Al-ion batteries. *J. Mater. Chem. A.* 2014, *2* (6), 1721.

[141] H. Lahan, R. Boruah, A. Hazarika, S. K. Das. Anatase TiO$_2$ as an anode material for rechargeable aqueous aluminum-ion batteries: remarkable graphene induced aluminum ion storage phenomenon. *J. Phys. Chem. C.* 2017, *121* (47), 26241.

[142] M. Kazazi, Z. A. Zafar, M. Delshad, J. Cervenka, C. Chen. TiO$_2$/CNT nanocomposite as an improved anode material for aqueous rechargeable aluminum batteries. *Solid State Ion.* 2018, *320*, 64.

[143] H. Lahan, S. K. Das. Al$^{3+}$ ion intercalation in MoO$_3$ for aqueous aluminum-ion battery. *J. Power Sources.* 2019, *413*, 134.

[144] H. Lahan, S. K. Das. Reversible Al$^{3+}$ ion insertion into tungsten trioxide (WO$_3$) for aqueous aluminum-ion batteries. *Dalton. T.* 2019, *48* (19), 6337.

[145] W. Tang, Y. Zhu, Y. Hou, L. Liu, Y. Wu, K. P. Loh, H. Zhang, K. Zhu. Aqueous rechargeable lithium batteries as an energy storage system of superfast charging. *Energ. Environ. Sci.* 2013, *6* (7), 2093.

[146] H. Kim, J. Hong, K. Y. Park, H. Kim, S. W. Kim, K. Kang. Aqueous rechargeable Li and Na ion batteries. *Chem. Rev.* 2014, *114* (23), 11788.

[147] N. P. Stadie, S. Wang, K. V. Kravchyk, M. V. Kovalenko. Zeolite-templated carbon as an ordered microporous electrode for aluminum batteries. *ACS Nano.* 2017, *11* (2), 1911.

[148] H. Wang, X. Bi, Y. Bai, C. Wu, S. Gu, S. Chen, F. Wu, K. Amine, J. Lu. Open-structured V$_2$O$_5$·nH$_2$O nanoflakes as highly reversible cathode material for mono-valent and multivalent intercalation batteries. *Adv. Energy. Mater.* 2017, *7* (14), 1602720.

[149] H. Tian, S. Zhang, Z. Meng, W. He, W.-Q. Han. Rechargeable aluminum/iodine battery redox chemistry in ionic liquid electrolyte. *ACS Energy. Lett.* 2017, *2* (5), 1170.

[150] J. W. Diggle, T. C. Downie, C. W. Goulding. Anodic oxide films on aluminum. *Chem. Rev.* 1969, *69* (3), 365.

[151] S. Choi, H. Go, G. Lee, Y. Tak. Electrochemical properties of an aluminum anode in an ionic liquid electrolyte for rechargeable aluminum-ion batteries. *Phys. Chem. Chem. Phys.* 2017, *19* (13), 8653.

[152] Y. Ito, T. Nohira. Non-conventional electrolytes for electrochemical applications. *Electrochim. Acta.* 2000, *45* (15), 2611.

[153] A. Kitada, K. Nakamura, K. Fukami, K. Murase. Electrochemically active species in aluminum electrodeposition baths of AlCl$_3$/glyme solutions. *Electrochim. Acta.* 2016, *211*, 561.

[154] P. Eiden, Q. Liu, S. Zein El Abedin, F. Endres, I. Krossing. An experimental and theoretical study of the aluminium species present in mixtures of AlCl$_3$ with the ionic liquids [BMP]Tf$_2$N and [EMIm]Tf$_2$N. *Chem-Eur. J.* 2009, *15* (14), 3426.

[155] S. Licht, G. Levitin, R. Tel-Vered, C. Yarnitzky. The effect of water on the anodic dissolution of aluminum in non-aqueous electrolytes. *Electrochem. Commun.* 2000, *2* (5), 329.

# 7 Calcium-Ion Batteries

*Ricardo Alcántara, Marta Cabello, Pedro Lavela and José L. Tirado*

## 7.1 INTRODUCTION

Multivalent ions could be at the forefront of the next-generation battery. Calcium-ion battery (CIB) is one of the emerging post-Li battery technologies [1, 2], but the first Ca-based energy storage device was described as early as the 1960s and it was a primary (non-rechargeable) system based on Ca as an anode and carbon as a cathode [3]. The use of a primary battery of calcium metal as an anode in $SOCl_2$ electrolytes was investigated by Staniewicz in 1980 [4]. In the 1990s, the reversible plating of Ca was considered impossible [5], limiting the interest in calcium batteries for many years. The proof of concept of reversible batteries based on calcium intercalation was published only about two decades back [6, 7]. The discovery of the ability to achieve reversible Ca plating/striping rapidly increased the research attention toward calcium metal batteries [8].

There are many reasons to use calcium in batteries instead of lithium [9, 10]. After three decades of commercialization, probably lithium-ion technology is reaching the limits of its capabilities. Calcium is the fifth most abundant element on the earth's crust (41,500 ppm), over 2,000 times more abundant than lithium. Calcium has a divalent oxidation state, which can counterbalance its heavier atomic weight compared to monovalent metals (Li and Na). The higher density of Ca (1.55 g cm$^{-3}$) compared to Na (0.968 g cm$^{-3}$) and Li (0.533 g cm$^{-3}$) is also an advantage in terms of volumetric energy density. A calcium metal anode can offer significantly higher volumetric (2073 mAh mL$^{-1}$) and gravimetric (1337 mAh g$^{-1}$) capacities than current carbonaceous anodes in Li-ion batteries. The standard reduction potential of calcium (−2.87 V vs. SHE) is only 0.17 V above lithium. Calcium is considered as much safer than lithium because various studies reported the possibility of Ca plating over dendritic growth on the electrode surface and also the melting point is higher (842°C (Ca) vs. 180.5°C (Li)) for Ca.

The ionic radius of calcium (100 pm) is equivalent to sodium (102 pm) and larger than those of lithium (76 pm) and magnesium (72 pm). The lower charge density of Ca$^{2+}$ (0.49 e Å$^{-3}$) compared to Mg$^{2+}$ (1.28 e Å$^{-3}$), Al$^{3+}$ (4.55 e Å$^{-3}$), and Zn$^{2+}$ (1.18 e Å$^{-3}$) could be an advantage for mobility, rapid desolvation, and rapid (de)intercalation. Calcium ion has a strong affinity for binding to oxygen and the preferred coordination numbers range from 6 to 8 [11]. We learned from the Li-ion battery that the

DOI: 10.1201/9781003208198-7

intercalation process must be rapid and reversed many times. Nature teaches that $Ca^{2+}$ is one of her preferred ions for signal transduction in living organisms, due to its rapid rate of ligand exchange ($3 \times 10^{-8}$ $s^{-1}$ for water exchange) and variability in the coordination sphere, which allows sudden changes in the ion concentration. Thus, the binding of calcium to the protein troponin-C is responsible for muscle contraction in vertebrate animals, and perhaps we should learn this lesson from nature.

Despite the aforementioned reasons to develop CIBs, the efforts of the scientific community on calcium seem to have been rather limited compared to magnesium and sodium batteries. So the advancement in the field of CIBs has been relatively smaller in the last decades. CIB is still at an early stage, but we are seeing the light at the end of the tunnel. The knowledge gained from Li-ion batteries can help to develop batteries based on the reversible intercalation, or other types of charge storage, of calcium in electrode active materials. We can highlight the progress for plating/stripping of Ca, the theoretical calculations on calcium intercalation, and the finding of promising electrode materials for the intercalation of $Ca^{2+}$. It seems that there is a chance for the future commercialization of a CIB.

## 7.2 ELECTROCHEMICAL CHARGE STORAGE MECHANISMS

Each $Ca^{2+}$ which is intercalated into a host compound results in a transfer of two electrons. This is an advantage because it could reduce the number of cations needed to transfer the same electrical charge by half as compared to univalent cations, such as lithium and sodium. Further, it could open the path to developing batteries with higher electrochemical capacities. However, the distribution of the guest species in the host and the interaction between the inserted cations and the host framework can be different in calcium as compared to lithium. The structural changes during the intercalation process and the voltage response could also be different.

The electrochemical charge storage in CIBs has been typically researched with no reversible plating/stripping of Ca in the electrochemical devices and using three-electrode cells. Irrespective of this experimental limitation, many authors found that calcium insertion indeed occurs. Besides that, theoretical calculations such as the density functional theory (DFT) modelling have been used to better understand the mechanisms of calcium diffusion in oxides as cathodic materials [12]. On contrary to the invention of the Li-ion battery in the 20th century, the progress in the calculations could be faster than the advancement in the laboratory experiments for CIB.

Organic compounds assembled by weak van der Waals forces can provide a flexible solid structure for calcium-hosting and a different charge storage behaviour [13]. Thus, organic crystals containing aromatic molecules and planar structures with $\pi$-$\pi$ stacked layers can be considered favourable materials for facilitating the storage and diffusion of $Ca^{2+}$. In addition, the negative charge of functional groups delocalized through the stacks, such as enolate, can contribute to stabilizing $Ca^{2+}$. Interestingly, the organic compounds with redox properties could be easily tuned.

The studies about the intercalation of calcium into previously well-known compounds often revealed diffraction peaks that could not be ascribed to any known phase. This fact could cast doubt on the interpretation of the results and the feasibility of calciation. There is a chance that new phases are formed upon calciation and, even, the new phases could be metastable. We have learned from lithium intercalation into oxides that some transition metal ions, such as vanadium and manganese, also have mobility [14]. Thus, it is expected that the intercalation of a larger cation ($Ca^{2+}$) also induces structure transformation, irrespective of the identification of the new Ca-based phases and their suitability for reversible charge–discharge cycling. Maybe the layered-type structures do not offer enough shielding for the electrostatic cation–cation repulsion, particularly for large calcium intercalation. Oxide with tunnel-type structures and polyoxoanions, such as phosphates, could offer better shielding.

A drawback of batteries based on multivalent cations is the sluggish diffusion into host structures, particularly oxides. To overcome the diffusion limitations, the typical method is to shorten the diffusion distance and increase the interfacial areas, and this can be achieved by using active materials with small particle sizes. However, the electrochemical reactions during charge and discharge could take place on the particle surface rather than the bulk of the material. It may drive erroneous interpretations of the experimental results. In that case, the faradic process at the particle surface or near-surface region, which is known as pseudocapacitance, must not be ignored. The relative contribution of the surface to the overall capacity is expected to increase proportionally to the imposed current density and when the solid-state diffusion decreases. The pseudocapacitance can be more relevant for solids with nanometric particles and sluggish diffusion rate, which could be the case for multivalent cation diffusion. Compared to magnesium, calcium is larger, and it reduces the charge polarization. Therefore, calcium has lower binding strength to oxygen atoms and could diffuse easily [15]. The enhancement of the pseudocapacitance by decreasing the particle size could overlap with other apparent mechanisms to improve the performance, such as vacancies and doping.

Theoretical calculations indicated that bidimensional materials, for example, $VO_2$ and $MoO_2$ [16], could provide high capacity and excellent calcium mobility, with low diffusion barrier: 0.306 eV ($VO_2$) and 0.22 eV ($MoO_2$). Moreover, the systems Ca-$VO_2$ and Ca-$WS_2$ could display metallic nature [17]. Another option to decrease the diffusion barrier could be selenium or sulphur instead of oxygen lattice [18]. On the other hand, the highly defective structures can favour the diffusion of the multivalent ions [19, 20]. A method to create crystallographic sites available for the intercalation of calcium is firstly to deintercalate other cations, such as sodium, before electrochemical calcium intercalation.

An alternative to the mechanism for the intercalation of the naked calcium cation could be molecule-assisted intercalation, particularly when the electrolyte solution contains water molecules. The addition of highly diluted water concentrations to the electrolyte solution could improve calcium intercalation. Nevertheless, the uncontrolled addition of water may provoke undesirable side reactions. In fact, it is very difficult to remove completely all traces of water from the electrolyte solutions. The nanosized particles of metallic oxides and other materials also can contain water adsorbed on their surface. These water molecules may generate protons, via oxidation to $O_2$. It can be promoted by the presence of ions in their high oxidation state. In certain cases, the intercalation of these protons could compete against calciation. On the other hand, the host material can also contain crystal water, which may improve calcium diffusion into the framework.

According to both theoretical calculations and electrochemical experiments, the alloying of calcium with metallic elements has been demonstrated to be a feasible way to store calcium. Unfortunately, the huge volume change during the alloying process makes the efficiency and reversibility of the CIB doubtful.

## 7.3 POTENTIAL ELECTRODES AND ELECTROLYTES

Similar to lithium-ion batteries, calcium intercalation compounds are expected to be crucial electrode materials for a potential CIB. The successful development of $LiCoO_2$ as cathode material points towards a similar finding in calcium-containing transition metal oxides. Having in mind the divalent valence of calcium, the compounds based on transition metals that exhibit multiple oxidation states, such as $V_2O_5$ and $MoO_3$, would be particularly interesting. On the other hand, the large radius of calcium would challenge finding host structures with large cavities and/or the ability for lattice expansion, such as layered 2D structures and tunnelled 3D structures. Although many structures have been reported for the reversible accommodation of calcium, no one of them has been yet selected as the most adequate.

$V_2O_5$ was one of the first materials to be researched for calcium intercalation. In a pioneering work, Amatucci et al. reported crystalline $V_2O_5$ for the intercalation of multivalent cations, including $Ca^{2+}$ [6]. Hayashi et al. proposed to use amorphous $V_2O_5$ to accelerate calcium diffusion. They found some new reflections emerging for $Ca_xV_2O_5$ in the XRD pattern, but these new phases could not be identified [21]. Contrarily, Verellli et al. did not find significant intercalation of calcium into $V_2O_5$, suggesting that the results of other authors could be due to parasitic reactions [22]. The known ability of $V_2O_5$ to intercalate water and protons can sow doubt about the validity of calcium intercalation. However, carefully driven studies showed that the presence of water can be even an advantage for veritable calcium intercalation. Thus, bilayered-type $Mg_{0.25}V_2O_5.H_2O$ was demonstrated to be a stable cathode for reversible $Ca^{2+}$ insertion, and the tiny variation of its interlayer spacing (10.76 Å) results in outstanding cycling stability [15]. It was proposed that $Mg^{2+}$ strongly interacts with the adjacent layer and oxygen in crystal water providing excellent structural stability for reversible calcium storage (ca. 120 mA h g$^{-1}$) for hundreds of cycles.

$A_xCoO_2$ compounds, with A = Li, Na, and K, exhibit layer-type structure. $Mg_xCo_2O_4$ prefers a spinel-type structure and $Ca_xCo_2O_4$ also possesses a layer-type structure. A CIB based on layered $CaCo_2O_4$ (s.g. P2/m) as the positive electrode and $V_2O_5$ (s.g. Pmmn) as the negative electrode was proposed [7]. The electrochemical cycling started with the oxidation of $Co^{3+}$ to $Co^{4+}$ at the positive electrode and calcium deintercalation from $CaCo_2O_4$. First-principles calculations found that layered $CaCo_2O_4$ exhibits both thermodynamic and kinetic properties suitable for topotactic Ca ion intercalation, and low energy barrier for calcium migration is comparable to lithium in conventional Li-based cathodes [23]. Another advantage of this electrode material is its high operating voltage (beyond 3 V vs. Ca). Nevertheless, the limited Ca extraction was attributed to the poor kinetics at the anode electrode [24].

Barde et al. patented the use of molybdenum oxides in calcium batteries [25]. The perovskite-type $CaMoO_3$ was found to be not suitable for calcium intercalation, because of the low mobility of calcium in this framework [26]. However, a more

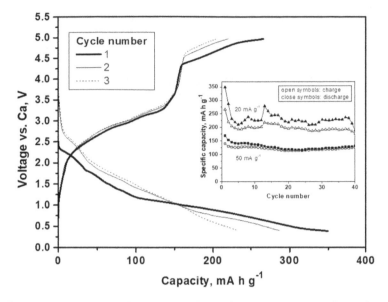

**FIGURE 7.1** Voltage vs capacity curves, and capacity vs cycle number (inset) for α-MoO$_3$ working electrode in calcium cell. Counter electrode: activated carbon. Electrolyte solution: 0.1 M Ca(TFSI)$_2$ in acetonitrile. The voltage of the working electrode was experimentally measured against activated carbon, the OCV voltage of the activated carbon vs. Ca was measured, and the voltage of the working electrode vs. Ca was extrapolated.

recent study has reported that the perovskite-type CaMnO$_3$ is a promising host for Ca intercalation [27]. We reported the intercalation of calcium into layered-type molybdite (α-MoO$_3$) in a nonaqueous electrolyte with an experimental reversible capacity of around 80–100 mA h g$^{-1}$ and accompanied by a relatively small change of the interlayer spacing [28]. We can confirm that the capacity of MoO$_3$ can be further improved (Figure 7.1). Chae et al. proposed that a way to improve the electrochemical behaviour of molybdenum trioxide is by using preinserted calcium and crystal water in Ca$_{0.13}$MoO$_3$·0.41(H$_2$O) [29]. This material could demonstrate a reversible calciation and a maximum capacity of 192 mA h g$^{-1}$. This value is even higher than the capacity of LiCoO$_2$ in lithium batteries. Thus, the mechanism of the charge/discharge process of molybdenum trioxide would deserve an in-depth study in future.

Although some studies reported that the A$_x$Mn$_2$O$_4$ spinels could be utilized as a suitable material for the reversible intercalation of divalent cations (A= Zn, Mg and Ca) and the theoretical calculations suggest that the migration energy of calcium through the diffusion path in Mn$_2$O$_4$ is relatively low [30], but the (de)intercalation of calcium from the polymorphs of CaMn$_2$O$_4$ has only been reported theoretically without any experimental backup [31].

Regarding sustainability, cost and toxicity, the Ca–Fe–O system would be particularly interesting. However, computational results reveal that perovskite-type CaFeO$_3$ and brownmillerite Ca$_2$Fe$_2$O$_5$ are not competitive due to poor calcium mobility. The attempts to electrochemically deintercalate Ca$^{2+}$ ions from the promising Ca$_4$Fe$_9$O$_{17}$

were unsuccessful. It is justified by the electrolyte solution instability at the predicted decalciation voltage (4.16 V vs $Ca^{2+}/Ca$) [32].

The relevance of chalcogenides in the early lithium batteries has prompted their research for calcium batteries. Early reports on calcium intercalation into $TiS_2$ and its solvation reactions can be found in the literature [33]. Later, the use of $TiS_2$ electrodes in calcium batteries was described [34]. More recently, Ca insertion was studied at temperatures around 100°C, reaching discharge capacities close to 500 mA h $g^{-1}$ at C/100 [35]. In 2019, Ca/$TiS_2$ cells were cycled at room temperature, performing reversible capacities of ca. 90 mA h $g^{-1}$ [36]. Recent work confirmed the important role of the electrolyte selection for the proper working of the cells at room temperature [37].

Regarding Chevrel phases, theoretical calculations by using DFT methods showed that the voltages of $Mo_6X_8$ (X = S, Se, Te) electrodes vs. Ca could be twice than those predicted for Mg. However, the diffusion of $Ca^{2+}$ ions would be more limited than that of $Mg^{2+}$ ions due to the different ionic radii [38].

Since the discovery of MXenes ten years ago, their potential applications have been widely explored, with particular emphasis on energy storage. These multilayer 2D solids of transition metals, carbon, and nitrogen have been theoretically evaluated for $Ca^{2+}$. The calculated capacity for Ca-MXenes (e.g. 487 mAh $g^{-1}$ for $Ti_2CO_2$) was lower than the value computed for Ca-defective graphene (913 mAh $g^{-1}$) [39].

Moving to polyanionic salts, we could find polyoxoanionic salts, basically phosphates, and Prussian blue analogs. Several phosphates have been experimentally considered so far: $NaV_2(PO_4)_3$ [40], $FePO_4$ [41], $VOPO_4 \cdot 2H_2O$ [42], and $NaFePO_4F$ [43], and many other oxysalts were studied theoretically [44]. The layered α-$VOPO_4$ was reported to possess the highest theoretical specific capacity, low energy barriers to overcome the migration of $Ca^{2+}$, and interesting voltage. Over 500 charge–discharge cycles have been reported for the intercalation of calcium into $Na_{0.5}VPO_{4.8}F_{0.7}$ [45]. In addition, it was found that the change in the polyanionic framework volume was smaller for calcium than for sodium intercalation.

Lipson et al. [46] demonstrated that manganese hexacyanoferrate (analogues to Prussian blue) could intercalate calcium reversibly. This cell possesses an operating potential of 3.4 V vs. Ca, making it an interesting candidate for cathode application in CIB. However, the reported capacity (80 mA h $g^{-1}$) was limited to the oxidation of Mn.

In analogy to sodium, graphite cointercalates calcium solvated by glyme molecules. Thus, Prabakar et al. [47] showed that graphite could act as an anode in CIB, thanks to the electrochemical co-intercalation of $Ca^{2+}$ and tetraglyme ($G_4$) solvent molecules, leading to a maximum composition of Ca-$G_4 \cdot C_{72}$. Also, a full-cell reversible potential of ca. 1.6 V and capacities around 80 mA h $g^{-1}$ were reported for graphite in combination with an organic cathode (perylene-3,4,9,10-tetracarboxylic dianhydride). Due to the electrochemical amphoterism of graphite, this material has also been reported as an anion insertion cathode in CIBs [48].

In Group 14 of the Periodic Table, the interesting Ca-alloying elements Si, Ge, and Sn are found. First-principles calculations have suggested that amorphous silicon electrodes could reach up to a $Ca_{2.5}Si$ composition, having a slightly higher capacity than the magnesium end composition, but penalized by a much higher volume expansion upon alloy formation and poorer kinetics [49]. Germanium has been experimentally shown to store calcium reversibly at 600°C in a molten salt electrolyte.

The electrochemical process demonstrated multiple successive discharge plateaus reaching a maximum specific capacity of 553 mA h g$^{-1}$ [50]. In tin-based compounds, experimental results showed a calcium uptake up to $Ca_7Sn_6$. A full CIB with a natural graphite cathode and tin foils as the anode rendered 4.45 V for more than 300 cycles [51]. Theoretical calculations have been used to compare Sn with other potential intermetallics, such as $Ca_xBi$ [52].

In principle, the ideal anode for the CIB would be Ca metal, but it is not without problems. Lithium metal in contact with many electrolyte solutions based on organic solvents spontaneously built a solid electrolyte interface (SEI). This layer blocks electron and conduct Li-ion, hence protecting the surface of Li against further irreversible reactions. Unlikely, it is believed that Ca usually forms a passivating layer in the presence of organic solvents, which will block the $Ca^{2+}$ diffusion, thus hindering the reversibility of Ca-electrochemistry. Unlike alkaline elements, the plating/stripping of multivalent metals is not easy. In primary Ca batteries, it was found that Ca continuously corrodes through reaction with $SOCl_2$, thus historically limiting the application of Ca batteries [4]. The poor reversibility of Ca plating in many electrolytes was also reported by Aurbach's group, and it was attributed to the formation of several calcium compounds in the SEI, which depends on the electrolyte composition [5]. Nowadays, it is generally accepted that the reversibility of calcium insertion is highly dependent on the electrolyte chemistry, and many recent studies are being conducted on improving the efficiency of Ca-electrode. Two other main issues for selecting the electrolyte solution are the solubility of the calcium salts and the mono/bi-dentate coordination of calcium to the solvents [53].

The electrochemical behaviour of the multivalent metals and, particularly, calcium is related to their coordination chemistry. The behaviour of Ca metal and $Ca^{2+}$ in different electrolyte solutions has started to be understood only recently. Although little was known about the solvation chemistries and the SEI of calcium, at least there was consensus on the very important role played by the coordination around calcium in the reversibility of these batteries [54]. The strong oxophilicity of calcium favours the effective coordination of oxygen atoms to $Ca^{2+}$. The double-charged cation can hinder desolvation and decrease ionic conductivity. In primary batteries, it was found that small molecules, such as $SO_2$, in the solvating sphere increase the ionic conductivity of $Ca^{2+}$-solvent [55]. The main component of the interface in Ca primary batteries containing chloride ions is the Cl-conductor ($CaCl_2$) [56]. When the electrochemical system involves different mobile ions, such as $Ca^{2+}$ and $Cl^-$, probably the SEI is reversibly precipitated/dissolved, and this dynamic mechanism of the SEI hinders the reversibility of the plating and enhances the corrosion.

The calcium salts currently used as electrolytes in nonaqueous secondary batteries are $Ca(ClO_4)_2$, $Ca(BH_4)_2$, $Ca(TFSI)_2$, $Ca(CF_3SO_3)_2$, $Ca(B(hfip)_4)_2$ and $Ca(PF_6)_2$. Some of these salts are not commercially available and must be synthesized in the research lab [46]. Therefore, it is important to untangle the properties of the prepared electrolytes such as water and air sensitivity, and thermal stability. For that purpose, techniques such as thermal analysis, X-ray diffraction (XRD), X-ray absorption spectroscopy (XAS) and nuclear magnetic resonance (NMR) are useful. It was reported that there is a strong connection between calcium coordination in solution and electrochemical efficiency. Thus, the evaluation of XAS at the Ca K-edge to understand the local coordination in calcium salts is very crucial [54]. Many solvents have been reported in this regard, such as tetrahydrofurane (THF), γ-butyrolactone (BL), acetonitrile (ACN), ethylene carbonate (EC), propylene carbonate (PC), ethyl methyl carbonate (EMC), diglyme (DGM), monoglyme or dimethoxiethane (DME), triglyme (TGM), tetraglyme (TEGDME) and their combinations. Thus, most of these nonaqueous solvents can be classified into two major classes: carbonates and ethers. Unfortunately, most of the electrolyte solutions are not suitable for the reversible plating of calcium. The interaction between $Ca^{2+}$ and solvent molecules is critical to achieving good electrochemical behaviour. The DFT calculations can be utilized as a useful tool to optimize the calcium-solvent geometries. Using such theoretical calculations and modeling, it was confirmed that monodentate molecules are preferred over strongly chelating solvents (e.g. TGM) [54]. The coordination strength of calcium for $Ca(TFSI)_2$ in several solvents increases according to the sequence THF < DME < DGM < TGM [57]. Strong coordination inhibits both desolvation and plating, leading to adverse side reactions on the electrode surface (e.g. decomposition of TFSI⁻) [54]. In addition, THF solvent leads to increased F-content in the SEI, while DME and DGM solvents lead to large O-content, and DGM shows the lowest cell polarization [58]. Solvent molecules coordinated to the cation can be decomposed during plating, and it seems that the TEGDME is more stable at the Ca/electrolyte interface [59].

Ponrouch et al. firstly reported the reversible plating/stripping of calcium on stainless steel at 75–100°C [8]. They used $Ca(BF_4)_2$ dissolved in (EC:PC) solvent, featuring a wide voltage window. The reduction of the electrolyte solution, in parallel with calcium plating, forms the SEI. The nature of this SEI, including composition and defects, is critical for allowing reversible plating. It was found that $Ca(BF_4)_2$ is more active than others salts, and that the SEI contains $CaF_2$ (F-conductor) and other compounds. The same group also found that Ca plating using $Ca(TFSI)_2$ in EC:PC solution is impossible. Computational simulations found that $Ca(TFSI)_2$ and $Ca(PF_6)_2$ are completely decomposed at 300 K on a Ca surface, forming products such as $CaF_2$, CaO, and CaS [59]. The reversibility of Ca plating/stripping at room temperature has been demonstrated only recently. Biria et al. reported that the reversible plating of Ca on copper is feasible using $Ca(BH_4)_2$ in EC:PC at room temperature and low current densities. In this study, they did not find any $CaF_2$ in the SEI, which is a promising attribute [9]. Bruce's group used $Ca(BH_4)_2$ dissolved in THF, which demonstrated a narrow voltage window, resulting in a limited upper attainable voltage because of the unstable nature of borohydrides towards oxidation at ca. 3 V vs. Ca [60]. In addition, the presence of $CaH_2$ on the Au substrate was detected, besides Ca metal. Interestingly, the electrolytes based on $Ca(B(hfip)_4)_2$ exhibited a reversible Ca deposition at room temperature, stability up to ca. 4.5 V, and high ionic conductivity [61,62]. It seems that the high electronegativity of fluorine and the delocalization of the charge in bulky anions (hfip) results in weak cation–anion interactions and, consequently high ionic conductivity, and superior anodic stability. The strong carbon–fluorine bond can also contribute to the high anodic stability (> 4.5 V). The XRD results on $Ca[B(hfip)_4]_2 \cdot 4DME$ reveal that $Ca^{2+}$ is solvated by four DME molecules in a slightly distorted square antiprismatic geometry, with an average Ca–O distance of 2.43 Å, which is longer than Mg–O distance, suggesting that the solvation energy of calcium could be lower than magnesium [61]. The weaker association of Ca–B(hfpi) compared to Ca–$BH_4$ and Ca–TFSI seems to be the key. The same group found that Ca metal soaked in $Ca[B(hfip)_4]_2$/DME electrolyte for weeks does not change the surface of Ca, while ca. 7% of $CaF_2$ is found in all the Ca deposits after electrochemical cycling. This $CaF_2$ would not inhibit plating and is coming from the reduction of the electrolyte at low voltage, similar to the formation of LiF in $LiPF_6$ electrolytes. According to computer simulations, only a few of the C–F bonds are cleaved from $Ca[B(hfip)_4]_2$ to form $CaF_2$ on Ca [59] and, thus $Ca[B(hfip)_4]_2$ on a Ca surface is more stable than other F-containing salts. The carborane-based salt $Ca[CB_{11}H_{12}]_2$ was proposed as a new electrolyte for CIB. Theoretical analyses suggest the superior stability of this anion on Ca surface as compared to other anions [59].

It is generally accepted that multivalent ions do not easily form dendrites. However, a recent study revealed that adverse dendritic growth could happen during calcium plating, and the characteristic branching fractal morphology was observed by in situ electron microscopy, particularly under high current density [63]. Upon oxidation/reduction cycling, calcium dendrites are detached from the electrode, resulting in electrically isolated Ca deposits or "dead calcium".

There are several reasons for the slow advancement of the Ca-battery. The preparation of new electrolytes and favouring reversible Ca plating are the two very important steps towards the development of CIB. In addition, the mechanical properties of Ca must be improved. It is known that the mechanical features (e.g. plasticity, modulus and hardness) of Li and Mg metals influence the electrodeposition process and, particularly, the dendrites formation [64]. The detachment of Ca particles from the electrode surface upon electrochemical cycling can decrease the apparent coulombic efficiency. Therefore, the deposits of Ca on the separators should be examined. The high sensitivity of Ca to the atmosphere can also influence its electrochemical behaviour. It has been proposed an initial five charge–discharge cycles to decrease the over-potential and improve the reversibility of Ca plating [61]. The initial deposition of Ca would facilitate the nucleation of crystalline Ca in the next cycles.

The compatibility between the electrolyte solution and the electrodes in calcium batteries is a serious concern, which will not get resolved easily. For example, $Ca(TFSI)_2$ is expected to be oxidatively decomposed at ca. 3.8 V vs. Ca. Lipson et al. reported that the corrosion of stainless steel used as a current collector must be considered in Ca-based electrolytes, particularly with chloride-containing electrolytes [65]. The traces of water in the electrode or the electrolyte can generate protons, and these protons could be intercalated into the working electrode instead of $Ca^{2+}$.

The best active material which should be used as negative electrode is not still chosen, and this fact limits the advancements towards a commercial CIB [66]. Since Ca metal as the counterelectrode is not a practical option with many electrolyte solutions, it is not employed in many of the calcium-based electrochemical cells reported in the literature. An affordable option is to use another metal, such as Ag [6], as counter and reference electrodes. After the calibration of the reference electrode, the working voltage can be referred to $Ca^{2+}$/Ca. Lipson et al. used Ca as the reference electrode and the alloy CaSn as the counter electrode. This alloy was previously prepared by calciation of Sn through discharging it to −1.5 V vs Ca metal [46]. Considering the failure of the Ca plating/stripping at the negative electrode with common electrolytes, activated carbon (AC) can be used as an alternative counterelectrode. The resulting electrochemical device is an asymmetric cell that can be described as a hybrid battery/supercapacitor system. It is expected to have a reversible adsorption/desorption of anions on the activated carbon. In order to achieve comparable results, the AC working electrode cell voltage must be converted into a voltage value referred to $Ca^{2+}/Ca$. For the validity of this procedure, it is desirable to use a large excess of the activated carbon mass, because the voltage of the activated carbon would increase linearly as anions are adsorbed in the double layer. As an example, the result using this procedure is depicted in Figure 7.1. In addition, the irreversible decomposition of the electrolyte solution at both electrodes and the resulting by-products can contaminate the reference electrode. Henceforth, the potential values reported in the literature for calcium intercalation should be carefully taken.

Aqueous solutions have been also employed as electrolytes for CIBs. The implementation of non-toxic, low-volatile and non-flammable aqueous electrolytes would bring safer batteries. In addition, the needlessness of air and moist-free manufacturing processes could diminish fabrication costs. However, the narrow window of water stability can be addressed as an important drawback for achieving competing devices. Nevertheless, recent research has evidenced that the margin of improvement in aqueous systems is noticeable, deserving the attention of numerous reports [67,68].

Early studies showed the electrochemical formation of cobalt(II) hexacyanoferrate films in $Ca(NO_3)_2$ solution as electrolyte [69]. Afterwards, several works reported the effective insertion of alkaline ions in related metal hexacyanoferrate (HCF) compounds [70,71]. Moreover, Mizuno et al. noted that the activation energy for $Li^+$ and $Na^+$ was considerably smaller in aqueous electrolytes as compared to organic ones [72], and it was explained as a result of the suppression of electrostatic repulsions at the electrode–electrolyte interface. These results encouraged Wang et al. to examine the reversible insertion of $Ca^{2+}$ in Prussian blue-like compounds. For this purpose, Ni-HCF was assembled with activated carbon as counterelectrode and 1 M $Ca(NO_3)_2$ acidic aqueous solution as electrolyte. They reported good capacity retention with an increased charge rate, which was attributed to the high ionic conductivity and the short diffusion pathways of the nanoparticulate Ni compound [73]. Moreover, their results showed that the increase of the charge–discharge voltage hysteresis is proportional to the radius of the cations $Mg^{2+}$, $Ca^{2+}$, $Sr^{2+}$, and $Ba^{2+}$, and this fact was related to the rate-limiting dehydration step. Thus, Mizuno et al. reported partial dehydration at the interfacial charge transfer for the insertion of alkaline ions. It would explain the ionic radius diminution of the inserting species, though preserving a screening effect and low coulombic repulsions [72]. Further studies have also revealed significant differences in the calcium solvation shell structure depending on the solvent polarity. Thus, XAS and DFT/MD calculations highlighted an inverse relationship between the $Ca^{2+}$ coordination number and both the electrolyte concentration and solvent steric hindrance [74].

Motivated by the early successful results, other authors reported the reversible insertion in aqueous cells based on the use of Ba-HCF [75] and Cu-HCF [76–79]. Lee et al. found that the use of a super-concentrated electrolyte ($Ca(NO_3)_2$: $H_2O$ = 1: 7.5) led to a stable cyclability, probably due to the suppression of the CuHCF structural crumbling [76]. The same authors in a later paper found that the activation energy for the $Ca^{2+}$ insertion into CuHCF decreases on increasing the electrolyte concentration, and it was ascribed to the diminution of the hydration number of $Ca^{2+}$ ions [77]. Adil et al. proposed a different approach for the copper compound in which a full cell was assembled with polyaniline-coated carbon cloth as the anode. It delivered a capacity of 130 mA h $g^{-1}$ at a 0.8 A $g^{-1}$ current rate and performed good cycling stability (95% capacity retention over 200 cycles) [78]. Alternatively, Gheytani et al. used the poly-imide poly[$N,N'$-(ethane-1,2-diyl)-1,4,5,8-naphthalenetetracarboxiimide] (PNDIE) as an anode exhibiting 54 W h $kg^{-1}$ at 1 C rate and excellent cycle life regardless of the current density [79].

Another group of inorganic host structures suitable for the insertion of large $Ca^{2+}$ ion is the layered materials. Particularly, the occurrence of water in the interlayer space of hydrated compounds involves an expansion of the flexible framework, which facilitates the reversible insertion of this alkaline-earth ion. In this way, Simon et al. reported a number of vanadium-containing oxides prepared by a molten salt route offering advantages to obtain nanoribbon-like morphologies [80]. The superior per-formance of $CaV_6O_{16}\cdot7H_2O$ was attributed to the 1D-nanostructured architecture and the pillaring effect exerted by the interlayer water molecules. The latter alleviates structural strains upon cycling and provides improved cycling stability. A further ex-situ XPS and XRD experiments on layered potassium vanadate ($K_2V_6O_{16}\cdot2.7$ $H_2O$) provided a direct proof of the reversible intercalation of $Ca^{2+}$ into the structures. Their nanowire-like particles, prepared by a hydrothermal procedure, were able to deliver a high initial capacity (113.9 mA h $g^{-1}$ at 20 mA $g^{-1}$) and good reten-tion of capacity (78.30 %) after 100 cycles at 50 mA $g^{-1}$ [81]. Potassium birnessite ($K_{0.31}MnO_2\cdot0.25H_2O$) is also categorized as a layered structure capable to host large $Ca^{2+}$ ions. It features a promising average discharge voltage (2.8 V vs. Ca) and initial discharge capacity (153 mA h $g^{-1}$ at 0.1 C). However, the material demonstrated an unstable cyclability, likely due to manganese dissolution into the electrolyte from the host material [82]. Another example is doublesheet vanadium oxide ($V_2O_5\cdot0.63$ $H_2O$). The mechanism for the storage of $Ca^{2+}$ ions was a combination of bulk insertion reactions and surface capacitive phenomena. Similar to previous studies, the revers-ible expansion of the lattice cell and energy binding band shifting was considered direct evidence of calcium intercalation [83].

Despite the references cited above, it is yet controversial whether the insertion of proton could be the origin of the capacity at some host materials. Thus, Palacin et al. reported the absence of any evidence of $Ca^{2+}$ extraction from $CaV_2O_5$ prepared by solid-state reaction when subjected to either electrochemical or chemical oxidation [22]. Nonetheless, the co-intercalation of the divalent cation and protons has been identified in aqueous $Zn/MnO_2$ cells, and its high reversibility and cyclability allowed to infer that the presence of protons did not exert any detrimental effects on the cell performance [84].

Besides inorganic compounds, organic materials feature interesting proper-ties to play the role of anodes in CIBs. Their molecular frameworks provide a flexible layout, while they are easily synthesized, less toxic and renewable [85]. Noticeable applicability of these electrode materials is to serve as an anode in full cells. Crystalline 3,4,9,10 perylenetetracarboxylic dianhydride (PTCDA) is a red pigment that could reversibly incorporate $Ca^{2+}$ ion. Although this material could deliver an initial capacity of 87 mA h g$^{-1}$, this capacity quickly faded in the subsequent cycle, most probably due to the electrode amorphization observed in the material using XRD analyses [86]. PNDIE delivered a capacity of 160 mA h g$^{-1}$ at −0.45 V versus Ag/AgCl. The reaction mechanism is characterized by a two-step two-electron enolization during $Ca^{2+}$ storage [79]. A full cell assembled with CuHCF as a positive electrode and PNDIE as a negative electrode revealed an outstanding cyclability regardless of the current density. Eventually, Cang et al. reported the reliability of a composite material consisting of poly(3,4,9, 10-perylentetracarboxylic diimide) (PPTCDI) supported on mesoporous silica SBA-15. Their synergistic combination provided excellent cycling stability with 95% capacity retention after 1,500 cycles [87]. Based on the assumption mentioned above, by which proton co-insertion exerts a beneficial effect on the overall cell reversibility, Han et al. reported proton-assisted $Ca^{2+}$ storage in aromatic 5,7,12,14-pentacenetetrone [88]. The weak π–π interaction between the stacked layers favoured a flexible host structure, while the accessible channels provided fast ionic diffusion for $Ca^{2+}$ storage. According to the authors, the contribution of proton insertion is not only significant to overall capacity (40–50%), but essential to the insertion of calcium. Thus, the attempt to insert $Ca^{2+}$ in an organic-based electrolyte solution failed due to the absence of protons [88].

## 7.4   BENEFITS AND CHALLENGES

M. Stanley Whittingham, Nobel Laureate in Chemistry 2019, said "calcium cells have a potential much closer to that of lithium cells, and the larger $Ca^{2+}$ should be able to diffuse faster" [89]. Despite the potential applicability of these cells, its transition towards practical batteries is quite challenging.

Regarding slurry preparation of the electrode materials, it is not usually more difficult than in the case of lithium cells, bearing in mind that product characteristics are strongly dependent on particle size. Therefore, common mixing equipment for slurry scaling-up can be used, such as dissolvers, kneaders, stirred media mills, three roller mills, disc mills, and ultrasonic homogenizers. At this point, electrode loadings are a key factor to meet energy density targets. These loadings are not usually the ones used in experiments at lab-scale, with a range that typically varies from 0.5 to 1 mAh cm$^{-2}$, electrode loadings of 5 mAh cm$^{-2}$ or even higher should be reached. For this purpose, a minimum of 0.5 kg of active material would be needed for slurry preparation.

For the realistic assessment of a Ca metal anode, we must consider and compare the capacity values. Since calcium is relatively heavy and dense, the volumetric capacity of Ca metal is comparable to that of Li metal, but the gravimetric capacity of calcium is significantly lower (1,337 mA h g$^{-1}$ against 3,861 mA h g$^{-1}$). Nevertheless, the gravimetric capacity of calcium surpasses other anodes, such as sodium, zinc and potassium. Besides the theoretical capacities, it is mandatory to consider the electrochemical performances. The enabled reversible plating/stripping of calcium is a technological breakthrough that has renewed the interest on calcium batteries and that can also impact some other areas. It seems nowadays it is quite possible to form a

$Ca^{2+}$-permeable SEI, contrary to previous thoughts. Thus, the experimental capacity of calcium has an opportunity to be competitive, but it needs to be explored thoroughly. Between the benefits already gained through the research about CIB are:

- Based on the available studies on the environmental impact of different metals, such as cumulative energy demand, global warming potential, and yearly global $CO_2$ emission, the most promising candidate to replace Li is Ca [90].
- Net advancements in the theoretical calculations about calcium intercalation.
- As a new strategy, different from Ca-metal batteries, full CIB have been tested: $CaCo_2O_4$ vs. $V_2O_5$ [7], and $MnFe(CN)_6$ vs. CaSn alloy [46].
- Discarding of some materials as electrodes for CIB due to their poor electrochemical behaviour.
- The relationships between the Ca-solvent coordination and the electrochemical performance have started to become clear.
- New electrolyte solutions which are very promising for reversible calcium plating have been developed, and the key factors for the optimum behaviour have become known.

However, some concerns prevent the development of commercial CIB, and even the proper electrochemical behavioural study of the material at the cell level. Unfortunately, the differences between calcium and lithium intercalation can result in a lower amount of reversibly intercalated calcium as compared to lithium. Further, for most of the Ca-based electrode materials, the experimental capacity was observed to be much lower than the theoretically calculated capacity. By considering all the aforementioned points, it is quite obvious that the road to developing CIBs must be paved with theoretical calculations/models, finding more suitable electrolyte solutions, and using stable electrode materials.

Several of the main challenges and opportunities for CIB are:

- Improving the understanding of the behaviour of calcium metal in the electrolyte solutions, the formation of the SEI layer and the influence of the SEI on the plating/stripping of Ca. Optimization of the SEI formation.
- Evaluation of calcium dendrites and their effect on practical applications.
- Improving the chemistry knowhow of $Ca^{2+}$ in the electrolyte solutions, including the coordination chemistry.
- Distinguishing side reactions from the reaction of calcium accommodation. Elucidating the exact role played by the traces of water and protons in the intercalation of calcium.
- Characterization of the new phases formed upon calciation of previously known compounds.
- Identification of the best structures that can accommodate reversibly a great number of $Ca^{2+}$.
- Discovering and developing high-voltage, high-capacity, and high-efficiency cathodes. There are still many opportunities to investigate new materials.
- Improving the cycling stability.

## 7.5   CURRENT DEVELOPMENTS AND PROSPECTS

The current methods to reduce the effect of poor multivalent diffusion, such as calcium diffusion, are [6]:

- The use of hydrated compounds. The oxygen atoms of water molecules can shield the coulombic interactions between the cationic guest and the anions of the host. The water molecules could be present in the crystal of the host compound, or initially in the electrolyte solution and then being intercalated in the host.
- Preparation of materials of small diffusion distances. The small particle size reduces the diffusion path length.
- To design host materials with structures with low activation energy for rapid diffusion. The theoretical calculations can help to find the materials with the most adequate structures for calcium diffusion.

To offer prospects, the next strategies could be followed:

- $^{43}$Ca NMR could help to understand the behaviour of calcium in batteries.
- Optimization of the microtexture, particle morphology, particle size and porosity of the electrode materials.
- Finding new calcium-containing compounds, which were not previously synthesized, with rapid Ca-diffusion rate and great energy density.
- Chemical or electrochemical deintercalation of Li, Na or Mg from well-known compounds and then calciation.
- Finding relationships between the mechanical properties of Ca metal and its electrochemical behaviour during cycling. A comprehensive study of the rigidity of Ca metal for battery applications.
- Enhancement of the reversibility and efficiency of calcium plating/stripping at room temperature. Suppression of the formation of a blocking layer on the surface of the Ca electrode. A deeper understanding of the SEI on Ca.
- Identification of the most adequate electrolytes. Ca-compatible electrolytes with a wide voltage window (> 4 V).
- Commercialization of new calcium salts which were recently reported by several research groups at laboratory scale or via computer simulation, such as $Ca[B(hfip)_4]_2$ and $Ca[CB_{11}H_{12}]_2$.
- Improved Ca resilience against oxidation by water and oxygen. Coating and alloying could allow the handling of calcium electrode under air atmosphere.
- Replacement of the Ca electrode by other types of anodes that could operate with common calcium-based electrolytes.
- Dual-ion batteries. Reversible calciation at the negative electrode and reversible accommodation of another cation or anion at the positive electrode.
- Adequacy of the stability window of the electrolyte solution to the voltages of the two electrode materials.
- Suppression of current collector corrosion. Optimization of the combination of current collector and electrolyte solution.
- Reversible redox process of two electrons per transition metal atom.

- Moving from three-electrode cell and half-cell to full cell.
- Engineering, scaling-up, and prototyping of commercial CIBs. Current equipment (welding, stacking machines…) for assembly of lithium-ion batteries operate under oxygen-containing atmosphere. However, strict control of the inert atmosphere is needed for handling Ca metal and assembling of the batteries, and it would increase the cost. The electrode of Ca could be modified to increase its resilience against corrosion. Due to the mass and density of Ca, thin films of Ca should be used to minimize the impact on the overall weight of the battery. It is needed to investigate the casting and rolling of Ca on a current collector. New battery architectures could be proposed.
- Optimization of the electrolyte concentration. The low solubility of many calcium salts could limit the concentration of some electrolytes. Since calcium is heavier than lithium, the overall weight of the battery could be raised by the weight of the electrolyte solution. Thus, using low-concentration electrolytes should be further explored.
- Economic modelling. The cost of the CIB should be calculated.
- Sustainability, cycle life, wastes and circular economy.

## REFERENCES

1. M.E. Arroyo-De Dompablo, A. Ponrouch, P. Johansson, M.R. Palacín. Achievements, challenges, and prospects of calcium batteries. *Chem. Rev.* 2020, 120 (14), 6331.
2. L. Stievano, I. de Meatza, J. Bitenc, C. Cavallo. S. Brutti, M.A. Navarra. Emerging calcium batteries. *J. Power Sources* 2021, 482, 228875.
3. S. M. Selis, J. P. Wondowski, R. F. Justus. A high-rate, high-energy thermal battery system. *J. Electrochem. Soc.* 1964, 111 (1), 6.
4. R. J. Staniewicz. A study of the calcium-thionyl chloride electrochemical system. *J. Electrochem. Soc.* 1980, 127 (4), 782.
5. D. Aurbach, R. Skaletsky, Y. Gofer. The electrochemical behavior of calcium electrodes in a few organic electrolytes. *J. Electrochem. Soc.* 1991, 138 (12), 3536.
6. G.G. Amatucci, F. Badway, A. Singhal, B. Beaudoin, G. Skandan, T. Bowmer, I, Plitz, N. Pereira, T. Chapman, R. Jaworskai. Investigation of yttrium and polyvalent ion intercalation into nanocrystalline vanadium oxide. *J. Electrochem. Soc.* 2001, 148 (8), A940.
7. M. Cabello, F. Nacimiento, J.R. González, G. Ortiz, R. Alcántara, C. Pérez-Vicente, J.L. Tirado. Advancing towards a veritable calcium-ion battery: $CaCo_2O_4$ positive electrode material. *Electrochem. Commun.* 2016, 67, 59.
8. A. Ponrouch, C. Frontera, F. Bardé, M.R. Palacín. Towards a calcium-based rechargeable battery. *Nat. Mater.* 2016, 15, 169.
9. S. Biria, S. Pathreeker, H. Li, I.D. Hosein. Plating and stripping of calcium in an alkyl carbonate electrolyte at room temperature. *ACS Appl. Energy Mater.* 2019, 2 (11), 7738.
10. R.J. Gummow, G. Vamvounis, M.B. Kannan, Y. He. Calcium-ion batteries: current state-of-the-art and future perspectives. *Adv. Mater.* 2018, 30 (39), 1801702.
11. A.K. Katz, J.P. Gusker, S.A. Beebe, C.W. Bock. Calcium ion coordination: a comparison with that of beryllium, magnesium, and zinc. *J. Am. Chem. Soc.* 1996, 118 (24), 5752.

12. T. Das, S. Tosoni, G. Pacchioni. Layered oxides as cathode materials for beyond-Li batteries: A computational study of Ca and Al intercalation in bulk $V_2O_5$ and $MoO_3$. *Comput. Mater. Sci.* 2021, 191, 110324.

13. C. Han, H. Li, Y. Li, J. Zhu, C. Zhi. Proton-assisted calcium-ion storage in aromatic organic molecular crystal with coplanar stacked structure. *Nat. Commun.* 2021, 12, 2400.

14. M.S. Whittingham, C. Siu, J. Ding. Can multielectron intercalation reactions be the basis of next generation batteries? *Acc. Chem. Res.* 2018, 51 (2), 258.

15. X. Xu, M. Duan, Y. Yue, Q. Li, X. Zhang, L. Wu, P. Wu, B. Song, L. Mai. Bilayered $Mg_{0.25}V_2O_5 \cdot H_2O$ as a stable cathode for rechargeable Ca-ion batteries. *ACS Energy Lett.* 2019, 4 (6), 1328.

16. Y. Wang, N. Song, X. Song, T. Zhang, Q. Zhang, M. Li. Metallic $VO_2$ monolayer as an anode material for Li, Na, K, Mg or Ca ion storage: a first-principle study. *RSC Adv.* 2018, 8, 10848.

17. G. R. Vakili-Nezhaad, A. M. Gujarathi, N. A. Rawahi, M. Mohammadi. Performance of $WS_2$ monolayers as a new family of anode materials for metal-ion (Mg, Al and Ca) batteries. *Mater. Chem.* Phys. 2019, 230, 114.

18. Y. Liang, H. Dong, D. Aurbach, Y. Yao. Current status and future directions of multivalent metal-ion batteries. *Nat. Energy* 2020, 5, 646.

19. S. Kang, K.G. Reeves, T. Koketsu, J. Ma, O.J. Borkiewicz, P-Strasser, A. Ponrouch, D. Dambournet. Multivalent $Mg^{2+}$-, $Zn^{2+}$-, and $Ca^{2+}$-ion intercalation chemistry in a disordered layered structure. *ACS Appl. Energy Mater.* 2020, 3 (9), 9143.

20. Y. Orikasa, K. Kisu, E. Iwama, W. Naoi, Y. Yamaguchi, Y. Yamaguchi, N. Okita, K. Ohara, T. Munesada, M. Hattori, K. Yamamoto. P. Rozier, P. Simon, K. Naoi. Noncrystalline nanocomposites as a remedy for the low diffusivity of multivalent ions in battery cathodes. *Chem. Mater.* 2020, 32 (3), 1011.

21. M. Hayashi, H. Arai, H. Ohtsuka, Y. Sakurai. Electrochemical characteristics of calcium in organic electrolyte solutions and vanadium oxides as calcium hosts. *J. Power Sources* 2003, 119–121, 617.

22. R. Verrelli, A. P. Black, C. Pattanathummasid, D. S. Tchitchekova, A. Ponrouch, J. Oró-Solé, C. Frontera, F. Bardé, P. Rozier, M. R. Palacín, On the strange case of divalent ions intercalation in $V_2O_5$. *J. Power Sources* 2018, 407, 162.

23. H. Park, P. Zapol. Thermodynamic and kinetic properties of layered-$CaCo_2O_4$ for the Ca-ion batteries: a systematic first-principles study. *J. Mater. Chem. A* 2020, 8 (41) 21700.

24. H. Park, Y. Cui, S. Kim, J.T. Vaughey, P. Zapol. Ca cobaltites as potential cathode materials for rechargeable Ca-ion batteries: Theory and experiment. *J. Phys. Chem. C* 2020, 124 (11) 5902.

25. F. Barde, M.R. Palacín, A. Ponrouch, M.E. Arroyo-de Dompablo. Molybdenum-based electrode materials for rechargeable calcium batteries. Patent WO2017/097437 A1, 15 June 2017.

26. M.E. Arroyo-de Dompablo, C. Krich, J. Nava-Avendaño, M.R. Palacín, F. Bardé. In quest of cathode materials for Ca ion batteries: the $CaMO_3$ perovskites (M = Mo, Cr, Mn, Fe, Co, Ni). *Phys. Chem. Chem. Phys.* 2016, 18, 19966.

27. S. Pathreeker, S. Reed, P. Chando, I.D. Hosein. A study of calcium ion intercalation in perovskite calcium manganese oxide. *J. Electroanal. Chem.* 2020, 874, 114453.

28. M. Cabello, F. Nacimiento, R. Alcántara, P. Lavela, C. Pérez-Vicente, J.L. Tirado. Applicability of molybdite as an electrode material in calcium batteries: a structural study of layer-type $Ca_xMoO_3$. *Chem. Mater.* 2018, 30 (17), 5853.

29. M.S. Chae, H.H. Kwak, S.T. Hong. Calcium molybdenum bronze as a stable high-capacity cathode material for calcium-ion batteries. *ACS Appl. Energy Mater.* 2020, 3 (6), 5107.

30. Z. Rong, R. Malik, P. Canepa, G.S. Gautam, M. Liu, A. Jain, K. Persson, G. Ceder. Materials design rules for multivalent ion mobility in intercalation structures. *Chem. Mater.* 2015, 27 (17), 6016.

31. M.E. Arroyo-de Dompablo, C. Krich, J. Nava-Avendaño, M.R. Palacín, F. Bardé. A joint computational and experimental evaluation of $CaMn_2O_4$ polymorphs as cathode materials for Ca ion batteries. *Chem. Mater.* 2016, 28 (19), 6886.

32. A.P. Black, A. Torres, C. Frontera, M.R. Palacín, M.E. Arroyo-de Dompablo. Appraisal of calcium ferrites as cathodes for calcium rechargeable batteries: DFT, synthesis, characterization and electrochemistry of $Ca_4Fe_9O_{17}$. *Dalton Trans.* 2020, 49, 2671.

33. A. Lerf, R. Schollhorn. Solvation reactions of layered ternary sulfides $A_xTiS_2$, $A_xNbS_2$, and $A_xTaS2$. *Inorg. Chem.* 1977, 16 (11), 2950.

34. Y.T. Jeong, S.K. Jeong. International Conference on Power Electronics and Energy Engineering (PEEE). Hong Kong, Peoples R. China 2015, 42.

35. D.S. Tchitchekova, A. Ponrouch, R. Verrelli, T. Broux, C. Frontera, A. Sorrentino, F. Bardé, N. Biskup, M.E. Arroyo-de Dompablo, M.R. Palacín. Electrochemical intercalation of calcium and magnesium in $TiS_2$: Fundamental studies related to multivalent battery applications. *Chem. Mater.* 2018, 30 (3), 847.

36. C. Lee, Y.T. Jeong, P. Maldonado Nogales, H.Y. Song, Y.S. Kim, R.Z. Yin, S.K. Jeong. Electrochemical intercalation of $Ca^{2+}$ ions into $TiS_2$ in organic electrolytes at room temperature. *Electrochem. Commun.* 2019, 98, 115.

37. R. Verrelli, A. Black, R. Dugas, D. Tchitchekova, A. Ponrouch, M.R. Palacin. Steps towards the use of $TiS_2$ electrodes in Ca batteries. *J. Electrochem. Soc.* 2020, 167 (7), 070532.

38. Ma. Smeu, M.S. Hossain, Z. Wang, V. Timoshevski, K.H. Bevan, K. Zaghib. Theoretical investigation of Chevrel phase materials for cathodes accommodating $Ca^{2+}$ ions. *J. Power Sources* 2016, 306, 431.

39. Y. Xie, Y. Dall'Agnese, M. Naguib, Y. Gogotsi, M.W. Barsoum, H.L. Zhuang, P.R.C. Kent. Prediction and characterization of MXene nanosheet anodes for non-lithium-ion batteries. *ACS Nano* 2014, 8 (9), 9606.

40. B. Jeon, J.W. Heo, J. Hyoung, H.H. Kwak, D.M. Lee, S.T. Hong. Reversible calcium-ion insertion in NASICON-type $NaV_2(PO_4)_3$. *Chem. Mater.* 2020, 32 (20), 8772.

41. S. Kim, L. Yin, M.H. Lee, P. Parajuli, L. Blanc, T.T. Fister, H. Park, B.J. Kwon, B.J. Ingram, P. Zapol, R.F. Klie, K. Kang, L.F. Nazar, S.H. Lapidus, J.T. Vaughey. High-voltage phosphate cathodes for rechargeable Ca-ion batteries. *ACS Energy Lett.* 2020, 5 (10) 3203.

42. J. Wang, S. Tan, F. Xiong, R. Yu, P. Wu, L. Cui. Q. An. $VOPO_4 \cdot 2H_2O$ as a new cathode material for rechargeable Ca-ion batteries. *Chem. Commun.* 2020, 56, 3805.

43. A.L. Lipson, S. Kim, B. Pan, C. Liao, T.T. Fister, B.J. Ingram. Calcium intercalation into layered fluorinated sodium iron phosphate. *J. Power Sources* 2017, 369, 133.

44. A. Torres, J.L. Casals, M.E. Arroyo-de Dompablo. Enlisting potential cathode materials for rechargeable Ca batteries. *Chem. Mater.* 2021, 33 (7), 2488.

45. Z.L. Xu, J. Park, J. Wang, H. Moon, G. Yoon, J. Lim, Y.J. Ko, S.P. Cho, S.Y. Lee, K. Kang. A new high-voltage calcium intercalation host for ultra-stable and high-power calcium rechargeable batteries. *Nat. Commun.* 2021 12, 3369.

46. A.L. Lipson, B. Pan, S.H. Lapidus, C. Liao, J.T. Vaughey, B.J. Ingram. Rechargeable Ca-ion batteries: a new energy storage system. *Chem. Mater.* 2015, 27 (24), 8442.

47. S.J.R. Prabakar, A.B. Ikhe, W.B. Park, K.C. Chung, H. Park, K.J. Kim, D. Ahn, J.S. Kwak, K.S. Sohn, M. Pyo. Graphite as a long-life $Ca^{2+}$-intercalation anode and its implementation for rocking-chair type calcium-ion batteries. *Adv. Sci.* 2019, 6 (24), 1902129.

48. M. Wang, C. Jiang, S. Zhang, X. Song, Y. Tang, H.M. Cheng. Reversible calcium alloying enables a practical room-temperature rechargeable calcium-ion battery with a high discharge voltage. *Nat. Chem.* 2018, 10, 667.

49. S. Lee, M. Ko, S.C. Jung, Y.K. Han. Silicon as the anode material for multivalent-ion batteries: a first-principles dynamics study. *ACS Appl. Mater. Interfaces* 2020, 12 (50), 55746.

50. H. Yin, D. Wang. Electrolytic germanium for calcium storage. *J. Electrochem. Soc.* 2016, 163 (13), E351.

51. M. Wang, C. Jiang, S. Zhang, X. Song, Y. Tang, H.M. Cheng. Reversible calcium alloying enables a practical room-temperature rechargeable calcium-ion battery with a high discharge voltage. *Nat. Chem.* 2018, 10, 667.

52. Z. Yao, V.I. Hegde, A. Aspuru-Guzik, C. Wolverton. Discovery of calciummetal alloy anodes for reversible Ca-ion batteries. *Adv. Energy Mater.* 2019, 9 (9), 1802994.

53. J.D. Forero-Saboya, E. Marchante, R.B. Araujo, D. Monti, P. Johansson, A. Ponrouch. Cation solvation and physicochemical properties of Ca battery electrolytes. *J. Phys. Chem. C* 2019, 123 (49), 29524.

54. D.M. Driscoll, N.K. Dandu, N.T. Hahn. T.J. Seguin, K.A. Persson, K.R. Zavadil, L.A. Cutiss, M. Balasubramanian. Rationalizing calcium electrodeposition behavior by quantifying ethereal solvation effects on $Ca^{2+}$ coordination in well-dissociated electrolytes. *J. Electrochem. Soc.* 2020, 167 (16) 160512.

55. C.W. Walker Jr, W. L. Wade Jr, M. Binder. Interactions of $SO_2$ with $Ca(AlCl_4)_2.SO_2Cl_2$ electrolyte: Raman studies. *J. Power Sources* 1989, 25, 187.

56. A. Meitav, E. Peled. Solid electrolyte interphase (SEI) electrode. Part V: the formation and properties of the solid electrolyte interphase on calcium in thionylchloride solutions. *Electrochim. Acta* 1988, 33 (8), 1111.

57. N. T. Hahn, D.M. Driscoll, Z.Yu, G.E. Sterbinsky, L. Cheng, M. Balasubramanian, K.R. Zavadil. Influence of ether solvent and anion coordination on electrochemical behavior in calcium battery electrolytes. *ACS Appl. Energy Mater.* 2020, 3 (9), 8437.

58. K.V. Nielson, J. Luo, T.L. Liu. Optimizing calcium electrolytes by solvent manipulation for calcium batteries. *Batteries Supercaps* 2020, 3 (8), 766.

59. S.S.R.K.C. Yamijala, H. Kwon, J. Guo, B.M. Wong. Stability of calcium ion battery electrolytes: predictions from ab initio molecular dynamics simulations. *ACS Appl. Mater. Interfaces* 2021, 13 (11), 13114.

60. D. Wang, X. Gao, Y. Chen, L. Jin, C. Kuss, P.G. Bruce. Plating and stripping calcium in an organic electrolyte. *Nat. Mater.* 2018, 17, 16.

61. Z. Li, O. Fuhr, M. Fichtner, Z. Zhao-Karger. Towards stable and efficient electrolytes for room-temperature rechargeable calcium batteries. *Energy Environ. Sci.* 2019, 12, 3496.

62. A. Shyamsunder, L.E. Blanc, A. Assoud, L.F. Nazar. Reversible calcium plating and stripping at room temperature using a borate salt. *ACS Energy Lett.* 2019, 4 (9), 2271.

63. S.D. Pu, C. Gong, X. Gao, Z. Nig, S. Yang, J.J. Marie, B. Liu, R.A. Houese, G.O. Hartley, J. Luo, P.G. Bruce, A.W. Robertson. Current-density-dependent electroplating in Ca electrolytes: from globules to dendrites. *ACS Energy Lett.* 2020, 5 (7), 2283.

64. C.D. Fincher, D. Ojeda, Y. Zhang, G.M. Pharr, M. Pharr. Mechanical properties of metallic lithium: from nano to bulk scales. *Acta Mater.* 2020, 186, 215.

65. A.L. Lipson, D. L. Proffit, B. Pan, T.T. Fister, C. Liao, A.K. Burrell, J.T. Vaugher, B.J. Ingram. Current collector corrosion in Ca-ion batteries. *J. Electrochem. Soc.* 2015, 162 (8), A1574.

66. D.S. Tchitchekova, D. Monti, P. Johansson, F. Bardé, A. Randon-Vitanova, M.R. Palacín, A. Ponrouch. On the reliability of half-cell tests for monovalent (Li⁺, Na⁺) and divalent (Mg²⁺, Ca²⁺) cation based batteries. *J. Electrochem. Soc.* 2017, 164 (7), A1384.

67. X. Yuan, F. Ma, L. Zuo, J. Wang, N. Yu, Y. Chen, Y. Zhu, Q. Huang, R. Holze, Y. Wu, T. van Ree. Latest advances in high voltage and high energy density aqueous rechargeable batteries. *Electrochem. Energ. Rev.* 2021, 4, 1.

68. Y. Liu, G. He, H. Jiang, I.P. Parkin, P.R. Shearing, D.J.L. Brett. Cathode design for aqueous rechargeable multivalent ion batteries: challenges and opportunities. *Adv. Funct. Mater.* 2021, 31 (13), 2010445.

69. S.M. Chen. Preparation, characterization, and electrocatalytic oxidation properties of iron, cobalt, nickel, and indium hexacyanoferrate. *J. Electroanal. Chem.* 2002, 521 (1–2), 29.

70. M. Pasta, C.D. Wessells, R.A. Huggins, Y. Cui. A high-rate and long cycle life aqueous electrolyte battery for grid-scale energy storage. *Nat. Commun.* 2012, 3 (1), 1149.

71. C.D. Wessells, S.V. Peddada, R.A. Huggins, Y. Cui. Nickel hexacyanoferrate nanoparticle electrodes for aqueous sodium and potassium ion batteries. *Nano Lett.* 2011, 11 (12), 5421.

72. Y. Mizuno, M. Okubo, E. Hosono, T. Kudo, H. Zhou, K. Oh-ishi. Suppressed activation energy for interfacial charge transfer of a Prussian blue analog thin film electrode with hydrated ions (Li⁺, Na⁺, and Mg²⁺). *J. Phys. Chem. C* 2013, 117 (21), 10877.

73. R.Y. Wang, C.D. Wessells, R.A. Huggins, Y. Cui. Highly reversible open framework nanoscale electrodes for divalent ion batteries. *Nano Lett.* 2013, 13 (11), 5748.

74. F. Yang, Y.S. Liu, X. Feng, K. Qian, L.C. Kao, Y. Ha, N.T. Hahn, T.J. Seguin, M. Tsige, W. Yang, K.R. Zavadil, K.A. Persson, J. Guo. Probing calcium solvation by XAS, MD and DFT calculations. *RSC Adv.* 2020, 10, 27315.

75. Md. Adil, P.K. Dutta, S. Mitra. An aqueous Ca-ion full cell comprising BaHCF cathode and MCMB anode. *ChemistrySelect* 2018, 3 (13), 3687.

76. C. Lee, S.K. Jeong. A novel superconcentrated aqueous electrolyte to improve the electrochemical performance of calcium-ion batteries. *Chem. Lett.* 2016, 45 (12), 1447.

77. C. Lee, S.K. Jeong. Modulating the hydration number of calcium ions by varying the electrolyte concentration: electrochemical performance in a Prussian blue electrode/aqueous electrolyte system for calcium-ion batteries. *Electrochim. Acta* 2018, 265, 430.

78. Md. Adil, A. Sarkar, A. Roy, M.R. Panda, A. Nagendra, S. Mitra. Practical aqueous calcium-ion battery full-cells for future stationary storage. *ACS Appl. Mater. Interfaces* 2020, 12 (10), 11489.

79. S. Gheytani, Y. Liang, F. Wu, Y. Jing, H. Dong, K.K. Rao, X. Chi, F. Fang, Y. Yao. An aqueous Ca-ion battery. *Adv. Sci.* 2017, 4 (12), 1700465.

80. L. Liu, Y.C. Wu, P. Rozier, P.L. Taberna, P. Simon. Ultrafast synthesis of calcium vanadate for superior aqueous calcium-ion battery. *Research* 2019, 6585686.

81. L.W. Dong, R.G. Xu, P.P. Wang, S.C. Sun, Y. Li, L. Zhen, C.Y. Xu. Layered potassium vanadate $K_2V_6O_{16}$ nanowires: A stable and high capacity cathode material for calcium-ion batteries. *J. Power Sources* 2020, 479, 228793.

82. J. Hyoung, J.W. Heo, S.T. Hong. Investigation of electrochemical calcium-ion energy storage mechanism in potassium birnessite. *J. Power Sources* 2018, 390, 127.

83. M.S. Chae, J.W. Heo, J. Hyoung, S.T. Hong. Double-sheet vanadium oxide as a cathode material for calcium-ion batteries. *ChemNanoMat* 2020, 6 (7), 1049.

84. W. Sun, F. Wang, S. Hou, C. Yang, X. Fan, Z. Ma, T. Gao, F. Han, R. Hu, M. Zhu, C. Wang. $Zn/MnO_2$ battery chemistry with $H^+$ and $Zn^{2+}$ coinsertion. *J. Am. Chem. Soc.* 2017, 139 (29), 9775.

85. B. Ji, H. He, W. Yao, Y. Tang. Recent advances and perspectives on calcium ion storage: Key materials and devices. *Adv. Mater.* 2021, 33 (2), 2005501.

86. I.A. Rodríguez-Pérez, Y. Yuan, C. Bommier, X. Wang, L. Ma, D.P. Leonard, M.M. Lerner, R.G. Carter, T. Wu, P.A. Greaney, J. Lu, X. Ji. Mg-ion battery electrode: An organic solid's herringbone structure squeezed upon Mg-ion insertion. *J. Am. Chem. Soc.* 2017, 139 (37), 13031.

87. R. Cang, C. Zhao, K. Ye, J. Yin, K. Zhu, J. Yan, G. Wang, D. Cao. Aqueous Calcium-ion battery based on a mesoporous organic anode and a manganite cathode with long cycling performance. *ChemSusChem* 2020, 13 (15), 3911.

88. C. Han, H. Li, Y. Li, J. Zhu, C. Zhi. Proton-assisted calcium-ion storage in aromatic organic molecular crystal with coplanar stacked structure. *Nat. Commun.* 2021, 12, 2400.

89. M. S. Whittingham. Lithium batteries: 50 years of advances to address the next 20 years of climate issues. *Nano Lett.* 2020, 20, 8435.

90. S. Ferrari, M. Falco, A. B. Muñoz-García, M. Bonomo, S. Brutti, M. Pavone, C. Gerbaldi. Solid-state post Li metal ion batteries: a sustainable forthcoming reality? *Adv. Energy Mater.* 2021, 2100785.

# 8 Metal Ion to Metal Batteries

*Wei Tang*

## 8.1 INTRODUCTION

The increasing demand for a low-carbon footprint economy has triggered significant changes in global energy consumption, driving us toward an imminent transition from hydrocarbon fuels to renewable and sustainable energy technologies. Electrochemical energy storage systems, such as batteries, are critical for enabling sustainable yet intermittent energy harvesting from sources including solar, wind, and geothermal. To date, various battery technologies have been employed for rapid and efficient energy storage/release, especially the prevailing lithium-ion batteries (LIBs), which fulfilled their promise for portable electronic devices and are being established in other niche markets such as stationary units and electric vehicles. The global market is projected to hit A\$200 billion by 2027, however, the cell-level energy density and pack-level costs of practical systems come short of the corresponding expectations with characteristics in the order of 350 W h kg$^{-1}$, 750 W h L$^{-1}$, and <A\$170 kW h$^{-1}$, respectively.

The need for better rechargeable batteries to enable high energy densities (around 500 W h kg$^{-1}$) with low cost has motivated researchers to replace the current metal ion batteries over the past few decades. The transition to metal-electrode-based battery chemistries, which have been known since the 1970s, can significantly enhance the figures of merit of batteries, and one of the possible alternatives such as Li metal anodes featuring a high specific capacity (3860 mA h g$^{-1}$) and electropositive potential (−3.04 V vs. standard hydrogen electrode, SHE). Unfortunately, most solid metal anodes, such as Zn and Li metal anodes, suffer from the possibility/risk of dendrite growth, which induces safety incidents and impedes security-critical applications. Consequently, finding alternatives with high energy densities, cost effectiveness, and good safety features is still an important part of metal-based battery research. This realization led to interest in liquid-to-liquid-based electrochemistry to gradually substitute for solid-to-solid chemistries. There is significant competitiveness for liquid systems beyond solid-state systems in terms of (i) the capability to operate with relatively high voltage efficiencies at high rates due to the superior kinetics provided by the liquid–liquid interfaces; (ii) the ability of liquid electrodes to render themselves

DOI: 10.1201/9781003208198-8

immune to microstructural electrode deformation and/or dendrite formation; (iii) the advantage of establishing flexible energy storage components; and (iv) the eventual potential of reduced cost because most of the active materials are earth-abundant and inexpensive.

The challenges associated with LIBs led to disappointment, but new findings about the prospects of this emerging battery technology have broadened today's ambitions. In this chapter, we first examine the strengths and weaknesses of metal batteries in terms of energy density and safety features. We note that the anticipated direct use of metals as the anode materials is central to the safety promises and potential advantage in energy density of the technologies, but both aspects cannot be taken for granted and require thorough analyses and further research. We then critically review the status of anode materials, electrolyte solutions, and cathode materials for the batteries with an emphasis on metal anode growth behavior, coulombic efficiency and cathode storage mechanism. We recommend characterization practices that could help overcome the rampant confusion seen in the research of these aspects. Finally, we highlight material design strategies that could eventually lead to high-performance metal batteries.

## 8.1.1  Introduction to Metal Batteries

Metal batteries generally refer to batteries that use lithium, sodium, potassium, calcium, multivalent metals, and alloys as the anode. The broader definition also includes liquid metal batteries and metal air batteries. Recently, metal batteries have attracted extensive attention of battery researchers due to their high theoretical specific capacity and low electrochemical potential. Taking lithium metal batteries for example, lithium–air (Li–air) battery and lithium–sulfur (Li–S) battery are the promising next-generation energy storage systems because of their attractively high energy densities ($\approx 3500\,\text{W h kg}^{-1}$ in Li–air battery, and $\approx 2,600\,\text{W h kg}^{-1}$ in Li–S battery) [1]. Compared with the traditional graphite and metal oxide/sulfide/phosphating composite anode, the metal anode possesses higher theoretical specific capacity and possesses lower electrochemical potential when paired with the cathode, which has very important research value for achieving higher energy density. In addition, the metal can be directly used as the anode, without additional synthetic materials and the use of current collector, which is also of great research significance for simplifying metal ion batteries. However, the metal materials used in metal batteries are mostly alkali metals, whose intrinsic high reactivity will lead to corrosive reactions in most organic solvents, making the solid electrolyte interface (SEI) formed is not strong and stable enough. In the process of charging and discharging, large volume expansion will be produced and the interfacial stability will be deteriorated. Furthermore, during electrochemical deposition, the Li dendrites will form and grow, which could penetrate through the separator and cause cell short circuit, imposing on the metal anode wide-ranging safety concerns. As a result, the coulomb efficiency of the metal batteries is low, the cycle life is short, and the safety is not enough for commercial use. As a consequence, the practical applications of the metal batteries remain far off, efforts must be made to resolve the fundamental challenges of the metal anode.

## 8.1.2 LI-METAL BATTERIES

### 8.1.2.1 Introduction

Li-metal batteries are considered to be the promising next-generation energy storage system. The Li-metal anode possesses an unrivaled specific capacity (3,860 mA h g$^{-1}$) and the lowest electrochemical potential (−3.04 V vs standard hydrogen electrode) [2–4]. However, since rechargeable Li-metal batteries emerged in the 1970s, the Li-metal anode has long been considered "unsafe" because dendritic Li formed during electrodeposition can cause internal short circuit and safety hazards. In contrast to the insertion reaction mechanism of host electrode materials such as graphite, the Li metal electrode is directly plated and stripped on the current collector during charging and discharging, in turn. Because of such a different mechanism of electrochemical plating and stripping, the Li metal electrode exhibits a different failure behavior from the host electrode materials [5]. The formidable challenges of Li metal anode that need to be overcome can be divided into three issues, as shown in Figure 8.1. The first one is the huge volume change of the electrode during Li deposition/dissolution, which is even more severe than that of the silicon anode because Li deposition can lead to virtually infinite volume expansion. The volume effect further deteriorates the interfacial stability. The second problem is the formation and growth of Li dendrites during electrochemical deposition, which could penetrate through the separator and cause cell short circuit, imposing on the Li-metal anode wide-ranging safety concerns. The dendritic Li could also lead to an electrical detachment of Li from the current collector and become "dead Li", significantly shortening the cycle life of the Li-metal battery. The third issue lies in the interfacial instability of Li metal in the organic electrolyte. Due to the intrinsic high reactivity of Li-metal in the organic electrolyte, corrosive reactions occur on the Li metal in most organic solvents. The side reactions on the Li-metal surface will deplete the electrolyte, increase the resistance, and further reduce the coulombic efficiency and cell life [1].

### 8.1.2.2 Strategies and Methods

In order to overcome these challenges in Li metal electrodes, a variety of strategies have been suggested, including architecture of Li host, electrolyte engineering,

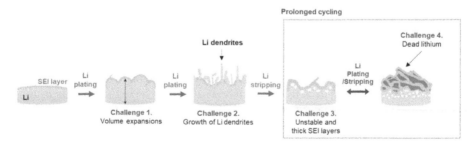

**FIGURE 8.1** Schematic illustration of the various challenges faced by Li metal anodes during prolonged cycling [6].

interface engineering for stabilizing Li metal anodes, and separator modification engineering [7]. They have shown promising electrochemical performance of Li metal electrodes.

The inhomogeneous Li-ion flux distribution usually leads to the non-uniform Li nucleation. The non-uniform Li nucleation on the planar Li foil electrodes further induces the inhomogeneous Li deposition and dendrite proliferation. Superior to simple Li foil anodes, Li-metal anodes with structure engineering, such as coated Li-powder electrode pressed on a current collector [8] and Li-metal anode with micro-structured surface (modified mechanically by microneedle or micropatterned stamp) [9] are beneficial to uniform Li deposition. Rational 3D electrode frameworks with novel architectures have been designed to control the Li plating/stripping behavior and improve the Li-metal anode stability. The ideal framework for Li-metal anode should hold a large specific surface area to reduce the local current density and create a homogeneous Li-ion flux, which is crucial for uniform Li-ion nucleation. Porous structure with sufficient pore volume in the Li substrate is desired to accommodate the Li volume change effectively upon cycling. In addition, the Li host substrate should possess mechanical and electrochemical stability, high electrical or ionic conductivity for fast electron/ion transfer, and a low gravimetric density for a high energy density [1]. Various 3D conduction skeletons based on porous Cu, foam Ni, and hybrid carbon nanomaterials as supports for Li-metal anodes have been addressed (Table 8.1), highlighting the significant influence of current collectors in inhibiting Li dendrite growth, and pushing forward the exploration of safe rechargeable LMBs. More importantly, taking the advantage of some particular 3D metal-based current collectors for uniform Li deposition, Li-metal anodes with largely enhanced cycle stability are achieved.

In addition to the work on the lithium anode, researchers have also done a lot of work on the regulation of electrolyte. The composition of the SEI film on the Li-metal anode relies greatly on the electrolyte (solvent and salts) and additives (or impurities). Rational choice of organic solvents, Li salts, and functional additives is desired to reinforce the SEI film on the Li-metal anode and improve its performance [23].

The conventional carbonate-based electrolytes show few merits in Li-metal batteries in terms of Li-dendrite suppression and cell life. Worse yet, studies have revealed that the carbonate-based electrolytes are not compatible with cathodes of Li–air and Li–S batteries, in which the carbonates decompose because of nucleophilic attack by reduced oxygen species or polysulfides. Recently, electrolytes using ethers, such as 1,3-dioxolane (DOL), 1,2-dimethoxyethane (DME), and tetraethylene glycol dimethyl ether (TEGDME), as solvents have received more attention. The ether-based electrolytes have low viscosity and high ionic conductivity. Moreover, unlike the carbonates, the ether-based electrolytes are compatible with high-energy Li metal batteries (Li–air and Li–S batteries). It is also found that the Li-metal anode has a high coulombic efficiency and low voltage hysteresis in the ether-based electrolytes. In addition to the carbonates and ethers, some other organic solvents such as dimethyl sulfoxide and acetonitrile have also been employed in Li-metal batteries. Recently, the room temperature ionic liquid (RTIL) has aroused tremendous interest in energy storage systems. The RTIL–ether hybrid electrolytes have been studied using $N$-propyl-$N$-methylpyrrolidinium TFSI (Py13TFSI) and DOL/DME [24]. Compared

**TABLE 8.1**
**Comparison of Electrochemical Performance Corresponding to Different Methods**

| Synthesis Methods | Anode Materials | Current Density | Initial Capacity (mAh g$^{-1}$) | Cycle Number | Final Capacity (mAh g$^{-1}$) | Capacity Retention | Refs.. |
|---|---|---|---|---|---|---|---|
| Mechanical rolling | 3D Cu/Li | 4 C | 109.6 | 500 | 87.4 | 79.70% | [10] |
| PECVD | Cu@VG@Li | 0.1 C | 102.1 | 400 | 73.5 | 72% | [11] |
| Chemical synthesis | Li-RCOFs | 2.1 mA cm- 2 | 137 | 500 | ~135.6 | 99% | [12] |
| Melting | CC/CNT@Li | 1 C | 160 | 100 | 157 | 99% | [13] |
| In situ gelation | 3D Li/CF | 0.2 C | 146.6 | 120 | 129 | 88% | [14] |
| Smelting reaction | Li-B-Mg | 0.5 C(1C = 274 mA g$^{-1}$) | 134.7 | 250 | 104.1 | 77.30% | [15] |
| Mechanical rolling | Li/Li22Sn5 | 5 C (4 mA cm$^{-2}$) | 132 | 500 | 120 | 91% | [16] |
| Thermal infusion | Li/CuZn | 1 C (170 mA g$^{-1}$) | 179 | 500 | 172 | 96% | [17] |
| Vacuum filtration | MgxLiy/ LiF-Li-rGO | 1 C | 134.1 | 400 | 107.1 | 80.80% | [18] |
| Seeds growth | Li-CC@ZnO | 0.5 C (1 C = 167 mA g$^{-1}$) | 142 | 300 | 117 | 82% | [19] |
| Graphene architecture | rAGA-Li | 2 C | 150 | 500 | 106.4 | 71% | [20] |
| Rolling-cutting | Composite Li anode | 2 C (1.28 mA cm$^{-2}$) | 125 | 100 | 77.5 | 62% | [21] |
| N-doped carbon | CC@CN–Co@Li | 5 C | 148 | 250 | ~145 | 98% | [22] |

Source: [7].

with the conventional ether-based electrolyte the Li plating and stripping reversibility is remarkably improved. The optimized electrolyte using Py13TFSI, DOL, and DME hybrid solvent with 2 M LiTFSI shows an excellent cycling stability and a high coulombic efficiency (99.1% after 360 cycles). These studies imply that developing novel electrolyte solvents is beneficial to mitigate the Li dendrite growth and anode interface problems.

The other key component of the electrolyte is the Li salt. The Li salts used for Li-metal batteries should not only meet essential requirements of the electrolyte (i.e., wide voltage window, high dissociation constant, high transference number, etc.) but also be chemically and electrochemically compatible with the Li metal anode and the cathodes. The ideal Li salt should also facilitate the formation of thin and stable SEI films. The frequently used Li salts in Li-metal batteries are $LiPF_6$, LiTFSI, $LiSO_3CF_3$, $LiClO_4$, and so on. The concentration of the salt in the electrolyte also has a significant effect on the ionic conductivity, Li-ion transferring number, and viscosity of the electrolyte. Compared with the standard 1 M molarity, a concentrated electrolyte provides sufficient Li-ion resources and high viscosity for suppressing the formation and proliferation of Li dendrites [25].

In Li-metal batteries, electrolyte additives have also been adopted to improve the Li–electrolyte interface, modify the properties of the SEI film on Li-metal anode, and minimize the side reactions at the Li interface. Various electrolyte additives that have been previously used in Li-ion batteries are employed in Li-metal batteries. Generally, these conventional additives have proven advantageous to stabilize the interface, improve the Coulombic efficiency, and reduce the interface impedance, but are not effective enough to eliminate the growth of Li dendrites. Novel functional additives have been developed for Li-metal batteries, such as $LiNO_3$, $Cs^+$ or $Rb^+$ cations, which based on a self-healing electrostatic shield mechanism, Li haloid salts, $LiPF_6$ and $H_2O$(a trace amount). Novel additives can be applied in Li-metal batteries for different functions beyond the electrochemical performance. For example, flame retardants such as phosphites have been adopted as additives to inhibit the flammability of organic electrolytes for safer Li-metal batteries [26].

The SEI film generated on a Li anode in organic electrolyte serves as a protecting layer for the Li-metal anode that can prevent side reactions between the Li metal and organic electrolyte. An ideal protective layer should be chemically and electrochemically stable, mechanically robust to inhibit Li dendrite penetration, and flexible to accommodate the volume changes during Li plating/stripping. The SEI film formed in the electrolyte in batteries (or in situ formed SEI) cannot fully meet these criteria because it is prone to cracking by Li dendrites. Constructing an additional stable and robust film as a protective layer (or so-called artificial SEI film) is worthwhile to reinforce the protection effect on the Li-metal anode. A facile method to form the artificial SEI film is to generate a passivation layer on the Li-metal surface by chemical pretreatments prior to electrochemical reactions [1]. As a favorable artificial SEI layer, the Guo group [27] introduced the uniform $Li_3PO_4$ protective layer obtained from the reaction between $H_3PO_4$ and Li metal surface. The $Li_3PO_4$ protective layer with a high Young's modulus (approximately 10–11 GPa) showed promising electrochemical performance because $Li_3PO_4$ is strong enough to mitigate the growth of Li dendrites. In contrast to the chemical formation of artificial SEI layers, Wang and his co-workers [28] recently demonstrated a new facile approach of in situ electrochemical formation of an interfacial protective layer. Methyl viologen, an additive in ether-based electrolytes, was electrochemically decomposed on the Li metal surface through in situ reduction during cycling, resulting in the formation of a stable and uniform interfacial coating layer. This protection layer showed excellent electrochemical performance, including 99.1% of columbic efficiency for 300 cycles. Moreover, the Cui group [29] proposed a new concept of smart protective layers, such as a dynamic polymer that can reversibly switch between its liquid and solid properties in response to the rate of lithium growth, because the static SEI cannot match the dynamic volume changes of the Li anode. The dynamically cross-linked polymer is primarily comprised polydimethylsiloxane (PDMS) cross-linked by transient boron-mediated crosslinks. The dynamic solid–liquid property is originated from the dynamic covalent bonds, where the mechanical properties of dynamic polymers vary with the stretching rate.

Li metal is one of the ideal candidates for the anode material of lithium batteries because of the highest theoretical specific capacity. However, Li metal electrodes have challenging issues such as poor cycle stability and safety due to Li metal dendrite growth and severe electrolyte decomposition on the Li metal surface. Moreover, their poor rate capability is inherent because the surface area of Li metal foil electrodes is much smaller than that of conventional composite electrodes comprising particles with a size of a few micrometers. Recently, much effort has been expended to overcome these problems, but the state-of-the-art technology of Li metal electrodes is still insufficient to meet the demands of the applications.

### 8.1.3 Na-Metal Batteries

#### 8.1.3.1 Introduction

The enormous demands for LIBs are dependent on the availability of Li resources, and Li is not regarded as an abundant element on the earth's crust. The cost of Li-containing materials has also rapidly increased in the past years, resulting in increased prices for LIBs. Because of the high abundance, low cost, and suitable redox potential of Na metal ($E_{Na+/Na}$ = 2.71 V vs. standard hydrogen electrode), rechargeable Na batteries are considered to be ideal alternatives to LIBs [30]. The current anode materials for Na-ion batteries (NIBs), including carbonaceous materials, titanium-based materials, alloy materials, chalcogen-based materials and organic materials are summarized in Figure 8.2(a), according to specific capacity (mAh g$^{-1}$) and voltage (V). Na metal can serve as the promising choice among all the anode materials because of its higher specific capacity of ~1,165 mA h g$^{-1}$ and lowest electrochemical potential.

The advantages of using Na metal as anode is very obvious; however, not many researches are conducted on the room temperature Na metal batteries due to its high reactivity, resulting in uncontrollable reaction with electrolyte, especially in the liquid electrolytes. In this way, Na metal can easily react with liquid electrolyte in a very short time to form a loose and fragile SEI layer in the process of battery assembly. During the Na plating, Na$^+$ prefers to deposit persistently on sites with the higher electric field, where this uneven deposition results in the Na plating crack. Further plating would cause failure of the SEI protective layer to induce the Na dendrite

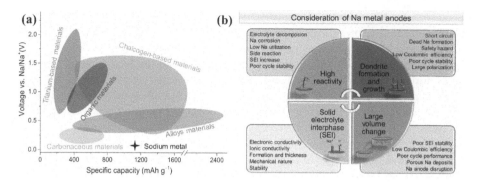

**FIGURE 8.2** (a) The available anode materials for Na-ion batteries (NIBs) according to specific capacity (mAh g$^{-1}$) and voltage (V). (b) The typical consideration for development of Na metal anodes [31].

formation and growth. When stripping, a part of dendrite may lose connection to the Na plate due to the Na$^+$ migration from the anode to cathode resulting in the formation of isolated Na or "dead" Na wrapped in SEI passivation layers; another part of dendrite will grow steadily and penetrate the separator to cause the short circuit of a battery. Even if the short circuit does not happen, lots of dead Na would no doubt produce a porous Na metal electrode and increase the thickness of SEI layer, fading the batteries' performance. Possible problems may happen with the Na metal anode illustrated in Figure 8.2(b).

### 8.1.3.2 Strategies and Methods

In order to overcome the above discussed challenges and considerations, deep understandings on the Na$^+$ plating and striping behavior need to be paid more attention. Fortunately, several general strategies (Figure 8.3) are proposed to suppress Na dendrite growth and large volume change and have yielded improved results.

1. *Interface engineering*: Though ex-situ or artificial ways to form a SEI protective layer on the surface of the Na anode by an additional chemical or physical treatments. This way won't cause Na metal anode to consume and/or decompose electrolytes, which can create a chemical, morphological, and thermodynamic stable passivation layer to suppress dendrite growth and limit volume change, e.g, Al$_2$O$_3$ thin layer [32].

2. *Electrolyte composition*: Electrolytes can be broadly categorized into liquid and solid states. For the former one, the Na salts, solvents and additives are the main parts for the formation and stability of SEI layers, while the latter can suppress the Na dendrite growth due to its high mechanical strength. Both the interface engineering and electrolyte composition are intended to produce a protective layer on the surface of the metal anode to prevent the growth of Na dendrite and enhance cycling stability, where the electrolyte composition

# Strategies for safe Na metal anodes

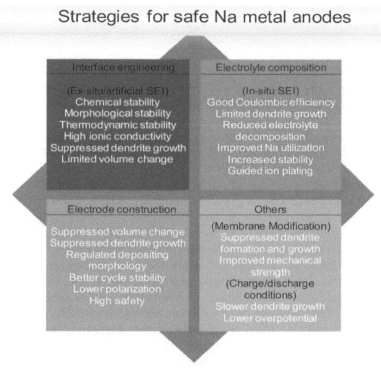

**FIGURE 8.3** General strategies for safe Na metal anodes: interface engineering refers to the formation of SEI layer via ex-situ or artificial ways through which an SEI protective layer is created by an additional treatment; electrolyte composition refers to the formation of SEI layer by in-situ ways during the operation of the battery (both the interface engineering and electrolyte composition are intended to produce a protective layer on the surface of the metal anode to prevent the growth of Na dendrite and enhance cycling stability); electrode construction is the recent development to construct a 2D or 3D matrix for Na accommodation (plating/stripping) suppressing the volume change of Na metal anode; some other strategies, including membrane modification, battery working conditions and parameters have also been applied to improve performance of Na metal anode to some extent [31].

     strategy tends to form an SEI layer by an in-situ way with consumption of electrolytes.

3. *Electrode construction*: This is a new developed method to suppress the volume change of Na metal anode and improve electrochemical behavior of a battery, which can construct a 2D or 3D matrix for Na accommodation. During Na plating/stripping, this matrix can sustain the changes of the structure and morphology without expansion and contraction.

4. *Other strategies*: Membrane can be seen as an important line of defense dendrite to cause short circuit, therefore, a membrane with good mechanical strength and higher stability can effectively suppress the growth of Na metal dendrite to some extent. In addition, the battery working conditions and

parameters, such as temperature, current density, charge capacity, etc., will also have considerable influence on the Na metal batteries.

## 8.1.4 K-METAL BATTERIES

Potassium batteries are currently considered a viable alternative option for substituting rechargeable lithium batteries in a number of applications, including large-scale energy storage, metal–sulfur, and metal–oxygen cells due to a number of beneficial properties. These include, in comparison to lithium: (i) higher abundancy of K metal on the earth's crust. (ii) Slightly lower standard potential of $K^+/K$ electrode in organic electrolytes (e.g., propylene carbonate and acetonitrile), promising higher voltage and high power operation. (iii) Potentially improved ion diffusion, cationic transference number and charge transfer resistance in liquid battery electrolytes due to the weaker Lewis acidity of $K^+$ ions, more disordered and flexible solvation structures, smaller Stokes radius of the solvated $K^+$ ions, lower electrolyte viscosity and the reduced values of desolvation energies. (iv) The fact that the best performing K cathode materials do not contain cobalt, associated with toxicity, ethical and sustainability problems. (v) Production costs reduction related to the current price of potassium carbonate and the possibility of using aluminum foils as current collectors on the anode side [33]. A prototype of potassium-ion batteries (KIBs) was successfully launched in 2004 [34], which has also been a research hotspot in recent years. Despite the increased interest in this type of batteries, a number of inherent detrimental properties remain, such as safety issues related to the high reactivity of K, low specific capacity of K compounds for a given electrode voltage window and their poor room temperature ionic conductivity, both issues being related to the larger size of the $K^+$ ion, as well as interface and interphase related problems when K metal is used as an anode material.

Similar to Li, the formation of dendrites can also be observed on the K-metal anode during battery charging and discharging processes. The formed dendrites can cause a decrease in coulombic efficiency (CE) and will pierce the separator and short circuit the battery, leading to severe thermal runaway. When the situation is serious, it may cause catastrophic fire hazards. Understanding K metal anode electrochemical behavior is scientifically exciting, providing a fresh palette for exploring a wide range of interfacial phenomena. However, potassium metal anodes are neither well-controlled nor well-understood.

The two strategies to prevent K metal dendrites growth includes modifying K metal–electrolyte interface and controlling the potassium metal-substrate [35]. The researches on the K metal–electrolyte interface mainly focus on the formation and growth of solid electrolyte interphase (SEI) due to its important role in the K metal dendrite growth process. Based on established descriptions for root causes of dendrites, the dendrite growth problem should be less severe for K than for Li or Na, while in fact the opposite is observed. The key reason that the K metal surface rapidly becomes dendritic in common electrolytes is its unstable solid electrolyte interphase (SEI). An unstable SEI layer is defined as being non-self-passivating. No SEI is perfectly stable during cycling, and all SEI structures are heterogeneous both vertically and horizontally relative to the electrolyte interface. The difference between a "stable" and an "unstable" SEI may be viewed as the relative degree to which during cycling it thickens and becomes further heterogeneous. The unstable SEI on K metal leads to a number of interrelated problems, such as low cycling Coulombic efficiency (CE), a severe impedance rise, large overpotentials, and possibly electrical shorting, all of which have been reported to occur as early as in the first 10 plating/stripping cycles. Many of the traditional "interface fixes" employed for Li and Na metal anodes, such as various artificial SEIs, surface membranes, barrier layers, secondary separators, etc., have not been attempted or optimized for the case of K. This is an important area for further exploration, with an understanding that success may come harder with K than with Li due to K-based SEI reactivity with both ether and ester solvents.

The second challenge with K metal anodes is that they do not thermally or electrochemically wet a standard (untreated) Cu foil current collector. Published experimental and modeling research directly highlights the weak bonding between the K atoms and a Cu surface. Existing surface treatment approaches that achieve improved K wetting includes employing new metallic, non-metallic, and composite substrate materials, constructing porous structure, doping impurity atoms, introducing adsorption sites, so as to increase the contact area of material and potassium, improve the efficiency of electrons and ion migration.

At present, the improvement of K metal batteries based on a single strategy is very limited. In the future, a comprehensive strategy should be adopted to consider all the inherent problems of K metal negative electrodes, so as to achieve as much improvement as possible. There are three interrelated thrusts to be pursued: First, K salt-based electrolyte formulations have to mature and become further tailored to handle the increased reactivity of a metal rather than an ion anode. Second, the K-based SEI structure needs to be further understood and ultimately tuned to be less reactive. Third, the energetics of the K metal–current collector interface must be controlled to promote planar wetting/de-wetting throughout cycling [36].

## 8.1.5 MULTIVALENT METAL (MG, ZN, CA, AND AL) BATTERIES

The emerging demand for electronic and transportation technologies has driven the development of rechargeable batteries with enhanced capacity storage. Especially, multivalent metal (Mg, Zn, Ca, and Al) and metal-ion batteries have recently attracted considerable interest as promising substitutes for future large-scale energy storage devices due to their natural abundance and multi-electron redox capability. These metals are compatible with nonflammable aqueous electrolytes and are less reactive when exposed in ambient atmosphere as compared with Li metals, hence enabling potential safer battery systems [37]. The Mg and Al anode will not induce metal dendrite growth during the cycle, which improves the safety performance of the battery. Although the zinc anode will produce dendrites, safety will be greatly improved in the aqueous electrolyte. Meanwhile, when the cathode material provides the same number of intercalation sites, the multivalent metal cation can carry more charge during the charging and discharging process, which means that the multivalent ion battery can provide higher capacity than the lithium-ion battery. Figure 8.4 clearly shows that volumetric capacities of Ca is on a par with Li, while Mg and Al are approximately two and four times more performant, respectively.

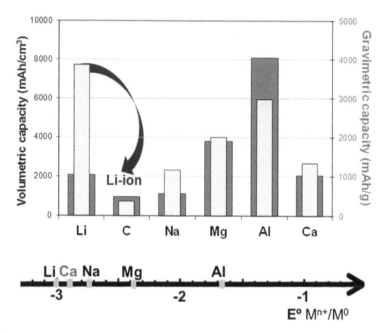

**FIGURE 8.4** Standard reduction potential and gravimetric/volumetric capacities of metal electrodes compared to values for graphite, typically used in the Li-ion technology [38].

However, due to the late development of multivalent metal ion batteries, there are still many problems to be solved. Taking positive electrode materials as an example, due to the small radius and large charge of multivalent ions, their polarization effect is extremely strong. When multivalent metal ions are intercalated in cathode materials during the battery charging and discharging process, they are easy to have a strong attraction effect with the anions in the materials, resulting in the collapse of the material structure, which leads to the deterioration of the electrochemical performance of the battery. In terms of electrolyte, it is always a great challenge to find the coordination negative ion groups that can make multivalent metal ions deposit on the metal anode smoothly. The development of multivalent ion battery is relatively slow due to the high demands of stability and conductivity of electrolyte.

Several steps were taken recently in the field of novel Mg-based electrolytes, cathode materials, and cell configurations. With respect to cathodes, both organic materials and also sulfur (under some specific conditions) have demonstrated cells with long-term cyclability being possible. While for the former there is a need to develop non-nucleophilic electrolytes with improved capability of polymer swelling, the Mg–S efforts should be directed toward suppressing the solubility of polysulfides. Inorganic cathodes are mostly chalcogenides, as use of oxides is still prevented by the lack of electrolytes with higher oxidative stability windows. Moreover, the (in) stability of interfaces is not fully yet explored and there are many open technological questions: anode composition, electrolyte additives, current collectors, separators, additives, cell casing, etc. Parallel efforts on several concepts in that direction by different groups should enable assessing the conditions for practical viability and commercialization.

For Ca batteries, despite reversible plating/stripping being demonstrated with two different concepts, improvements are highly necessary for kinetics/efficiency and/or widening the electrochemical stability window of the electrolyte. On top of that, the road toward the development of a positive electrode material operating at high potential and enabling decent capacities with fast kinetics is clearly very long and winding. For the moment, only inorganic materials have been considered, with organic materials and sulfur remaining interesting pathways to follow. Overall, a deeper understanding of the nature of cation complexes formed in the electrolyte, its transport properties and the interfacial processes such as adsorption, (de-)solvation, etc, will be key for further developments.

For Al batteries, there are, in principle, two main (very) large obstacles to overcome. One is to develop functional electrolytes free from $Cl^-$ and capable of fast cationic $Al^{3+}$ transport, as this would render Al batteries both more performant and more stable including solving the corrosion problems, the latter today affect the reliability of tests made, but will in the future cause problems for encapsulation. The second is that there are no cathodes capable of either efficiently coordinating or even accommodating the three $e^-/Al^{3+}$ in the framework, and this means to target different hosts and host–guest interactions than those employed at present. Here the path of organic cathodes seems to be the more promising with the large amount of flexibility in design.

## 8.2 THE ADVANTAGES OF METAL BATTERY

Alkali metal anodes mainly include Li, Na, and K. Among them, metal Li anode with minimum ionic radius (0.76 A°), density (0.53 g $cm^{-3}$), lowest redox potential (−3.04 V), and highest theoretical specific capacity (3,860 mAh $g^{-1}$) possesses tremendous potential for high-energy-density $Li–O_2$ and Li–S batteries [39]. According to the report, state-of-the-art LIBs with graphite anodes and lithium transition metal oxide (LMO) cathodes can achieve a specific energy of 300 Wh $kg^{-1}$. However, if the graphite anode is replaced by the metal Li, Li–LMO batteries would deliver a higher specific energy of 440 Wh $kg^{-1}$ [40]. Moreover, if the cathode is replaced by S or $O_2$ to assemble Li–S and $Li–O_2$ batteries, both of which have been intensively studied as next-generation energy storage devices over the past few decades. The realistic specific energy of Li–S batteries is ~600 Wh $kg^{-1}$, which is twice that of LIBs. As for $Li–O_2$ batteries, whose realistic specific energy may reach up to 950 Wh $kg^{-1}$, are regarded as an ideal battery system [41]. As for Na and K metal, metal Na anode possess a high theoretical specific capacity of 1,166 mAh $g^{-1}$ and low redox potential (−2.71 V), and metal K anode possesses a theoretical specific capacity of 685 mAh $g^{-1}$ and low redox potential (−2.93 V) [42] although the theoretical specific capacity and electrochemical potential of metal Na and K are inferior to those of Li. The abundance of metal Li, Na and K is 0.00017%, 2.36%, and 2.09%, respectively, and the price for each is $5,000 $ton^{-1}$, $150 $ton^{-1}$, and $216 $ton^{-1}$, respectively [43]. The high abundance and low cost of metallic Na and K are still attracting the intense interest of researchers, because Na metal batteries and K metal batteries can

**TABLE 8.2**
**Physical Parameters and Electrochemical Properties of Li, Na, and K Metal Anodes**

| Alkali Metal | Abundance (wt %) | Ionic Radius (A) | Density (g cm$^{-3}$) | Redox Potential (V) | Theoretical Gravimetric Specific Capacity (mAh g$^{-1}$) | Price ($ ton$^{-1}$) |
|---|---|---|---|---|---|---|
| Li | 0.0017 | 0.76 | 0.53 | −3.04 | 3860 | 500 |
| Na | 2.36 | 1.02 | 0.97 | −2.71 | 1166 | 150 |
| K | 2.09 | 1.38 | 0.86 | −2.93 | 685 | 216 |

partly replace Li metal batteries in some areas. Thus, Na–S, Na–O$_2$, K–S, and K–O$_2$ batteries as next-generation energy storage devices are also under intense research (Table 8.2).

Currently, the commercial LIBs with transition metal oxide/phosphate cathodes such as LiCoO$_2$, LiMn$_2$O$_4$, and LiFePO$_4$ and graphite anode (372 mAh g$^{-1}$) are close to their limited theoretical energy density, which cannot meet the ever-growing demand of the highly developed electric vehicle. A very promising approach to solve this problem is to move from the traditional insertion chemistry to an innovative conversion chemistry. The lithium–sulfur (Li–S) system is a good example of this (Figure 8.5), since the sulfur cathode could deliver a high theoretical capacity of 1,675 mAh g$^{-1}$ based on the following electrochemical process: $16Li + S_8 \rightarrow 8Li_2S$. Specifically, during the discharge process, sulfur first becomes lithiated to form a series of intermediate, long-chain lithium polysulfide species ($S_8 \rightarrow Li_2S_8 \rightarrow Li_2S_6 \rightarrow Li_2S_4$), which dissolve readily into ether-based electrolytes and contributes 25% of the theoretical capacity of sulfur (418 mAh g$^{-1}$). Upon further lithiation, the dissolved long-chain polysulfides form short-chain sulfide species ($Li_2S_4 \rightarrow Li_2S_2 \rightarrow Li_2S$), which reprecipitate onto the electrode as solid species. This contributes the remaining 75% of the theoretical capacity (1,255 mAh g$^{-1}$) [44]. Hence, the Li–S batteries offering an extraordinarily higher specific energy density than that provided by conventional LIBs, 2,600 Wh kg$^{-1}$ versus 300 Wh kg$^{-1}$. Besides the natural abundance and environmental benignity of elemental sulfur help Li–S batteries gain more advantages. Similarly, the metal Li anode can be replaced by other alkali metal such as Na and K to form Na–S and K–S batteries. The specific energy density of Na-S is 1,274 Wh kg$^{-1}$. Na–S batteries is based on the similar electrochemical process: $S_8 + 2Na^+ + 2e^- \rightarrow Na_2S_8$. The specific process can be divided into four stages: stage I, solid-state ring-like $S_8$ is converted into dissolved chain-like $Na_2S_8$; stage II: further reduction of $Na_2S_8$ into a series of other soluble long-chain polysulfide intermediate species ($Na_2S_6$, $Na_2S_5$ and $Na_2S_4$) via liquid–liquid transformation; stage III: precipitation reactions from soluble $Na_2S_4$ to a series of insoluble short-chain polysulfide intermediate species ($Na_2S_3$ and $Na_2S_2$); stage IV: solid–solid conversion between $Na_2S_2$ and $Na_2S$ [45]. As for K–S batteries, the specific energy density of K–S batteries is 1,023 Wh kg$^{-1}$, and the electrochemical process can be simply described as: $K+ S_8 \rightarrow K_2S_6 \rightarrow K_2S_5 \rightarrow K_2S_4 \rightarrow K_2S_3$ [46].

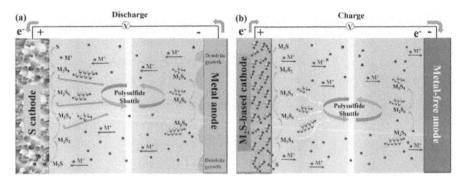

**FIGURE 8.5** (a) Schematic diagram of M–S (M = Li, Na, K) batteries and (b) $M_2S$ cathode-based M–S batteries with polysulfides present in the electrolyte, separately [44].

Despite the M–S batteries possessing higher specific energy density than conventional LIBs, S as cathode is much cheaper than market mainstream cathodes such as $LiFePO_4$ and NCM811. M–S batteries are still difficult in commercial applications. The major problem is the polysulfide shuttle effect. Due to the M polysulfide (MPS) intermediates generate, dissolve into the electrolyte, and shuttle to the anode, which makes it much more complicated to achieve dendrite inhibition in the existence of MPS intermediates, especially under the high loading sulfur cathode [44]. It is widely believed that polysulfides are able to penetrate the passivation layer and corrode the fresh metal underlining the surface layer, leading to capacity loss. However, it is critical to use high loading sulfur cathode and limited amount of metal (low N/P ratio) to achieve a high specific energy density in practical M–S batteries, which means a more severe shuttle effect and capacity loss in real M–S batteries. Extensive researches focus on inhibiting the shuttle effect and many methods have been put forward, including constructing cathode host, electrolyte modification, and membrane design [47–49]. All these methods aim at confining and adsorbing M polysulfide intermediates and then promoting its conversation. These methods have been proved succeesful in suppressing polysulfide shuttle effect. However, these still exist a certain gap with the commercial application of M–S batteries.

Apart from M–S (M = Li, Na and K) batteries, $M–O_2$ (M = Li, Na and K) batteries also captured worldwide attention as next-generation energy storage system due to its high theoretical energy densities and the use of the environmentally friendly and unlimited source of oxygen. The energy density of $M–O_2$ batteries is primarily determined by the type of discharge products, because its discharge/charge reaction mechanism differs from those of the conventional Li-ion or Na-ion batteries [43]. Instead of the typical topotactic storage of working ions, the cathode of an $M–O_2$ battery provides surface sites only for nucleation and growth of the solid metal–oxide discharge products during discharge. Among the various $M–O_2$ battery systems, their differences in energy density depend largely on the volume and weight of the discharge products and their redox potentials. For instance, The $Li–O_2$ batteries have higher gravimetric and volumetric energy densities than $Na–O_2$ batteries and $K–O_2$ batteries (3,500 Wh kg$^{-1}$ versus 1,602 Wh kg$^{-1}$ versus 935 Wh kg$^{-1}$), far exceeding

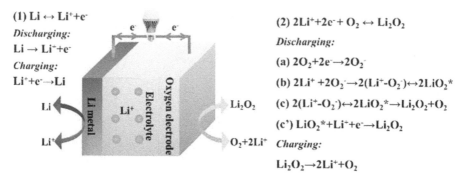

(1) $Li \leftrightarrow Li^+ + e^-$

*Discharging:*
$Li \rightarrow Li^+ + e^-$

*Charging:*
$Li^+ + e^- \rightarrow Li$

(2) $2Li^+ + 2e^- + O_2 \leftrightarrow Li_2O_2$

*Discharging:*

(a) $2O_2 + 2e^- \rightarrow 2O_2^-$

(b) $2Li^+ + 2O_2^- \rightarrow 2(Li^+ - O_2^-) \leftrightarrow 2LiO_2^*$

(c) $2(Li^+ - O_2^-) \leftrightarrow 2LiO_2^* \rightarrow Li_2O_2 + O_2$

(c') $LiO_2^* + Li^+ + e^- \rightarrow Li_2O_2$

*Charging:*

$Li_2O_2 \rightarrow 2Li^+ + O_2$

**FIGURE 8.6**   Schematic illustration of a Li–O$_2$ battery and the corresponding reactions at the Li-metal (left side) and oxygen electrodes (right side) [50].

commercial M-ion batteries. This high energy density can be attributed to its higher redox potentials, the lower atomic weight, and the higher specific capacity of lithium. Li–O$_2$ batteries have a simple reaction between oxygen and lithium ions. The ideal operation of Li–O$_2$ batteries is based on the electrochemical formation (discharge) and decomposition (charge) of lithium peroxide (Li$_2$O$_2$). To be specific, at the beginning of the discharge, oxygen is reduced on the electrode and successively combines with a Li ion in the electrolyte, forming the metastable LiO$_2$ (O$_2$ + Li$^+$ + e$^-$ → LiO$_2$). LiO$_2$ may subsequently undergo two different reaction pathways: either a chemical disproportionation involving two LiO$_2$ molecules (2LiO$_2^-$ → Li$_2$O$_2$ + O$_2$) or a continuous electrochemical reduction of LiO$_2$ with an additional electron and Li ion (LiO$_2^-$ + Li$^+$ + e$^-$ → Li$_2$O$_2$), both of which result in the formation of Li$_2$O$_2$ (Figure 8.6). The kinetics of each reaction step and the stability of the intermediates are greatly affected by the surrounding conditions, leading to distinct electrochemical properties, including specific capacity, rate capability, and energy efficiency [51]. However, Li–O$_2$ batteries research is still facing a lot of challenges. The discharge process involves the reduction of oxygen to superoxide (O$_2^-$), the formation of LiO$_2$ and its disproportionation into Li$_2$O$_2$ and O$_2$. While the charge process is the direct oxidation of Li$_2$O$_2$ into O$_2$, and the discharge product Li$_2$O$_2$ has a low conductivity. As a result of the asymmetric reaction mechanism, the battery charging process has a much higher overpotential (1–1.5 V) than that of discharge (0.3 V), which renders the system with a low round-trip energy efficiency around 60%. For lower charging overpotentials to be achieved, lithium ion has been replaced by larger cations such as Na$^+$ and K$^+$ to stabilize the superoxide (O$_2^-$) discharge product. Hence, Na–O$_2$ batteries and K–O$_2$ batteries have attracted increasing attention from the researchers owing to the fast charge-transfer kinetics with the one-electron reaction, which exhibits a low charging overpotential with a high round-trip energy efficiency (more than 90%) [52]. Na–O$_2$ batteries have a high theoretical specific energy density of 1,602 Wh kg$^{-1}$, the discharge reaction routes of Na–O$_2$ batteries and Li–O$_2$ batteries may be similar, but a relative stable NaO$_2$ compound exists in the Na–O phase diagram, which means a relative stable superoxide (NaO$_2$) could be produced (O$_2$ + Na$^+$ + e$^-$ → NaO$_2$) and compete with the formation of peroxides (Na$_2$O$_2$) during discharge (O$_2$ + 2Na$^+$ + 2e$^-$ → Na$_2$O$_2$)

for Na–$O_2$ batteries. Nevertheless, the cells that emit $NaO_2$ as a discharge product exhibit a very low charging potential, whereas the overpotential of cells with $Na_2O_2$ is comparable to that of the Li–$O_2$ batteries, indicating that it is crucial to control the composition of discharge products. Besides, $NaO_2$, as the reaction product of Na–$O_2$ batteries, is not thermodynamically stable enough, and it spontaneously deteriorates at room temperature into poor reversible $Na_2O_2$, which is very difficult to suppress this parasitic reaction. Consequently, the reversibility of the Na–$O_2$ batteries is still relatively low. Although K–$O_2$ batteries have the relatively low theoretical specific energy density of 935 Wh $kg^{-1}$, but based on the hard–soft acid–base (HSAB) theory, $O_2^-$ is more stable with cation of a lower charge density while heavier $K^+$ cations can effectively stabilize the superoxide form with the steric repulsion between cations in the crystal product [53]. Thus, $KO_2$ is more thermodynamically stable and commercially available in contrast with the Li–$O_2$ and Na–$O_2$ batteries. The formation and decomposition reactions of $KO_2$ in K–$O_2$ batteries are significantly more reversible than those of $LiO_2$ in Li–$O_2$ and $NaO_2$ in Na–$O_2$ batteries, leading to a high round-trip energy efficiency [54]. In addition, $KO_2$ is easily decomposed during charging with a low overpotential (approximately 50 mV lower than that of $NaO_2$), which is attributed to a small reorganization energy and a lower energy barrier for the $KO_2$ to $O_2$ conversion based on Marcus theory [55]. Among the M–$O_2$ batteries, therefore, the K–$O_2$ batteries have the highest energy efficiency without novel catalysts and additives. But, M–$O_2$ batteries still face a severe challenge: the formation of metal dendrites. The metal anodes could show the formation and growth of dendrites on the anode surface during cycling. The electrochemically grown dendrites result in further capacity loss and could penetrate the separator and then cause internal short circuits. Therefore, there is still a long way to realize the commercialization of M–$O_2$ batteries.

Despite the metal batteries own higher specific capacity and wider voltage operation window versus conventional metal-ion batteries, there is still a major problem that due to the low redox potential of Li, Na, and K metal. They will inevitably react with electrolytes to form solid-state electrolyte interface. What's worse, the hostless nature of Li, Na, and K metal makes them suffer from severe volume variation during cycling, leading to the fresh metal constantly exposed in the electrolytes to formed new SEI. The electrolytes and metal are further consumed, causing metal batteries capacity faded rapidly. Apart from this, conventional liquid organic electrolytes also face several safety issues such as the leakage and flammability [40]. A series of electric vehicle accidents have aroused people's concern. Replacing liquid electrolytes by solid-state electrolytes is a possible path toward safer batteries since most SSEs are nonflammable. More importantly, these SSEs possess high modulus can efficiently suppress the dendrites growth. Additionally, the SSE is convenient to achieve an ultrathin thickness, which can hopefully improve the energy density and apply to flexible and wearable devices. Therefore, when metal anode paired with the SSEs, the solid-state metal batteries can potentially deliver a high energy density and excellent safety assurance. Due to these benefits, a rapidly increasing trend of investigations on solid-state electrolytes (SSEs) for use in lithium batteries has emerged in recent years.

In lithium metal batteries, generally, the widely investigated SSEs can be divided into two main aspects: inorganic ceramic electrolytes and organic polymer electrolytes (Figure 8.7). The former is commonly based on oxides, sulfides, and halides, where the

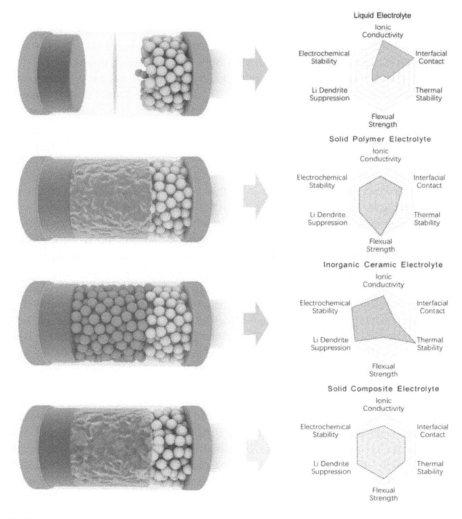

**FIGURE 8.7**    The performance comparisons of liquid electrolytes, SPEs, ICEs, and SCEs [56].

sulfide-based electrolytes and halide-based electrolytes, such as $Li_6PS_5Cl$, $Li_{10}GeP_2S_{12}$, $Li_{10}SnP_2S_{12}$, and $Li_3InCl_6$ show significantly high ionic conductivities at room temperature ($10^{-4}$–$10^{-2}$ S cm$^{-1}$) and good mechanical strength and flexibility. However, they have some drawbacks including low oxidation stability, sensitivity to $H_2O$, and poor compatibility with cathode materials [56]. Oxide-based solid electrolytes such as garnet $Li_7La_3Zr_2O_{12}$ (LLZO), perovskite $Li_{3.3}La_{0.56}TiO_3$ (LLTO), NASICON (lithium superionic conductor) $LiTi_2(PO_4)_3$, and LISICON (sodium superionic conductor), and $Li_{14}Zn(GeO_4)_4$ are widely studied due to their good conductivity ($10^{-4}$ S cm$^{-1}$) and outstanding stability. Nevertheless, some of them are still sensitive to $H_2O$, and the nature of ceramics is rigid, thus, the corresponding electrolytes are difficult to manufacture. In brief, the inorganic SSEs possess the excellent ionic conductivity, whereas their interfacial contact with electrodes are undesirable due to their nature of inorganic ceramic particle. In addition to this, most of inorganic ceramic electrolytes cannot process in the air atmosphere. Thus, it remains a challenge to achieve inorganic ceramic electrolytes large-scale production for further applications. The latter type of SSEs is organic polymer electrolytes, which usually consists of a polymer as the matrix such as poly(ethylene oxide) (PEO), poly(vinylidene fluoride) (PVDF), polyacrylonitrile (PAN), and poly(methyl methacrylate) (PMMA)) and some lithium salts as the internal ionic conductor to improve the polymer ionic conductivity, such as $LiClO_4$, LiTFSI ($LiN(CF_3SO_2)_2$), $LiAsF_6$, and $LiPF_6$. Different from crystalline inorganic electrolytes, organic polymer electrolytes are light, flexible, and scalable. Thus, they are compatible with electrodes, which can overcome the large interfacial impedance between electrodes and SSEs. Also organic polymer SSEs are more compatible with the state-of-the-art manufacturing process [57]. However, they have the drawbacks of limited thermal stability and poor mechanical strength. For example, the Young's modulus of typical pristine PEO/LiTFSI is only 0.4 MPa [58], which is not sufficient to inhibit the dendrites growth. Hence, the dendrites can form at the electrolyte–electrode interface. Moreover, the ionic conductivity of organic polymer organic polymer SSEs in the range of $10^{-6}$ to $10^{-5}$ S cm$^{-1}$ typically at room temperature, which is too poor for application in commercial batteries. Although some methods have been proposed to improve the ionic conductivity of polymer-based SSEs, such as the lithium salt-based additives and the introduction of liquid plasticizers into the polymer matrix, but it still cannot meet the basic requirement of ionic conductivity in practical applications.

It is obvious that both of inorganic and polymer SSEs have their advantages and disadvantages. Therefore, it is so far hard for individual inorganic or polymer electrolytes to satisfy the practical applications. To overcome the shortcomings of inorganic and polymer SSEs and make use of their advantages, it is a good option to combine inorganic and polymer SSEs so that one plus one is greater than two. Several recent studies have combined them by using polymer SSEs as skeleton to make sure the good contact between Li metal and SSEs, whereas the inorganic SSEs nanoparticles as filler inside the polymer SSEs skeleton to improve its ionic conductivity. The formed composite solid-state electrolytes (CSSEs) have demonstrated that they can result in varying degrees of performance improvement, which are considered as one of the most promising electrolytes for application in next-generation lithium batteries [59–62]. Moreover, the CSSEs are also suitable for the development of flexible electronic devices.

The widely investigated SSEs of Na and K metal batteries can also be divided into two aspects: inorganic ceramic electrolytes and organic polymer electrolytes. The research contents of organic polymer electrolytes in Na and K metal batteries are almost the same as that of Li metal batteries. Except the lithium salts is replaced by the sodium and potassium salts as the internal ionic conductor to improve the polymer ionic conductivity. As for inorganic ceramic electrolytes in Na metal batteries, it could be divided into β-alumina electrolytes, sodium superionic conductor (NASICON) electrolytes, and sulfide-based solid-state electrolytes. However, due to the large ionic size of K ions (1.38 Å), they have difficult migrating behavior in solid-state electrolytes. This leads to less possible structures for K solid-state electrolytes. Thus, it still remains a challenge for K metal batteries to choose a suitable inorganic ceramic electrolyte. Even though metal batteries can lift up the energy density two or three times than conventional metal ion batteries, none of metal batteries have been successfully commercialized so far. In spite of many strategies have been proved to improve the performance of metal batteries, but it's just lab-level testing. Hence, there is a still a long way to improve the performance of metal batteries in practical application.

## 8.3 CHALLENGES OF THE METAL BATTERIES

### 8.3.1 DENDRITE GROWTH

#### 8.3.1.1 Li Dendrite

Lithium, sodium, and potassium metals have extremely high chemical and electrochemical reactivity, which leads to the metal being easily corroded by electrolyte, thus reducing the utilization rate of lithium metal anode. At present, lithium metal battery is limited by the formation of lithium dendrite, which leads to insufficient safety and is difficult to be applied in practice [63–66]. The known problems caused by lithium dendrite growth are as follows (Figure 8.8):

1. *The battery short circuited.* Dendrites can penetrate the separator to contact the cathode, resulting in direct contact and short circuit of the battery [67]. Battery short circuit is often accompanied by thermal runaway, and further by electrolyte burning and battery explosion. The industrialization of MBs used to be cased because of the potential safety hazard caused by the dendrite short circuit. The short circuit caused by dendrite growth and its potential safety is the key to prevent the practical application of metal anodes.
2. *Aggravated adverse reactions.* On the one hand, the growth of dendrites destroys the SEI film on the anode surface, exposing more metals to the electrolyte to form new SEI; on the other hand, the growth of dendrites will increase the surface area of lithium metal, resulting in more contact areas

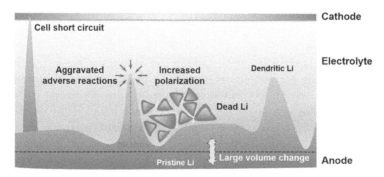

**FIGURE 8.8**   Scheme of dilemma for Li metal anode in rechargeable batteries. Reproduced with permission from Ref. [70].

and parasitic reactions between fresh metals and the electrolyte. Since these side reactions will irreversibly consume the active material lithium metal and the electrolyte and make no contribute to the capacity, they will significantly reduce the coulomb efficiency of the batteries.

3. *Evolution of dead (no-active) metal in existing dendrites.* Due to the activity of lithium, sodium, and potassium, the metal and electrolyte will form a non-conductive film. In the discharge process, due to the high current density at the dendrite root, the electrons will be lost rapidly, leading to the fracture of the dendrite root. Therefore, the lithium dendrites formed will continue to react with the electrolyte to generate large lithium dendrites wrapped in the SEI film, so they cannot contact the collector and electrons, thus converting the lithium dendrites into electrochemically inert dead lithium, greatly reducing the coulomb efficiency.

4. *Intensified polarization.* The lithium metal anode with dendrite has a porous and uneven structure composed of numerous dead lithium [68, 69]. Compared with the uniform and dense dendrite-free lithium coating, the diffusion path of lithium ions and electrons increases, and the resistance increases, resulting in greater polarization and undesirable energy efficiency.

5. *Huge volume change.* As a conversion electrode without matrix, the "hostless" metal anode has an infinite volume variation during the plating/stripping process, which is much higher than the existing intercalated and alloy anodes, such as graphite (10%) and silicon (400%). The growth of dendrite leads to porous structure of metal anode, larger volume, and lower filler density. This as-formed porous structure will lead to more severe volume changes, which will have a fatal impact on the safe and efficient operation of the metal anode.

### 8.3.1.2  Na Dendrite

Sodium dendrite growth is not a direct Li analog. The low oxidation–reduction potential of sodium metal −2.71 V makes it possible to grow dendrites in the electrolyte [70, 71]. The sodium metal anode is also hindered by problems similar to lithium anode [72]. The electrodeposition moss and dendritic growth of sodium reduce the coulomb efficiency due to the formation of SEI film on its surface, the electrolyte is depleted and the battery impedance rises, which ultimately leads to the reduction of battery life. As the reaction mode of metal sodium and carbonate electrolyte is slightly different [73], the morphology of sodium dendrite is slightly different from that of lithium dendrite, usually showing needle or moss [74]. The ionic radius of sodium is 55% larger than that of lithium, and the combination with solid carbon is obviously weak [75], resulting in less exothermic adsorption/insertion [76]. Although this makes Na unable to embed into graphite (standard LIB anode) [77], it can still embed into amorphous carbon reversibly. Importantly, it seems that it is not easy to form under potential deposition on the carbon surface inside nanopore [78]. It is worth noting that due to the different chemical/electrochemical reactivity of sodium and potassium metals, solubility of decomposition products, mechanical properties, and volume flow ratio, it is more difficult to form an adequate passivation layer than on lithium metals [79]. The SEI layer is well-recognized to be chemically and geometrically heterogeneous. The underlying Al or Cu current collectors are also geometrically and structurally non-uniform, containing mechanical scratches from processing, dislocation terminations in the form of steps, grain boundaries of various orientations, etc. This in turn leads to heterogeneous nucleation and growth kinetics of the plated/stripped Na. Over a number of cycles, the result is poor CE due to additional SEI growth and "dead Na". This leads to an increase in the plating–stripping overpotential, SEI-induced electrolyte depletion [80, 81] and in some cases dendrite growth through the battery separator.

### 8.3.1.3   Potassium Dendrite

As metal K has strong reducing activity to organic electrolyte [82], it is difficult to form a passivation layer on its surface. Potassium metal anodes are unstable in commonly used electrolytes for lithium and sodium metal anode [83]. The low mechanical modulus of K dendrite leads to continuous dissolution and damage of the dendrite, which results in new SEI, low efficiency of K metal anode, and shorter service life than Li and Na metal anode. The calculated metal volume per mole of K is almost four times that of Li. This will undoubtedly lead to more fragile and unstable SEI in Na and K systems.

### 8.3.1.4   Mg Dendrite

In previous reports, Mg metal anodes did not form dendrites [84, 85]. However, with the further research in recent years, Mg dendrites can be observed on the Mg metal anode [86, 87]. The formation of Mg dendrites will be affected by many factors, such as the distribution of electric field, surface diffusion rate, deposition rate, diffusion barrier, temperature, pressure, and various liquid electrolytes [88]. In fact, the formation mechanism of Mg dendrites in metal anodes is similar to that of other dendrites. For example, the uneven electric field on the electrode will lead to uneven distribution of $Mg^{2+}$ flux, which will lead to uneven deposition of Mg and promote the formation of Mg dendrites [89]. The weak magnesium affinity of the substrate will lead to a large Mg nucleation barrier, thus accelerating the formation of Mg dendrites.

### 8.3.1.5   Zn Dendrite

Disposable zinc metal batteries are widely used, but secondary zinc batteries are still in the laboratory stage due to the same problem of dendrite growth [90, 91]. By removing the used zinc and replacing it with a new zinc anode, the poor reversibility and instability of the bifunctional air electrode in the zinc oxygen battery can be solved, but this measure is difficult to apply in practice [92]. Unlike lithium, sodium, and potassium, zinc is relatively stable in electrolyte (water system and non-water system) because the low coulomb efficiency is not caused by metal anodes and electrolytes. The performance of zinc metal battery is mainly limited by the following four reasons: (i) dendrite growth; (ii) shape change; (iii) passivation and internal resistance; and (iv) hydrogen evolution.

### 8.3.2   Unstable Solid Electrolyte Interphase (SEI)

Due to the low electrochemical potential of metal anodes, liquid electrolytes can easily corrode metal electrodes and form a passive layer on the surface. This passive layer is the so-called natural SEI, which can act as a protective layer to resist the continuous corrosion of metal anodes by liquid electrolytes to some extent. However, the strength of the natural SEI formed in common liquid electrolytes is low and a small stress change on the surface of electrode can result in its damage. The breakdown of the SEI on metal anodes would cause the exposure of fresh metal and thus bring about endless side reactions. Nevertheless, the structure of SEI plays an important role in determining the cycling of metal anodes [93, 94]. SEI has many functions, including limiting further side reactions between metals and electrolytes, and promoting uniform metal deposition by regulating solid ion flux. The ideal SEI layer has the following characteristics: (1) appropriate thickness, enough to completely prevent electron transfer to the electrolyte, but not too thick; increase lithium-ion diffusion resistance; (2) high ion conductivity to reduce lithium-ion diffusion resistance; (3) strong mechanical properties to adapt to shocks and uneven volume changes in repeated deposition/stripping; (4) insoluble in electrolyte; and morphology under a wide range of operating temperatures and voltages Structural and chemical stability.

### 8.3.2.1  SEI Formation Mechanism

Practically, due to the most negative nature of the electrochemical potential of Li, the redox reactions between Li and nonaqueous electrolyte cannot always be avoided. When bare Li is exposed to solution, there are immediate reactions between Li, electrons, and solution species in time constants of milliseconds or less [95]. SEI growth is initially triggered by electrons. Endo et al. [96] investigated the reductive decomposition of electrolytes with electron spin resonance as well as molecular orbital calculation method. Anions in electrolytes are difficult to reduce through the calculations of enthalpy changes between anion positive reduction and free solvent, with $\Delta H_r \approx -1$ kcal mol$^{-1}$. Li$^+$ can coordinate with these solvents and help to decrease the reduction enthalpy to $\Delta H_r \approx -10^{-2}$ kcal mol$^{-1}$, rendering the process thermodynamically favorable. SEI grows with the continued electron/electrolyte-participated reactions and stops when lacking one or both of them. The growth rate of Li dendrites is in essence determined by the Li-ion migration direction and rate, which are driven by the electronic field and concentration gradient in electrolytes. These factors are mainly directed by external applied charging conditions (current density, charging time, etc.). Besides, the properties of the SEI layer (electronic and ionic conductivity) and electrolytes (viscosity, ionic mobility, etc.) also impact the Li-ion migration.

**Electronic Field**. The electronic field is the first driving force in Li dendrite growth. Several models have been proposed to describe the Li dendrite growth based on the electronic field. Chazalviel found that the growth of a ramified metallic electrodeposit like Li dendrites in dilute salt solutions is essentially driven by the space charge that tends to develop upon anion depletion near the anode [97]. The front of the dendrites has been shown to advance at a velocity $v_a = -\mu_a E_0$, which is just the velocity of the anion mobility, determined by the mobility $\mu_a$ and the electric field $E_0$ in the neutral region of an electrolyte. However, Chazalviel's model has to be improved because it does not take into account the variations of parameters such as the diffusion parameters associated with ionic concentration. Brissot et al. observed that individual dendrites exhibit different velocities. Nevertheless, these velocities are not very far

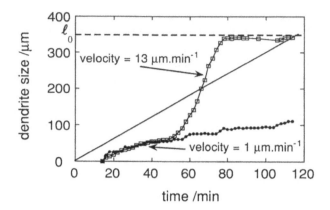

**FIGURE 8.9** Evolution of size of two typical Li dendrites under 0.7 mA cm$^{-2}$. The solid straight line exhibits the calculated length predicted by Chazalviel's model for a cationic transport number tc = 0.15. Reproduced with permission from Ref. [98].

from the velocity predicted by Chazalviel's model. In particular, they seem to be almost proportional to the current density [98]. After several polarizations of the cell, dendrites seem to be unable to grow beyond a "barrier" at a given distance from the negative electrode. This distance is about the size attained by dendrites during the first polarizations (Figure 8.9). The establishment of these models is mainly based on the electric field as the driving force, giving the growth rate and impacting factors of Li dendrites when the plating process is electronic field controlled. However, a practical system is more complicated than that, where many aspects act together.

**Diffusion Flux.** The electric field plays the main driving force for lithium-ion migration only at the beginning stage of the Li plating process, which will then become a diffusion-controlled process afterward. Therefore, an accurate model must take into account both the effect of the electric field and the effect of concentration gradients on the growth of dendrites. Akolkar developed a mathematical model to describe the dendritic growth process during Li electrodeposition near the Li surface along with electrochemical reactions (deposition) at the flat surface and at the dendrite tip (Figure 8.10) [99]. This model incorporated transient diffusional transport of Li ions in a diffusion boundary layer. Generally, Li dendrite growth is strongly affected by an applied current density. When a cell is operated well below a limited current density, dendritic growth with a relatively low dendrite growth rate can be also observed. The numerical diffusion–reaction model proposed by Akolkar [99] estimated a dendrite growth rate of about 0.02 mm s$^{-1}$ at an operating current density of 10 mA cm$^{-2}$. This result is very comparable with experimentally observed dendrite propagation rates from Nishikawa et al. [100].

The order of magnitude lower hardness of Na vs Li indicates the propensity for plastic flow under the mildest shear conditions [101], while its significantly lower modulus implies a greater elastic compliance of the dendrites. In liquid battery electrolytes, Na-metal dendrite growth is qualitatively recognized as a series of interrelated steps, involving an interplay between metal–support interactions and SEI

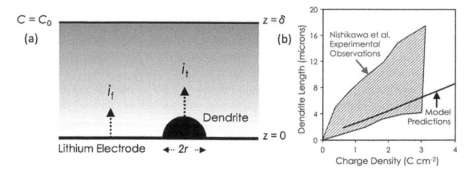

**FIGURE 8.10** (a) Scheme of Li surface. A hemispherical dendrite "precursor" of radius $r$ is presented on the Li surface. The dendrite tip grows at it, and the flat electrode surface grows. (b) Comparison of model prediction with experimental results on Li dendrite propagation at 10 mA cm$^{-2}$. The cross-hatched region indicates the spread in the experimental observation. Reproduced with permission from Ref. [99].

growth/stability. During the first charge, there is the formation of SEI layer at approximately below 1 V vs Na/Na$^+$. The SEI layer is well-recognized to be chemically and geometrically heterogeneous. In carbonate-based solvents, Na metal is more reactive than Li metal when tested under identical conditions. These include various combinations of ethylene carbonate (EC), propylene carbonate (PC), diethyl carbonate (DEC), and dimethyl carbonate (DMC) [102]. It may be immediately observed that at identical current density, the overpotential for the Na–Na case is significantly higher and noisier from the onset, indicating unstable SEI growth. Employing glyme-based solvents, either by themselves or combined with carbonates, the SEI structure is stabilized. The layer possess a more inorganic-based composition of Na$_2$O and NaF, versus the usual sodium carbonate-rich content in EC/PC/DEC/DMC, etc. [103]. It is essential to emphasize that the plating–stripping scenario with Na-metal cells is not just a Li analog that needs updating, but rather a process that is fundamentally different in every respect: (i) The Na-metal nucleation thermodynamics and kinetics differ since Na would bond differently with inorganic (e.g, Na$_2$O and NaF), metallic (e.g, cathode species crossover), and host carbon structures; (ii) Na-metal growth kinetics differ since at room temperature the Na cells are at a higher homologous temperature than Li, e.g, T/Tm = 0.8 vs 0.65; (iii) Dendrites of Na metal are an order magnitude more plastically complaint than ones of Li metal and possess roughly half the elastic rigidity. For instance, Chen et al. examined Na dendrite stability employing operando microscopy under quasi-zero electrochemical fields [104]. It was observed that in a quasi-zero electric field, the Na dendrites were gradually dissolved in a standard battery organic electrolyte (EC/DMC) and broken under external stress [105]. Conversely, under the same conditions, the Li dendrites were relatively stable. In addition, (iv) SEI chemistry, structure, and stability in Na-based systems do not follow the known trends observed with Li and do not respond to well-known Li additives. For instance, there are surprising reports that the well-accepted Li SEI stabilizing additive fluoroethylene carbonate (FEC) will induce significant voltage polarization when employed with Na metal [73, 105]. The SEI composition of a Na-metal anode tested in the same carbonate (EC/DMC) electrolyte with analogous salt as Li metal (1 M $LiPF_6$ and 1 M $NaPF_6$) displayed a much higher inorganic (Na–F to Na–O) content, implying core differences in the electrochemical/chemical passivation behavior.

The dominant presence of $Na_2O$ (shear modulus 49.7 MPa) and NaF (31.4 MPa) is encouraging since both are potentially elastically stiff enough to block Na dendrites. Yet, in general, SEI growth with Na is relatively unstable. (v) The low melting point of Na and the possibility of suppressing it further with eutectic and off-eutectic solutions [106] offer a unique possibility to fully prevent dendrite growth through alloy design.

### 8.3.3 CURRENT COLLECTORS

In addition, with the current Li metal production technology, it is hard to reduce the thickness of lithium foil to less than 50 μm [107]. This means that an excess amount of Li metal is not utilized and, therefore, reduces the gravimetric and volumetric capacities and can lead to severe safety issues. It has been shown that with 200% excess lithium, the theoretical volumetric capacity of the lithium metal battery would decrease from 1,060 mAh $L^{-1}$ to about 690 mAh $L^{-1}$, which is lower than that of a graphite based cell (719 mAh $L^{-1}$) [108]. Although lithium foil is avoided in the case of the PMLB assembly process, production of pre-deposited Li on the copper electrode and the assembly of a full battery still leads to potential exposure of free lithium.

Planar Cu foil has been used as the CC at the anode side since the first commercialization of LIB by Sony owing to its multiple outstanding properties, including good conductivity, ductility, and stability at low potentials [109]. Alternative materials, such as carbonaceous-based collectors face a number of challenges for commercialization in the LIBs. As an indispensable component which bridges the anode material and the external circuit, the Cu CC has a significant influence on the capacity, and long-term stability of batteries [110]. For Li-metal based batteries, the Cu CC not only serves as the connection between the negative electrode active material and the external circuit, but acts as the substrate for lithium plating, and, therefore, plays an especially important role in the nucleation and growth of lithium and accordingly the battery capacity and stability performance. However, commercially available Cu foils are generally not ideally defect-free and they contain numerous cracks and pits at the micro to nano length scales. These defect sites have lower charge-transfer resistance than the other surface locations, and, thus, act as "hot spots" for fast lithium nucleation and growth. In addition, planar Cu foil has a low specific surface area (SSA), which results in high local current densities, and accordingly, fast formation and growth of Li dendrites. All these drawbacks for the Cu CC contribute to the dramatic capacity fade in Li-metal based batteries and hinder their commercialization. During the past decade, the modification of the Cu CC has attracted considerable attention and effort to improve the mechanical/chemical properties of the Cu CC. An ideal material for CC for lithium should possess multiple outstanding properties including those discussed for LIB applications. Firstly, the CC should be defect-free and have a very low and uniform overpotential for Li nucleation, in order to obtain homogenous lithium deposition. Secondly, considering the volume changes during the lithium plating/striping process, the CC should provide excellent structural flexibility as well as mechanical stability. More importantly, it is known that the SEI plays a critical role in the lithium deposition and battery performance. An ideal CC should be able to allow the formation of a very thin, stable, and homogenous SEI layer during the lithium plating process, so that the plated lithium can be isolated from the electrolyte and consumption of active lithium can be prevented. Moreover, the fabrication of an ideal CC should be easy and low-cost.

## 8.4 DIFFERENT APPROACHES TO APPLY METAL BATTERIES IN PRACTICAL APPLICATIONS

Recently, metals (such as Li, Na, K and Zn) are directly employed instead of those anodes (such as graphite, silicon carbon, and hard carbon) with low capacity limits to build so-called metal batteries (MBs) to increase the energy density of the battery. Nevertheless, it should be pointed out that all these MBs face some severe problems, such as dendrite growth and cycle instability. Dendrite growth will cause short circuits, resulting in thermal runaway, thus triggering potential hazards such as fires and explosions. To this end, researchers have put great efforts into the studies of dendrite growth in the past decades, and various new strategies have been proposed and developed to inhibit the dendrite growth of metal anodes and improve the safety and service life of the battery, and we will show some of the strategies that researchers have proposed in recent years to address metal dendrites in this part. Although most of these strategies focus on solving lithium dendrites, they can also be applied to other metal dendrites (such as sodium and potassium) due to the commonalities of metals [88].

### 8.4.1 LITHIUM MBS

#### 8.4.1.1 Temperature

According to the coarse-grained-dynamical Monte Carlo (CGMC) theoretical calculation model for complex thermodynamic of electrolyte, the dendrite growth at high temperature is less obvious than that at low temperature, and high temperature also has a great influence on the growth of dendrite and the formation of SEI film. In addition, Akolkar et al. [111] and C.T. Love et al. [112] have proved the effect of temperature on the growth of dendrite from mechanism and experiment, respectively. Therefore, adjusting temperature is a promising method to control the formation of dendrite internal structure. Based on these models, researchers proposed some modification strategies for lithium metal anode. For instance, Li et al. [113] proposed a dendrite self-healing strategy to control electrochemical deposition at high current density. The current passing through the dendrite can generate enough Joule heat to promote the lithium atoms to migrate from the dendrite tip to the valley region between adjacent dendrites and make the dendrite smooth. Moreover, this strategy was applied on Li–S battery, which show excellent performance, and the batteries with normal current density cycling and repeated high current density healing also show better cycling performance, high coulomb efficiency, and low thermal run-away risk.

Temperature can not only directly affect the formation of dendrites, but also inhibit the growth of dendrites by affecting the electrolyte and SEI film. For example, Yan et al. [114] achieved improved electrochemical effect in ether electrolyte than at room temperature (20°C) by raising the temperature to 60°C, as shown in Figure 8.11. Based on the low surface area lithium particles deposited at high temperature (60°C), the contact between lithium and electrolyte was reduced, leading to a decrease of the occurrence of side reactions, and at 60°C, a mechanically stable and ordered multilayer structure SEI formed in the ether-based electrolyte, which also effectively passivates the anode and inhibits the formation of lithium dendrites.

To sum up, considering the thermodynamic parameters including temperature and energy factors is an effective strategy to construct a dendrite-free anode.

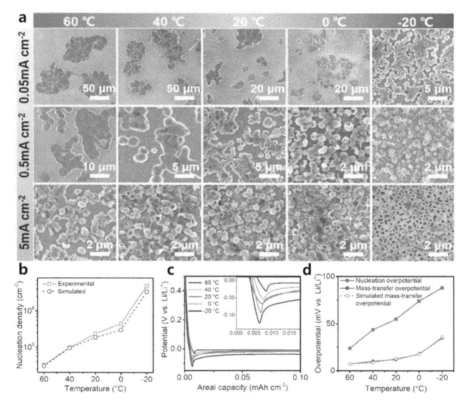

**FIGURE 8.11** Effect of temperature on dendrite formation and battery performance. (a) SEM images of Li nuclei layers at varied temperature and current density conditions with the capacity limitation of 0.1 mAh cm⁻². (b) Change in Li nucleation densities with temperature. (c) Chronoamperometric comparison of Li nucleation at varied temperature at 0.05 mA cm⁻². (d) Change in nucleation overpotentials and experimental and simulated mass-transfer overpotentials with temperature.

## 8.4.1.2 Pressure

Since various strategies have shown beneficial effects in button batteries, their application to the large-scale soft pack battery has been poor. The researchers found that the button cell is in the high stress state caused by the steel shell, while the soft pack cell is in the low stress state caused by the aluminum plastic film. Thus, the compression state plays an important role in this gap.

Zhang et al. constructed a force electrochemical phase field model [115]. They found that the lithium deposition reaction will trigger the stress response caused by the intrinsic strain and the local deformation will cause the redistribution of ions, which affects the growth of lithium dendrites. In addition, Zhang et al. [115] also studied the effect of external pressure on dendrite growth. They found that the whole changes from the expansion state to the compression state of the batteries when the applied external pressure is greater than the electrochemical stress, resulting in a maximum hydrostatic pressure at the dendritic tip to inhibit the growth of the dendritic tip and promote lateral growth (Figure 8.12a–e). It is worth noting that the increase of external pressure will bring the risk of material failure, which can be avoided by predicting an optimum external pressure based on the von Mises yield criterion. Moreover, researchers have taken both external pressure and electrolyte modulus variation into account. They found that the effect of external pressure on dendrite growth is more significant at lower elastic modulus of electrolyte under a fixed external pressure, and in the extreme case of pure liquid electrolyte (the elastic modulus is close to zero), the external pressure can always suppress the dendrite. Furthermore, Chen et al. [116] proposed a new strategy that applied external pressure to shape the dendrites before using it further for the development of stable metal anode, and the reuse of Li metal anode showed extraordinary electrochemical performance at high current density (Figure 8.12f–h).

It can be seen from the above work that the external pressure also plays a very important role in the electrochemical performance of MBs, and the relationship between pressure and other variables in the battery is also critical, which needs to be further explored.

## 8.4.1.3 Current Density and Substrate

A research group [116] studied the nucleation and growth process of Li deposited on copper foil electrode in the early stage, and elaborated the relationship between the size, morphology, area density, and current density of Li crystal nucleus. They found that the size of lithium decreases and dendrite growth increases with the increase of current density and overpotential. Liu et al. [117] found that dendrite growth also increases with the increase of exchange current density. In addition, the execution of the electroplating/stripping cycle for lithium depends largely on the substrate. Different substrates show different behaviors with lithium such as some are slightly soluble in lithium (platinum, aluminum, magnesium, zinc, silver, gold), and some are completely insoluble in lithium such as silicon, tin, carbon, and nickel [118]. For example, $Li_{22}Sn_5$ alloy is a better substrate for $Li^+$ deposition than Li and Cu [118], which can continuously change the exchange current density, thereby deposit metal ions uniformly and avoid dendrite formation.

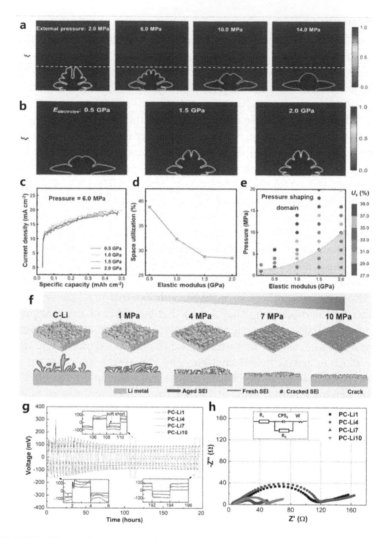

**FIGURE 8.12** External pressure and Electrolyte modulus variation. (a) The simulated results of Li dendrites under the external pressure ranging from 2.0 to 14.0 MPa, the snapshots of dendritic morphology. (b) The pressure shaping effect on the Li dendrite growth in electrolytes of elastic moduli ranging from 0.5 to 2.0 GPa. The applied external pressure is fixed at 6.0 MPa, the snapshots of dendritic morphology at a plating capacity of 0.40 mAh cm$^{-2}$. (c) The current density evolution with the proceeding of electroplating. (d) The space utilization in different electrolytes. (e) The phase diagram based on the applied external pressure and the elastic modulus of electrolyte at a plating capacity of 0.4 mAh cm$^{-2}$. The white region indicates enhanced performance of Li metal anodes, whereas the gray portion denotes that the external pressure fails to work, and the pre-cycled metal anode and external pressure strategy. (f) Schematic drawings of the surface (top row) and cross-sectional (bottom row) morphologies of the PC-Li samples. (g) Voltage hysteresis of the PC-Li foils with different compression pressure at the current density of 1 mA cm$^{-2}$ with the areal capacity of 2 mAh cm$^{-2}$. (h) Nyquist plots of the PC-Li symmetric cells.

### 8.4.1.4 Electrolyte Solvent and Electrolyte Additive and Concentration

Electrolyte, as an important part of battery, is also related to the growth of lithium dendrites. SEI is a passivation film formed on the surface of anode during the first charge and discharge of lithium metal battery, which is obtained by the reaction of metal ion with solvent, trace water, HF, etc. The stable SEI can ensure uniform distribution of lithium ions and reduction of dead lithium, thus inhibiting the growth of lithium dendrites. To this end, electrolyte modification is also one of the strategies for suppressing lithium dendrites.

Lithium metal often shows different morphologies and electrochemical properties in different electrolytes. For example, the Coulombic efficiency in the DMC electrolyte is only 20–30%, which can be increased to about 80% with the addition of EC. In common EC/DMC-based electrolyte, a large number of lithium dendrites are still produced. Fortunately, if EC is substituted with FEC, not only the growth of lithium dendrites can be inhibited, but also the coulombic efficiency (~96%) and high voltage electrochemical stability can be achieved. For 1 M $LiPF_6$ FEC/FEMC/HFE electrolyte, the potential of the electrolyte is raised to about 5.8 V. After replacing EC with FEC, the dendrite is completely inhibited and the lithium metal appears as a spherical shape. The formation of this morphology should be related to the large amount of inorganic SEI mainly composed of LiF. Fluoride with the strongest binding energy, the lowest electronic conductivity, and the highest thermodynamic stability can greatly inhibit the continuous side reactions between electrolyte and positive and negative electrode materials, expand the positive and negative electrode windows of electrolyte, and realize the efficient cycling of lithium metal and high voltage cathode materials [119].

In addition to electrolyte solvent, electrolyte additives have been also used to prevent dendrite formation and passive film formation on metal anodes, thus obtaining high-performance batteries. To this end, Meng et al. [120] proposed a simple and effective interface in situ catalytic grafting strategy to achieve high-efficiency stability and dendrite suppression of lithium metal anode. In this work, liquid polydimethylsiloxane ($PDMS-OCH_3$) terminated by $-OCH_3$ group was used as a grafting additive. Under the action of charge transfer, the dissociation reaction of $PDMS-OCH_3$ can be catalyzed by the thin "film" of $Li_2O$ and LiOH on the surface of lithium metal. The broken macromolecules can be grafted onto the surface of lithium metal, and smaller molecules can be densified into inorganic $Li_xSiO_y$ fast ion conductors. Such an organic–inorganic hybrid interface phase (i.e., grafted SEI) was further enhanced by the high concentration of LiF injected during the electrochemical process. The combination of hard inorganic components of LiF and $Li_xSiO_y$ can provide fast ion channel and interface, realize the homogenization effect of ion current, and act as a barrier to hinder the growth of lithium dendrite. At the same time, the soft PDMS branch can also enhance the flexibility and buffering effect of the whole SEI. Such liquid $PDMS-OCH_3$ as the additive of the carbonate system could give Li/Li symmetric battery a stable cycle of 1,800 h and achieve a small potential polarization of about 25 mV, and the CE of Li/Cu asymmetric battery is up to 97% under the condition of high current density and high capacity (Figure 8.13a and 8.13b). Furthermore, Wang et al. [121] proposed to use molecules containing s-conjugated

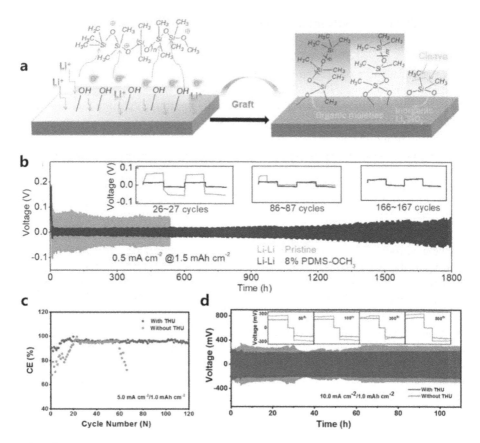

**FIGURE 8.13** (a) Schematic of in situ grafting of PDMS-OCH3 on thin skins of LiOH and Li₂O on Li anode and its evolution into fragmented organic and inorganic moieties. (b) Li plating-stripping performance comparison of Li/Li symmetric cells with and without PDMS addition at 0.5 mA cm⁻² with an areal capacity of 1.5 mAh cm⁻². Insets: corresponding voltage profiles at different cycling stages. (c) The cycling performance of Cu–Li half cells in 1 M LiTFSI DOL/DME electrolyte with or without THU additive at and 5.0 mA cm⁻²/1.0 mAh cm⁻² and (d) the cycling performance of symmetrical Li–Li cells in 1 M LiTFSI DOL/DME electrolyte with or without THU additive 10.0 mA cm⁻²/1.0 mAh cm⁻². Inset: Detailed analysis of voltage profiles at different cycles.

structure as electrolyte additives, in which thiourea is the representative to realize uniform lithium deposition. Because thiourea can promote the deposition of lithium metal and effectively avoid the formation of lithium dendrite, copper lithium battery shows high cycle stability at up to 5 mA cm⁻², and under different current densities, the Li/Li symmetric battery exhibited low overpotential with flat voltage profile and improved cycle stability at 10 mA cm⁻²/1.0 mAh cm⁻² (Figure 8.13c and 8.13d). What is more remarkable is that the electrochemical cycling test was carried out on the lithium metal electrode, which had already produced lithium dendrite in the presence of thiourea. It can be found that after 20 cycles, the morphology of lithium metal surface has been greatly improved, and dendrites are effectively eliminated. The coin cell coupled with $LiFePO_4$ shows obvious advantages in both cycle and rate performance. It can stably cycle more than 600 cycles at 5 C, and still maintain more than 94% discharge capacity. In conclusion, electrolyte additives are the simple and most promising application in practical batteries to improve the uneven deposition of lithium and inhibit the growth of lithium dendrites [88].

### 8.4.1.5 Interphase

In addition to the above-mentioned strategies, the inhibition of lithium dendrites can also be achieved by regulating the formation and growth of dendrites. Dendrite control is to allow the formation and growth of dendrites, but to control the nucleation position, electric field distribution and ion transport direction of dendrites to obtain a smooth large layer. Generally, two typical dendritic growth strategies are introduced: (i) adjusting nucleation sites and (ii) controlling the growth path and direction. Interface modification of metal anodes has recently become a popular research strategy.

The three-dimensional metal electrode can improve the stability and cycle performance of the electrode, such as carbon, porous copper, and polymer carrier [122–124]. Among them, carbon scaffold composites are mainly concerned. For example, Niu et al. [125] reported an amino functionalized mesoporous carbon fiber Li–C anode material, which utilizes the strong interaction between lithium and functional groups on the surface of carbon materials to achieve a uniform distribution of lithium ions. The constructed reversible and self-smooth lithium deposition system obtained an energy density up to 380 W h $kg^{-1}$ and cycle life up to 200 cycles, and combining such self-smoothing Li–C film as a negative electrode combined with commercial NMC622 cathode, it gives excellent cycling performance at 1 C for 1,000 cycles. In addition, a research group [126] prepared a lithium diffusion interface layer based on lithium multi-walled carbon nanotubes (Li-MWCNT) to inhibit the formation of lithium dendrites. Based on the difference of Fermi energy levels between lithium and MWCNT, the lithium MWCNT layer was constructed in situ by a simple physical contact method. The protective lithium anode with Li-MWCNT interface layer is applied to lithium–sulfur batteries, which could realize high current pulse discharge and 450 cycles with a different rate, and the CE was close to 99.9%. Li-MWCNT shows stable and smooth voltage profile for 2000 h at 2 mA $cm^{-2}$ as well as at different current densities up to 5 mA $cm^{-2}$. Furthermore, it has been reported that adjusting the structure of metal electrodes can achieve the homogenized electrodeposition (Figure 8.14a–c). Zhang et al. studied how current density affects the initial nucleation and electrodeposition kinetics of Na/Li. The results show that the current density determines the initial size of Na/Li nucleation and that the structure of electrodes with the high specific surface area can effectively reduce the local current density. Thus, the 3D deposition framework with a micro/nanostructure can achieve uniform nucleation and dendrite suppression by adjusting the local current density of the deposition, which is consistent with the above example.

Furthermore, the uniform metal ion infiltration method is also used in the suppression of lithium dendrites. Based on the metal with no nucleation barrier or low nucleation overpotential has a certain solubility in lithium, Yan et al. [127] designed a

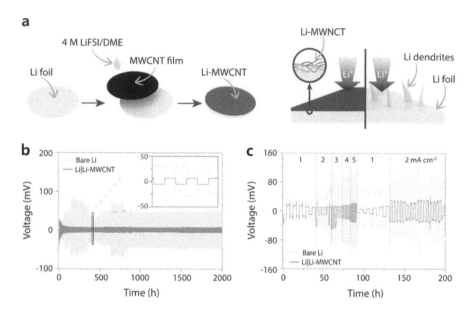

**FIGURE 8.14** Carbon composite scaffold method. (a) Schematic illustration of MWCNT-Li for the suppression of Li dendrite formation, representing the role of MWCNT-Li in suppressing of Li dendrite as compared with bare Li; (b) voltage profile of symmetric cells of bare Li and MWCNT Li at a current density of 2 mA cm$^{-2}$ and (c) voltage profile of bare Li and MWCNT-Li at different current densities.

nanostructure of lithium metal anode composed of hollow carbon. Gold nanoparticles are uniformly distributed on the inside of the spherical shell. During the deposition process, lithium metal mainly grows inside the hollow carbon sphere. The selective deposition and stable lithium metal coating of lithium metal slow down the formation of lithium dendrite and improve the cycling performance. Even in corrosive alkyl carbonate electrolyte, the lithium metal can be cycled at 98% coulomb efficiency for more than 300 cycles. Tantratian et al. [128] introduced lithium-impregnated titanium dioxide nanotubes with ultra-uniform surface curvature and uniform arrangement as lithium metal anode. A uniform local electric field is generated on the surface of the ultra-uniform nanotube, which adsorbs lithium ions to the surface uniformly, resulting in a uniform current density distribution, and the metal lithium can be uniformly deposited on the nanotube wall so that each lithium atom grows on the circumference of the nanotube without obvious evidence of lithium dendrite formation. Notably, the spaced Li tubes show high cycle stability and low overpotential than pristine Li and closed Li tubes and show a high specific capacity of 132mAh g$^{-1}$ at 1 C and the CE was about 99.85% in 400 cycles.

Importantly, the SEI mentioned above plays an important role in the dendritic growth of metal anodes. The correlation between SEI characteristics (composition and structure) and the behavior of interfacial kinetics of metal ions is a key factor in determining metal ion deposition behavior, dendrite growth, and battery

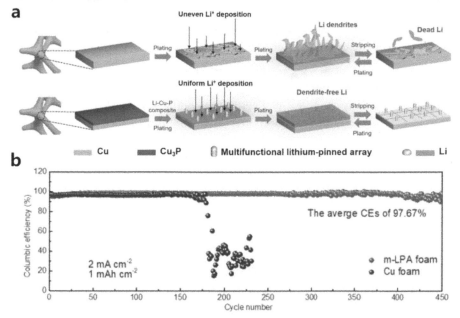

**FIGURE 8.15** Artificial SEI and ion/e- Infiltration methods. (a) Schematic illustration of Li–Cu–P and Li pin formation to inhibit dendrite formation and (b) voltage profiles of symmetric cells at a current density of 2 mA cm⁻².

rate performance. Based on this potential relationship, researchers have done a lot of exploration. For example, Xu et al. [129] constructed a lithiophilic multifunctional layer on 3D copper foam, Cu@Cu$_3$P, by a mild chemical vapor phosphating synthesis method (Figure 8.15a and 8.15b) and used it as lithium collector. The Cu$_3$P reacts with Li ion to form a multifunctional "solid lithium lattice" (Li$^{2+}_x$Cu$_1$-XP), which is the mentioned artificial SEI and can induce the uniform deposition of lithium as a lithium affinity site, thus inhibiting the growth of dendrites, and the average Coloumbic efficiency of m-LPA coating 3D Cu is 98.61% and 97.41% after 600 and 450 cycles with a fixed capacity of 1 mAh cm⁻² at a current density of 1 and 2 mA cm⁻², respectively. The above results provide an important basis for the commercial application of LMBs, and also have certain guiding significance for the research of other metal anodes (Na, K, Zn, etc.). Moreover, Tu et al. [130] prepared mixed materials (Sn–Li, etc.) artificial SEI by rapid and simple ion displacement deposition of active electrochemical metals (such as Sn, In, or Si) on alkali metal electrode materials. The realization of SEI film can protect the alkali metal electrode from the influence of side reactions and provide a good interface to inhibit the volume change of electrode materials, alloying and coating, and inhibit the formation of the dendrite, and in the detection of the electrochemical performance of Sn–Li, the hybrid material has a smooth voltage profile compared with the original Li metal after the protection of the Sn layer at a current density of 3 mA cm⁻²compared with the original Li metal. Moreover, the Sn–Li electrode assembled with LiNi$_{0.8}$Co$_{0.15}$Al$_{0.05}$O$_2$ (NCA) has excellent cycle stability with 3 mAh cm$^{-2}$ at 0.5 C. Compared with the original Li surface, the growth rate of Sn–Li electrode material is significantly slower.

### 8.4.1.6  Electrolyte Type

Dealing with lithium dendrites by electrolyte modification is also one of the hot topics in recent years. Various inhibition strategies have been proposed for this system, such as adjusting electrolyte composition or concentration as described in the above sections. In addition, based on the high capacity of lithium (Li) metal and the intrinsic safety of solid-state electrolyte, solid-state Li-metal batteries are regarded as a promising candidate for next-generation energy storage. Solid electrolytes, mainly divided into polymer electrolytes and inorganic solid electrolytes, were studied.

Polymer electrolytes have been widely studied because of their ion conductivity and certain flexibility. Rosso et al. [131] studied the formation mechanism of lithium dendrites in polymer electrolytes, showing that polymer electrolytes with high ion diffusion coefficient, high Li ion transference number, and high ion concentration can inhibit the formation of lithium dendrites. More importantly, Khurana et al. [132] reported a cross-linked ethylene/PEO-based electrolyte with high ionic conductivity at room temperature, which is one of the most widely studied polymer electrolyte. Poly(ethylene oxide)(PEO)/lithium bis-(-trifluoromethanesulfonyl) imide (LiTFSI) has been well proved by electric vehicle. This electrolyte in these LMBs work at 40–70°C. It shows high ionic conductivity and uniform deposition-dissolution of lithium (Figure 8.16a and 8.16b). Nevertheless, the low mechanical strength of PEO is still at risk of being pierced by growing lithium dendrites, thereby the recombination of polymer electrolytes (such as inorganic fillers) is also one of the strategies currently attracting attention. Su et al. [133] proposed a design idea of all-solid-state polymer electrolyte with high entropy micro-region interlocking. This all-solid-state polymer electrolyte with a special structure is formed by introducing a newly synthesized multifunctional ABC arm star-shaped ternary polymer into the PEO matrix. The problem of inconsistency between mechanical strength and ionic conductivity of all-solid-state polymer electrolytes needs to be solved. In the full battery test, it showed a long cycle stability (more than 4,000 h) (Figure 8.16c and 8.16d).

In order to prevent dendrite from penetrating the membrane, a high modulus solid electrolyte is studied to ensure the safe cycles of MBs [134]. For example, Gao et al. designed an asymmetric solid electrolyte (ASE) to inhibit the dendrite growth for SSLBs [135]. The initial discharge capacity of LFP = ASE| Li solid-state battery is 155.1 mAh g$^{-1}$, and the capacity retention is of up to 90.2% at 0.2 C for 200 cycles. The coulombic efficiency can exceed 99.6% during the whole cycling process. Compared with SPE, ASE shows a wide electrochemical window larger than 5.2 V. The specific capacity of LCO = Li battery and SSLBs is 130 mAh g$^{-1}$ and the capacity retention is of 78.9% after 110 cycles. Although the research of inorganic solid electrolyte has made a big step forward in the application of metal batteries, its high mechanical strength still has the problem of poor interfacial wettability. Therefore, perhaps the focus can be placed on the composite electrolyte in future research.

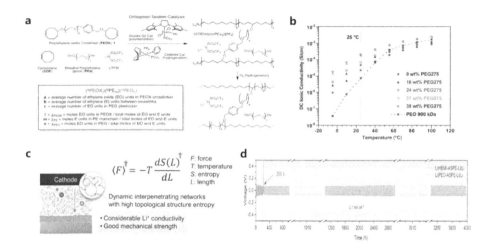

**FIGURE 8.16** (a) Polyethylene/poly(ethylene oxide) solid polymer electrolyte (SPE) synthesis and nomenclature and (b) plot of DC ionic conductivity as a function of temperature for 70PEOX electrolytes having different weight percent of PEG275 plasticizer. All films had [COE]:[1] ratio of 15:1 and [EO]:[Li] composition of 18:1. The conductivity of a PEO 900 kDa sample with [EO]:[Li] ratio of 18:1 is also shown for comparison purposes. (c) The illustration the design strategy of the work of Su et.al.; high entropy electrolyte with self-assembled dynamic interpenetrating polymer networks is designed, which promises high cationic conductivity and suppressed Li dendrites. (d) Cycling performance of the Li|HEMI-ASPE-Li|Li and Li|PEO-ASPE-Li|Li symmetrical cells at a current density of 0.1 mA cm$^{-2}$ at 70 °C.

## 8.4.2 SODIUM AND POTASSIUM MBs

Although lithium metal is the most studied alkali metal anode, the dendrite problem is not unique to lithium batteries. Other MBs, such as sodium and potassium batteries, also have safety problems caused by dendrite growth and low cycle stability. Thus, strategies for coping with lithium dendrites as mentioned above are also mostly applicable to sodium and potassium metal batteries. However, it is worth noting that although there are similarities among Li, Na, and K, it is more difficult to form sufficient passivation layers on Na and K metals than on Li metals due to their different chemical/electrochemical reactivity, solubility of decomposition products, mechanical properties, and volume flow ratio [79].

For the sodium MBs, strategies such as electrolytes, nanocarbon matrix, and membrane modification are also used to suppress sodium dendrites. For instance, Xu et al. [136] studied the electrochemical performance of Na–S battery under different solvent combinations and solute concentrations (such as 1 M, 1.5 M and 2 M NaTFSI in PC and PC-FEC). They proved that the concentrated electrolyte with polar part (2 M PC-FEC) can inhibit the dendrite formation and show excellent cycling performance. Moreover, Sun et al. [137] introduced functionally co-doped carbon nanotubes (CNTs) to control the nucleation behavior of dendritic nano-metal anode sodium. The N and S functional groups on the CNTs induce the

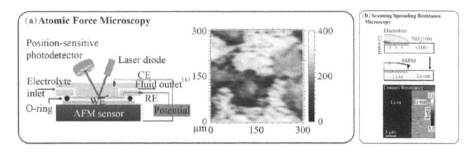

**FIGURE 8.17** Qualitative, 2D imaging techniques used for detection of Li. (a) Schematic of the cell used for AFM-based characterization (adapted with permission [142]. Copyright 2015, Elsevier.) and the topography changes due to heterogeneity of Li across the electrode surface (adapted with permission [142]. Copyright 2015, Springer Nature). (b) Surface map of contact resistance using SSRM for Li plated on a $TiO_2$, surface Adapted with permission.[143] Copyright 2019, American Chemical Society).

### 8.5.1.3  Neutron Depth Profiling for Metal Batteries

Neutron depth profiling (NDP) uses the absorption of cold neutrons by Li nuclei and the subsequent emission and measurement of charged particles (*He and 3H) to calculate the subsurface position of the absorption event at the Li atom, thus generating depth profiles of Li concentration. Nagpure et al. used NDP to observe the concentration gradient of Li in LIBs with a graphite electrode up to =12 μm below the electrode surface by bombardment of an area of =0.8 cm$^2$ with cold neutrons [148]. Lv et al. used NDP to investigate the subsurface heterogeneity of the plating and stripping processes of Li (to 25 um below the surface) on a Cu substrate and effect of the current density on the morphology of Li microstructures [144] with a liquid electrolyte, as shown in Figure 8.18a. Han et al. investigated the role of ionic conductivity in the formation of Li dendrites in Li solid-state batteries, using a time-resolved operando NDP measurement [149].

### 8.5.1.4  X-Ray Based Tomography for Metal Batteries

X-ray based tomography is an important technique to obtain volumetric reconstruction of the cell (or particular cell components) from the microscale to the nanoscale [150]. In traditional absorption-based X-ray-tomography, the primary challenge in characterizing Li in Li-ion batteries lies in distinguishing the graphite (typically the anode), electrolyte, and Li deposits, which have similar X-ray mass attenuation coefficients [151]. This is made even more difficult if there are imaging artifacts from high X-ray attenuating materials such as the copper current collector. Phase-contrast X-ray tomography has shown promise in separating cell components which have similar mass attenuation coefficients. Different phase-contrast X-ray tomography methods such as propagation-based imaging and differential phase-contrast X-ray imaging can measure both attenuation and phase-shift of X-rays that can be combined together for complete 3D visualization of cell components.

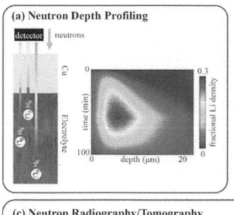

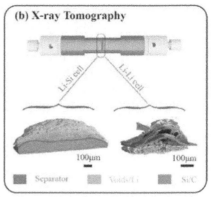

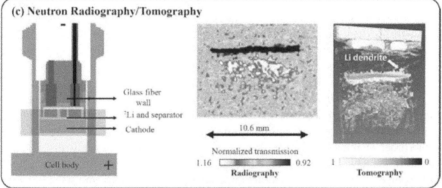

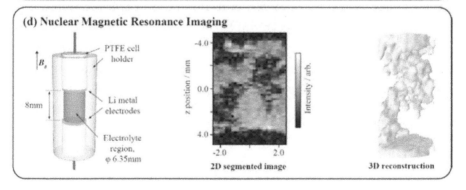

**FIGURE 8.18** Quantitative imaging techniques used for visualization of Li. (a) Using neutron depth profiling to obtain depth and time (SOC) resolved maps of Li (Adapted with permission [144]. Copyright 2018, Springer Nature). (b) 3D reconstructions of Li growth in a Li–Si and a Li–Li cell using phase-contrast X-ray tomography (adapted with permission [145]. Copyright 2016, American Chemical Society). (c) Use of neutron radiography (2D) and tomography (3D) on a custom-built Li-ion cell, to visualize Li dendrites (adapted with permission [146]. Copyright 2019, American Chemical Society). (d) Nuclear magnetic resonance imaging (nMRI) to produce 2D segmented images, which can be reconstructed into a full volume of Li growth between substrates (adapted with permission.[147] Copyright 2016, National Academy of Sciences).

Eastwood et al. demonstrated the 3D microstructures of deposited mossy Li on a Cu substrate in custom-built cells with in situ capabilities using a phase contrast filtered back projection algorithm [152]. They were also able to detect other Li-containing species (Li salts) due to the reaction of the mossy Li with the electrolyte. Sun et al. studied the morphology of plated Li in Li–Li symmetric and Li–Si half cells to infer the role of Li plating and stripping in cell failure using phase contrast X-ray tomography [145], as seen in Figure 8.18b. Frisco et al. used Zernike phase contrast to study the effect of current densities and temperatures on the morphology and nature of Li deposits on a Cu substrate at the nanoscale [150]. Harry et al. used X-ray tomography to visualize the formation of subsurface Li dendrites, below the Li metal anode and polymer (solid-state) electrolyte interface [153].

### 8.5.1.5 Neutron Tomography and Radiography for Metal Batteries

Neutron-based imaging (radiography/tomography) uses the interaction of the neutron beam with the nuclei of materials to visualize Li deposition in regions of plating. The main difference between X-ray and neutron-based characterization comes from their interaction with matter. The scattering cross-sections are dominated by electron interaction and nuclear interaction for X-ray and neutron, respectively. Accordingly, neutrons are particularly sensitive to some low Z elements such as Li. Radiography is a quick way to obtain quick (order of seconds) 2D snapshots of the cell while tomography involves 3D reconstructions from various 2D slices and takes significantly longer (order of minutes). Unlike X-rays, neutrons interact strongly with Li. Same et al. used this technique to observe Li plating in coin cells under in situ conditions [154]. Song et al. used a combination of operando 2D radiography and static 3D tomography to study the role of Li dendrites in shorting an LMB [146]. A schematic of the cell used for the imaging as well as an example 3D reconstruction as presented by Song et al. is shown in Figure 8.18c.

### 8.5.1.6 Nuclear Magnetic Resonance Imaging for Metal Batteries

NMR spectroscopy, as described in Section 2.1.3, is a quantitative, global, non-destructive method of detecting surface deposits during plating and stripping of Li in cells. NMR imaging (nMRI) is an application of NMR, which combines NMR with imaging, furnishing additional local spatial information (along one dimension) of Li [155]. Chandrashekar et al. used "bag" cells with Li electrodes and chemical-shift imaging to obtain spatial information on the various microstructures of Li (dendritic, mossy) versus SEI products [65]. Chang et al. used a similar chemical mapping to detect the point of onset of Li dendrite growth and separate it from regular Li microstructure growth in custom-built Li metal cells [156].

Traditionally, NMR and nMRI rely on the signals from Li and $^7$Li to quantify the amount of Li dendrites in the cell during cycling. However, this approach often leads to limits on the spatial and temporal resolution of the technique due to the low sensitivity of the Li nucleus. Iliott et al. proposed an indirect nMRI technique, which instead monitored the interactions of the Li dendrites with the surrounding electrolyte through monitoring of the H (proton) signal, which enhanced the sensitivity of the technique [147]. An example 2D reconstructed slice as well as the 3D reconstructed volume obtained by Iliott et al. on an LMB is shown in Figure 8.2d. An important point to be noted while using nMRI lies in the orientation of the electrochemical cell with respect to the radio frequency (RF) coil used for exciting the cell. The difference in signal propagation through the cell based on its orientation with respect to the coil can be resolved by calibrating the direction of the RF signal to be parallel to the electrodes.

### 8.5.1.7 X-Ray Diffraction for Metal Batteries

X-ray diffraction-based methods rely on the detection of the crystalline peaks from metallic Li and other Li-containing species (such as $Li_xC_6$). X-rays are non-destructive; thus, X-ray diffraction (XRD) scans can be conducted on most cell geometries, under in situ or operando conditions. Additionally, XRD can be used to conduct both global and local measurements of Li, by using a small X-ray beam to obtain local information and a spatial rastering scheme to cover the entire cell. Finally, a distinct feature of XRD is that owing to the non-destructive nature of X-rays, XRD can furnish structural information on all crystalline components of the cell simultaneously, such as the cathode or anode, apart from the Li-containing species such as plated Li or $Li_xC_6$ allowing for a simultaneous study of the degradation of all cell components in a correlative manner, thus facilitating a link between localized Li-plating and the SOC of active materials in the same region.

All of the studies described below employ synchrotron X-rays, which allow for fast characterization due to a high X-ray flux. Thus, realistic cells can be analyzed both globally and locally in a reasonable amount of time (order of hours for a 14 cm pouch cell with a sub-millimeter spatial resolution in [158]). Additionally, the wide variety of focusing optics available at synchrotron sources enable XRD to have a wide range of spatial resolution from sub-micrometer using micro Laue diffraction to millimeter-scale resolution using a standard powder diffraction scheme.

Out of the three spatial ($x$, $y$, $z$, the length, breadth, and depth) and one temporal ($t$) dimensions, different cell orientations with respect to the X-ray beam are used to resolve two of the four dimensions. Spatially resolved (along $x$, $y$) XRD has been utilized to study the heterogeneity of Li plating and associated degradation mechanisms across the cross-sectional area of a pouch cell. Figure 8.19a shows the cell components and a typical spatial map of Li from such an XRD scan. Such scans typically quantify the heterogeneity of Li across the cross-section of the cell at a fixed SOC, and by averaging over the thickness of the electrode. Additionally, integration of the spatially resolved Li across the entire cell also helps to directly correlate the exact contributions of various mechanisms of Li inventory loss to the cell performance (measured through electrochemistry). On the other hand, depth-resolved ($z$, $t$) XRD has been used to study the heterogeneity of Li plating and Li intercalation into a graphite anode under operando conditions, by averaging over the cross-sectional area of the cell [159] (see Figure 8.19b for a schematic of the orientation of the cell with respect to the X-ray beam and a spatial map of Li). Finally, another implementation of XRD: energy-resolved X-ray diffraction (EDXRD) has also been used to quantify similar heterogeneities across the depth of electrode, under operando conditions [160, 161]. EDXRD provides insights into the concentration gradient of Li across the depth

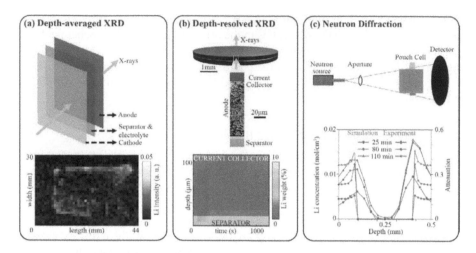

**FIGURE 8.19** Diffraction-based local characterization methods to generate 2D plots showing the distribution of plated Li. (a) Using spatially resolved X-ray diffraction (XRD) to obtain depth-averaged Li distribution over the cross-sectional area of the anode at a particular time (SOC) (adapted with permission [147]. Copyright 2020, Elsevier). (b) Using depth-resolved XRD to obtain spatial maps of Li over the anode, as a function of time (SOC) (adapted with permission [17]. Copyright 2020, The Royal Society of Chemistry). (c) Using depth-resolved neutron diffraction to obtain spatial maps of Li over the anode, as a function of time (adapted with permission [157]. Copyright 2016, American Chemical Society). On University, Wiley Online.

of the anode, which is the likely cause for the nucleation of Li plating at the anode–separator interface [162].

Despite several advantages as a characterization tool for Li, XRD-based methods can only probe the crystalline components of the cell. Thus, nanocrystalline or amorphous species (such as Li-containing SEI products) are unable to be determined by XRD, since they vanish into the background signal. Additionally, compared to imaging-based methods, another major limitation of diffraction techniques is their inability to distinguish the morphology of Li (e.g, dendritic vs mossy). Therefore, XRD is best combined with techniques that are sensitive to the Li morphologies, such as X-ray tomography. XRD computed tomography (XRDCT) involves taking a series of measurements at different positions to reconstruct tomograms with 3D spatial crystallographic information. XRDCT has recently been demonstrated with 1 μm resolution to spatially resolve sub-particle crystallographic detail within LIB electrodes [163, 164]. With the increased X-ray flux of recently upgraded synchrotron sources, operando XRDCT may be a promising method to provide sub-particle lithiation information simultaneously to detect the onset of Li-plating, facilitating both spatial crystallographic and 3D reconstructions.

### 8.5.1.8 Neutron Diffraction for Metal Batteries

Neutron-based diffraction for Li detection is similar to XRD-based methods, in terms of being a nonintrusive, in situ and quantitative technique. The difference in the interaction between X-rays and neutrons with matter, along with the challenges associated with generating a high flux for neutrons manifests in shorter data acquisition time and superior spatial resolution of XRD compared to neutron diffraction. Zinth et al. used diffraction to investigate Li plating and stripping on a graphite anode at sub-ambient temperatures (–20°C) and moderate cycling rates (C/5 to C/30) [165]. Von Lüders et al. used in situ neutron diffraction on commercial 18,650 cylindrical cells to study the dependence of C-rate on the amount of Li plating and intercalation of deposited Li into the graphite anode at sub-ambient temperatures (–2°C) [166]. Zhou et al. brought together the advantages of various neutron-based methods by combining neutron radiography and diffraction to develop a non-invasive, in situ characterization tool for Li-ion pouch cells [167], as illustrated in Figure 8.19c. Neutron diffraction can provide a spatial resolution at the sub-millimeter range. Additionally, the acquisition of neutron patterns to discern the Li peak takes considerably longer than in X-rays, which limits the C-rate of cycling during operando studies [165].

## 8.5.2 COMMERCIAL ASPECT

Li-ion batteries are transforming the transportation and grid sectors. Their scale up is truly historic: Li-ion is now the only rechargeable battery other than lead-acid produced at >5 GWh y$^{-1}$, with a worldwide manufacturing expansion reaching hundreds of GWh yr$^{-1}$ over the next five years. Li-ion battery packs achieve long cycle life (in the thousands), high charge/ discharge rates (>1 C), high energy content (specific energy of ~150 Wh kg$^{-1}$ and energy density of 250 Wh L$^{-1}$), and low capital costs (<\$300 kWh$^{-1}$) [168]. However, the present Li-ion material platform (a graphite negative electrode coupled with a metal oxide positive electrode) is not expected to reach the US Department of Energy's (DOE) electric vehicle pack goals of 235 Wh kg$^{-1}$, 500 Wh L$^{-1}$ and \$125 kWh$^{-1}$ [169]. The intercalation mechanism that fundamentally enables the excellent cycling of Li-ion also places an upper limit on energy content because of the weight and volume of the hosts into which Li* intercalates. Thus, there remains an acute need for higher-energy alternatives to the Li-ion material platform [169]. Replacing the graphite electrode with lithium metal, which results in a ~35% increase in specific energy and ~50% increase in energy density at the cell level, provides a path to reach those goals, especially if the introduction of lithium metal is combined with reduction of the liquid electrolytes, which impose both safety and thermal management mass and volume requirements at the pack level [170].

### 8.5.2.1 Necessity of Using Limited Lithium in Test Cells

The use of a much higher loading of lithium metal in a research cell than would be used in a commercial cell has a practical origin: the ready availability and low cost of lithium foils that are hundreds of micrometers thick at laboratory supply companies. However, we believe the lithium metal research community should shift to the use of limited lithium (defined here as <30 um, or < –6 mAh cm$^{-2}$) even in experimental test cells, for reasons described below. This shift can be made easily, as lithium foils with a thickness of 20 um are now available commercially, although surface passivation may need to be handled carefully.

First, we believe that some of the results from Li/Li symmetric cells in Figure 8.20 (for example, points 13, 18, and 26) reflect mixed ionic/electronic conduction in the separator, leading to deceptively "good" cycling results. In many research publications with Li/Li symmetric cells, the cell potential does not fall to 0 V during extended cycling but rather stabilizes, often after a transition period of tens to hundreds of hours, which we believe could indicate soft shorts. What may be taking place in these cases is the growth of lithium metal through the separator to contact the positive electrode and thus form a direct electronic pathway between the electrodes. If this growth occurs slowly, with significant ionic current taking place even as an electronic pathway is established, the cell potential may change only slowly, and never fall to zero as it would for a "hard" short. Once an electronic pathway is established, extremely stable cycling may be observed because lithium is no longer being passed between the electrodes. The use of a limited amount of lithium metal can conclusively identify the presence of electronic shorting in the following way. When a known amount of lithium is removed fully from one electrode, the potential should rise, thereby indicating full depletion; Faraday's law can be used to compare the charge passed with the amount of lithium present. Lithium with a thickness of 20 um, corresponding to ~4 mAh cm$^{-2}$, can be fully stripped at a meaningful current challenges density in a matter of hours, and a full strip can be done periodically during a long-term cycling test. In addition to the use of limited lithium and a full strip, three additional experiments are helpful to check for the presence of electronic shorts: (1) measure the activation energy of the conductivity (low or negative values indicate electronic conductivity); (2) check for inductive loops, or the absence of an imaginary loop, with impedance spectroscopy [171–173], and (3) monitor the open-circuit voltage of a full cell to check for electronic shorting.

Second, the use of limited lithium in research cells will more closely reflect viable commercial designs with high energy density and ensure that technical challenges for those designs are identified at an early point. Several groups are now reporting current densities, per-cycle areal capacities, and cumulative capacities approaching DOE goals; however, Figure 8.20 shows that these efforts use far more lithium than acceptable for the targeted energy density. As one example of a technical challenge, cycling 15 um (~3 mAh cm$^{-2}$) of a 500 μm lithium foil results in far less volume change (on a fractional basis) than cycling 15 um out of 19 μm (the 80% target shown in Figure 8.20). Coupling among the parameters shown in Figure 8.20 is certain, and additional technical challenges may emerge as limited lithium is used more widely.

Third, the limited lithium paradigm allows the Coulombic efficiency another critical electrochemical parameter to be assessed readily through a limited plate and a full strip. While long-term cycling tests rarely use limited lithium, the use of a limited plate and strip is well-established and practiced by some groups today, with copper or stainless steel typically used as the working electrode [174–176]. A full strip may be carried out on each cycle, or periodically, to assess the Coulombic efficiency. With 80% of the initial lithium passed per cycle (the goal), reaching 1,000 cycles requires a Coulombic efficiency of 99.98%—a significantly higher value than that reached by present approaches using liquid [174, 176] or solid electrolytes [177]. Notably, LiPON thin-film cells in which all of the lithium and cycled hundreds of

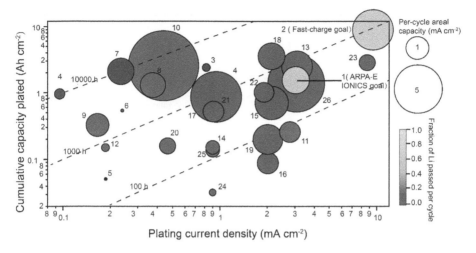

**FIGURE 8.20** Status of published efforts on the cycling of lithium metal. The data is analyzed in terms of four parameters: cumulative areal capacity plated, per cycle areal capacity, plating current density, and the fraction of the initial lithium metal present that is plated per cycle. The size of each point indicates the per-cycle areal capacity (see the key in the upper-right corner), and the color indicates the fraction of Li metal initially in the cell that is plated on each cycle (see the key in the lower-right corner). Points 1 and 2 are goals, are for LiPON thin-film cells, are PEO-based solid polymer electrolytes, are solid inorganic separators, 13 and 14 are custom nanostructures, and are liquid electrolytes.

times, thereby proving a Coulombic efficiency above 99.9% is possible in practice with lithium metal.

## 8.5.2.2 Hitting Aggressive Cost Goals

In addition to meeting aggressive performance goals, lithium metal batteries for transportation or the grid must also meet challenging cost goals to achieve mass adoption. It is important to note that the "cost" goal is really the price goal for cell buyers (for example, vehicle manufacturers) and includes the manufacturing cost, business overhead, warranty, and profit. With the DOE cell goal of $100 kWh$^{-1}$, and assuming the cost breakdown reflects that of an existing Li-ion energy cell, the areal cost goal for the cell repeat layers is $10–12 m$^{-2}$. With the cathode and current collectors retaining a combined cost of –$7 Scheme of Li surface. m$^{-2}$, the cost goal for the separator and any lithium beyond that present in the cathode is <$5 m$^{-2}$ [178]. Lowering the cathode or current collector costs (or using Li metal as the current collector rather than Cu), or reducing manufacturing costs, would allow a higher cost for the separator plus lithium metal sheets. The critical question is how this challenging cost goal can be achieved. Each of the major contributions to cost depends on manufacturing volume, and each needs careful consideration. We analyze the cost in the following three categories: (1) cost of the material inputs in the appropriate purity and form; (2) cost of the processing required to form the separator from the material inputs; and (3) cost

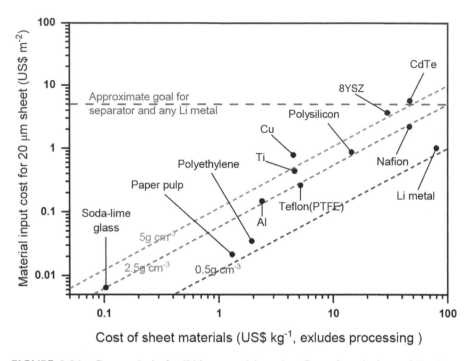

**FIGURE 8.21** Cost analysis for lithium metal batteries. Cost of producing a fully dense 20 μm sheet as a function of the cost of materials alone (diagonal lines represent constant densities) and approximate sheet product prices at the thickness at which they are sold, again as a function of the cost of the material inputs.

of integrating a separator, lithium metal (if needed) and other cell components into a cell. Figure 8.21 addresses costs (1) and (2); cost (3) will be addressed at the end of this section. Figure 8.21 shows material costs assuming full density and a 20 um thickness as a function of material cost (in US$ $kg^{-1}$); diagonal lines correspond to constant densities. Lithium metal, because of its low density and molecular weight, costs ~$1 $m^{-2}$ for 20 um (corresponding to ~4 mAh $cm^{-2}$) despite its cost (as an ingot) of ~$80 $kg^{-1}$. The lithium layer offers an opportunity for cost saving: if only a seed layer is required (for example, 5 um), with the cathode supplying the remaining lithium, material costs drop to ~$0.25 $m^{-1}$. Presently, lithium foil at these costs and thicknesses is not available, but remains of critical importance in the event of scale-up. Figure 8.21 shows the importance of reaching a separator thickness of 20 um, not only to ensure a high energy content for the repeating cell layers but also to enable per-kilogram costs that reflect the current prices of battery-grade materials such as electrolytes (–$20 $kg^{-1}$) and electrode materials ($15–30 $kg^{-1}$). The role of density is also evident, with significant density differences among leading Li conducting materials [179, 180]. Figure 8.21 shows that several commodity sheet materials (such as glass, Al, polyethylene, and paper pulp) have material costs under $0.10 $m^{-2}$, while more specialized materials like CdTe and 8 mol% $Y_2O_3$-stabilized $ZrO_2$ (8YSZ) have costs well above $1 $m^{-2}$.

The development of a reversible lithium metal electrode has eluded researchers for decades, but remains of great interest due to its potential to reach performance and cost goals. In this perspective, we identified critical parameters against which to assess progress and provided a summary of published efforts. We also encouraged the lithium metal battery community to use limited lithium (<30 um) rather than thick lithium foils for numerous reasons, including the ability to detect soft shorts. Finally, we established a cost target for a separator plus lithium sheet to enable a cell price of \$100 kWh$^{-1}$, and showed that there are several manufacturing and material classes that can reach or approach that target, although none to date with the performance properties required by a lithium metal battery. If the lithium metal electrode can be proven to cycle in small research cells using the four parameters identified here, with material and processing costs consistent with the cost target at scale, additional challenges and opportunities will be evident. A dense lithium foil cycled with high per-cycle utilization is required and will result in significant volume changes in large-format cells; the resulting stresses and shape change may limit cycling and have deleterious side effects. Lithium metal electrodes with minimal volume change, or novel types of large-format cells or packs containing them, are possible solutions. Determining the safety and rate capability of large-format cells, especially after extended cycling and during charge at low temperatures, will be another priority. In terms of manufacturing, lithium metal technologies that integrate easily into existing lithium-ion production lines will achieve the most rapid growth. The need to address these challenges is motivated by the definitive shift toward an electrified future, and the fact that high-energy rechargeable batteries are perhaps the single most important enabler for that transformation to be widely successful.

## REFERENCES

1. C. Yang, K. Fu, Y. Zhang, E. Hitz, and L. Hu. Protected lithium-metal anodes in batteries: From liquid to solid. *Adv. Mater.* 2017, *29* (36), 1701169.
2. P. G. Bruce, S. A. Freunberger, L. J. Hardwick, and J.-M. Tarascon. Li–O$_2$ and Li–S batteries with high energy storage. *Nature Mater.* 2012, *11* (1), 19.
3. X.-B. Cheng, R. Zhang, C.-Z. Zhao, and Q. Zhang. Toward safe lithium metal anode in rechargeable batteries: A review. *Chem.Rev.* 2017, *117* (15), 10403.
4. H. Ye, S. Xin, Y.-X. Yin, and Y.-G. Guo. Advanced porous carbon materials for high-efficient lithium metal anodes. *Adv. Energy Mater.* 2017, *7* (23), 1700530.
5. D. Koo, S. Ha, D.-M. Kim, and K. T. Lee. Recent approaches to improving lithium metal electrodes. *Curr. Opin. Electrochem.* 2017, *6* (1), 70.
6. J. I. Lee, G. Song, S. Cho, D. Y. Han, and S. Park. Lithium metal interface modification for high-energy batteries: Approaches and characterization. *Batteries & Supercaps* 2020, *3* (9), 828.
7. M. Qi, L. Xie, Q. Han, L. Zhu, L. Chen, and X. Cao. An overview of the key challenges and strategies for lithium metal anodes. *J. Energy Storage* 2022, *47*, 103641.
8. J. Heine, S. Krüger, C. Hartnig, U. Wietelmann, M. Winter, and P. Bieker. Coated lithium powder (CLiP) electrodes for lithium-metal batteries. *Adv. Energy Mater.* 2014, *4* (5), 1300815.

9. J. Park, J. Jeong, Y. Lee, M. Oh, M.-H. Ryou, and Y. M. Lee. Micro-patterned lithium metal anodes with suppressed dendrite formation for post lithium-ion batteries. *Adv. Mater. Interfaces* 2016, *3* (11), 1600140.

10. Q. Li, S. Zhu, and Y. Lu. 3D porous Cu current collector/Li-metal composite anode for stable lithium-metal batteries. *Adv. Funct. Mater.* 2017, *27* (18), 1606422.

11. Z. Hu, Z. Li, Z. Xia, T. Jiang, G. Wang, J. Sun, P. Sun, C. Yan, and L. Zhang. PECVD-derived graphene nanowall/lithium composite anodes towards highly stable lithium metal batteries. *Energy Storage Mater.* 2019, *22*, 29.

12. R. Li, J. Wang, L. Lin, H. Wang, C. Wang, C. Zhang, C. Song, F. Tian, J. Yang, and Y. Qian. Pressure-tuned and surface-oxidized copper foams for dendrite-free Li metal anodes. *Mater. Today Energy* 2020, *15*, 100367.

13. F. Liu, R. Xu, Z. Hu, S. Ye, S. Zeng, Y. Yao, S. Li, and Y. Yu. Regulating lithium nucleation via CNTs modifying carbon cloth film for stable Li metal anode. *Small* 2019, *15* (5), 1803734.

14. Y. Zhang, Y. Shi, X.-C. Hu, W.-P. Wang, R. Wen, S. Xin, and Y.-G. Guo. A 3D lithium/carbon fiber anode with sustained electrolyte contact for solid-state batteries. *Adv. Energy Mater.* 2020, *10* (3), 1903325.

15. C. Wu, H. Huang, W. Lu, Z. Wei, X. Ni, F. Sun, P. Qing, Z. Liu, J. Ma, W. Wei, L. Chen, C. Yan, and L. Mai. Mg Doped Li–LiB Alloy with in situ formed lithiophilic LiB skeleton for lithium metal batteries. *Adv. Sci.* 2020, *7* (6), 1902643.

16. M. Wan, S. Kang, L. Wang, H.-W. Lee, G. W. Zheng, Y. Cui, and Y. Sun. Mechanical rolling formation of interpenetrated lithium metal/lithium tin alloy foil for ultrahigh-rate battery anode. *Nat. Commun.* 2020, *11* (1), 829.

17. S.-S. Chi, Q. Wang, B. Han, C. Luo, Y. Jiang, J. Wang, C. Wang, Y. Yu, and Y. Deng. Lithiophilic Zn sites in porous CuZn alloy induced uniform Li nucleation and dendrite-free Li metal deposition. *Nano Lett.* 2020, *20* (4), 2724.

18. Q. Xu, X. Yang, M. Rao, D. Lin, K. Yan, R. Du, J. Xu, Y. Zhang, D. Ye, S. Yang, G. Zhou, Y. Lu, and Y. Qiu. High energy density lithium metal batteries enabled by a porous graphene/MgF2 framework. *Energy Storage Mater.* 2020, *26*, 73.

19. X. Wang, Z. Pan, Y. Wu, X. Ding, X. Hong, G. Xu, M. Liu, Y. Zhang, and W. Li. Infiltrating lithium into carbon cloth decorated with zinc oxide arrays for dendrite-free lithium metal anode. *Nano Res.* 2019, *12* (3), 525.

20. L. Dong, L. Nie, and W. Liu. Water-stable lithium metal anodes with ultrahigh-rate capability enabled by a hydrophobic graphene architecture. *Adv. Mater.* 2020, *32* (14), 1908494.

21. Z. Liang, K. Yan, G. Zhou, A. Pei, J. Zhao, Y. Sun, J. Xie, Y. Li, F. Shi, Y. Liu, D. Lin, K. Liu, H. Wang, H. Wang, Y. Lu, and Y. Cui. Composite lithium electrode with meso-scale skeleton via simple mechanical deformation. *Sci. Adv.* 2019, *5* (3), eaau5655.

22. T. Zhou, J. Shen, Z. Wang, J. Liu, R. Hu, L. Ouyang, Y. Feng, H. Liu, Y. Yu, and M. Zhu. Regulating lithium nucleation and deposition via MOF-derived Co@C-modified carbon cloth for stable Li metal anode. *Adv. Funct. Mater.* 2020, *30* (14), 1909159.

23. X.-B. Cheng, R. Zhang, C.-Z. Zhao, F. Wei, J.-G. Zhang, and Q. Zhang. A review of solid electrolyte interphases on lithium metal anode. *Adv. Sci.* 2016, *3* (3), 1500213.

24. N.-W. Li, Y.-X. Yin, J.-Y. Li, C.-H. Zhang and Y.-G. Guo. Passivation of lithium metal anode via hybrid ionic liquid electrolyte toward stable Li plating/stripping. *Adv. Sci.* 2017, *4* (2), 1600400.

25. D. Aurbach, and A. Zaban. Impedance spectroscopy of lithium electrodes: Part 1. general behavior in propylene carbonate solutions and the correlation to surface chemistry and cycling efficiency. *J. Electronal. Chem.* 1993, *348* (1), 155.

26. J. Wang, F. Lin, H. Jia, J. Yang, C. W. Monroe and Y. NuLi. Towards a safe lithium–sulfur battery with a flame-inhibiting electrolyte and a sulfur-based composite cathode. *Angew. Chem. Int. Ed.* 2014, *53* (38), 10099.

27. N.-W. Li, Y.-X. Yin, C.-P. Yang, and Y.-G. Guo. An artificial solid electrolyte interphase layer for stable lithium metal anodes. *Adv. Mater.* 2016, *28* (9), 1853.

28. H. Wu, Y. Cao, L. Geng, and C. Wang. In situ formation of stable interfacial coating for high performance lithium metal anodes. *Chem. Mate.* 2017, *29* (8), 3572.

29. K. Liu, A. Pei, H. R. Lee, B. Kong, N. Liu, D. Lin, Y. Liu, C. Liu, P.-c. Hsu, Z. Bao, and Y. Cui. Lithium metal anodes with an adaptive "solid-liquid" interfacial protective layer. *J. Am. Chem. Soc.* 2017, *139* (13), 4815.

30. Y. Zhao, K. R. Adair, and X. Sun. Recent developments and insights into the understanding of Na metal anodes for Na-metal batteries. *Energy Environ. Sci.* 2018, *11* (1), 2673.

31. C. Zhao, Y. Lu, J. Yue, D. Pan, Y. Qi, Y.-S. Hu, and L. Chen. Advanced Na metal anodes. *J. Energy Chem.* 2018, *27* (6), 1584.

32. W. Luo, C.-F. Lin, O. Zhao, M. Noked, Y. Zhang, G. W. Rubloff, and L. Hu. Ultrathin surface coating enables the stable sodium metal anode. *Adv. Energy Mater.* 2017, *7* (2), 1601526.

33. J. Popovic. Review—Recent advances in understanding potassium metal anodes. *J. Electrochemi. Soc.* 2022, *169* (3), 30510.

34. A. Eftekhari. Potassium secondary cell based on prussian blue cathode. *J. Power Sources* 2004, *126* (1), 221.

35. W. Bao, R. Wang, B. Li, C. Qian, Z. Zhang, J. Li, and F. Liu. Stable alkali metal anodes enabled by crystallographic optimization—A review. *Journal of materials chemistry. A, Materials for energy and sustainability* 2021, *9* (37), 2957.

36. P. Liu, and D. Mitlin. Emerging potassium metal anodes: Perspectives on control of the electrochemical interfaces. *Acc. Chem. Res.* 2020, *53* (6), 1161.

37. J. Xie, and Q. Zhang. Recent progress in multivalent metal (Mg, Zn, Ca, and Al) and metal-ion rechargeable batteries with organic materials as promising electrodes. *Small* 2019, *15* (15), 1805061.

38. A. Ponrouch, J. Bitenc, R. Dominko, N. Lindahl, P. Johansson, and M. R. Palacin. Multivalent rechargeable batteries. *Energy Storage Mater.* 2019, *20*, 253.

39. B. Li, Y. Wang, and S. Yang. A material perspective of rechargeable metallic lithium anodes. *Adv. Energy Mater.* 2018, *8* (13), 1702296.

40. X. B. Cheng, R. Zhang, C. Z. Zhao, and Q. Zhang. Toward safe lithium metal anode in rechargeable batteries: A review. *Chem. Rev.* 2017, *117* (15), 10403.

41. X. Zhang, Y. Yang, and Z. Zhou. Towards practical lithium-metal anodes. *Chem. Soc. Rev.* 2020, *49* (10), 3040.

42. E. Olsson, J. Yu, H. Zhang, H. M. Cheng, and Q. Cai. Atomic-scale design of anode materials for alkali metal (Li/Na/K)-ion batteries: progress and perspectives. *Adv. Energy Mater.* 2022, *12* (25), 2200662.

43. H. Wang, D. Yu, C. Kuang, L. Cheng, W. Li, X. Feng, Z. Zhang, X. Zhang, and Y. Zhang. Alkali metal anodes for rechargeable batteries. *Chem* 2019, *5* (2), 313.

44. H. Yang, B. Zhang, Y. Wang, K. Konstantinov, H. Liu, and S. Dou. Alkali-metal sulfide as cathodes toward safe and high-capacity metal (M= Li, Na, K) sulfur batteries. *Adv. Energy Mater.* 2020, *10* (37), 2001764.

45. X. Huang, Y. Wang, S. Chou, S. Dou, and Z. Wang. Materials engineering for adsorption and catalysis in room-temperature Na–S batteries. *Energy Environ. Sci.* 2021, *14* (7), 3757.

46. J. Ding, H. Zhang, W. Fan, C. Zhong, W. Hu, and D. Mitlin. Review of emerging potassium–sulfur batteries. *Adv. Mater.* 2020, *32* (23), e1908007.

47. J. Zhou, T. Wu, Y. Pan, J. Zhu, X. Chen, C. Peng, C. Shu, L. Kong, W. Tang, and S. Chou. Packing sulfur species by phosphorene-derived catalytic interface for electrolyte-lean lithium–sulfur batteries. *Adv. Funct. Mater.* 2021, *32* (4), 2106966.

48. W. Liu, K. Zhang, L. Ma, R. Ning, Z. Chen, J. Li, Y. Yan, T. Shang, Z. Lyu, Z. Li, K. Xie, and K. P. Loh. An ion sieving conjugated microporous thermoset ultrathin membrane for high-performance Li-S battery. *Energy Storage Mater.* 2022, *49*, 1.

49. N. Zhong, C. Lei, R. Meng, J. Li, X. He, and X. Liang. Electrolyte solvation chemistry for the solution of high-donor-number solvent for stable Li-S batteries. *Small* 2022, *18* (16), e2200046.

50. P. Zhang, Y. Zhao, and X. Zhang. Functional and stability orientation synthesis of materials and structures in aprotic Li-O2 batteries. *Chem. Soc. Rev.* 2018, *47* (8), 2921.

51. H. D. Lim, B. Lee, Y. Bae, H. Park, Y. Ko, H. Kim, J. Kim, and K. Kang. Reaction chemistry in rechargeable Li-O$_2$ batteries. *Chem. Soc. Rev.* 2017, *46* (10), 2873.

52. K. Song, D. A. Agyeman, M. Park, J. Yang, and Y. M. Kang. High-energy-density metal-oxygen batteries: Lithium-oxygen batteries vs sodium-oxygen batteries. *Adv. Mater.* 2017, *29* (48), 1606572.

53. J. Park, J. Y. Hwang, and W. J. Kwak. Potassium-oxygen batteries: Significance, challenges, and prospects. *J. Phys. Chem. Lett.* 2020, *11* (18), 7849.

54. N. Xiao, X. Ren, M. He, W. D. McCulloch, and Y. Wu. Probing mechanisms for inverse correlation between rate performance and capacity in K-O$_2$ batteries. *ACS Appl. Mater. Interf.* 2017, *9* (5), 4301.

55. W. Wang, and Y. Lu. The Potassium–air battery: Far from a practical reality. *Accounts Mater. Res.* 2021, *2* (7), 515.

56. J. C. Bachman, S. Muy, A. Grimaud, H. H. Chang, N. Pour, S. F. Lux, O. Paschos, F. Maglia, S. Lupart, P. Lamp, L. Giordano, and Y. Shao-Horn. Inorganic solid-state electrolytes for lithium batteries: Mechanisms and properties governing ion conduction. *Chem. Rev.* 2016, *116* (1), 140.

57. S. Li, S. Q. Zhang, L. Shen, Q. Liu, J. B. Ma, W. Lv, Y. B. He, and Q. H. Yang. Progress and perspective of ceramic/polymer composite solid electrolytes for lithium batteries. *Adv. Sci.* 2020, *7* (5), 1903088.

58. L. Liu, J. Lyu, J. Mo, P. Peng, J. Li, B. Jiang, L. Chu, and M. Li. Flexible, high-voltage, ion-conducting composite membranes with 3D aramid nanofiber frameworks for stable all-solid-state lithium metal batteries. *Sci. China Mater.* 2020, *63* (5), 703.

59. N. Wu, P. H. Chien, Y. Qian, Y. Li, H. Xu, N. S. Grundish, B. Xu, H. Jin, Y. Y. Hu, G. Yu, J. B. Goodenough. Enhanced surface interactions enable fast Li$_+$ conduction in oxide/polymer composite electrolyte. *Angew. Chem. Int. Ed.* 2020, *59* (10), 4131.

60. M. B. Dixit, W. Zaman, N. Hortance, S. Vujic, B. Harkey, F. Shen, W.-Y. Tsai, V. De Andrade, X. C. Chen, N. Balke, and K. B. Hatzell. Nanoscale mapping of extrinsic interfaces in hybrid solid electrolytes. *Joule* 2020, *4* (1), 207.

61. S. Liu, L. Zhou, J. Han, K. Wen, S. Guan, C. Xue, Z. Zhang, B. Xu, Y. Lin, Y. Shen, L. Li, and C. W. Nan. Super long-cycling all-solid-state battery with thin Li6PS5Cl-based electrolyte. *Adv. Energy Mater.* 2022, *12* (25), 2200660.

62. K. K. Fu, Y. Gong, J. Dai, A. Gong, X. Han, Y. Yao, C. Wang, Y. Wang, Y. Chen, C. Yan, Y. Li, E. D. Wachsman, and L. Hu. Flexible, solid-state, ion-conducting membrane with 3D garnet nanofiber networks for lithium batteries. *Proc. Natl. Acad. Sci. USA* 2016, *113* (26), 7094.

63. D. Aurbach, E. Zinigrad, H. Teller, and P. Dan. Factors which limit the cycle life of rechargeable lithium (Metal) batteries. *J. Electrochem. Soc.* 2000, *147* (4), 1274.

64. R. Bhattacharyya, B. Key, H. Chen, A. S. Best, A. F. Hollenkamp, and C. P. Grey. In situ NMR observation of the formation of metallic lithium microstructures in lithium batteries. *Nat. Mater.* 2010, *9* (6), 504.

65. S. Chandrashekar, N. M. Trease, H. J. Chang, L. S. Du, C. P. Grey, and A. Jerschow. $^7$Li MRI of Li batteries reveals location of microstructural lithium. *Nat. Mater.* 2012, *11* (4), 311.

66. K. N. Wood, E. Kazyak, A. F. Chadwick, K. H. Chen, J. G. Zhang, K. Thornton, and N. P. Dasgupta. Dendrites and pits: Untangling the complex behavior of lithium metal anodes through operando video microscopy. *ACS Centr. Sci.* 2016, *2* (11), 790.

67. D. A. Dornbusch, R. Hilton, S. D. Lohman, and G. J. Suppes. Experimental validation of the elimination of dendrite short-circuit failure in secondary lithium-metal convection cell batteries. *J. Electrochem. Soc.* 2015, *162* (3), A262.

68. D. Lu, Y. Shao, T. Lozano, W. D. Bennett, G. L. Graff, B. Polzin, J. Zhang, M. H. Engelhard, N. T. Saenz, W. A. Henderson, P. Bhattacharya, J. Liu, and J. Xiao. Failure mechanism for fast-charged lithium metal batteries with liquid electrolytes. *Adv. Energy Mater.* 2015, *5* (3), 1400993.

69. C. M. LóPez, J. T. Vaughey, and D. W. Dees. Morphological transitions on lithium metal anodes. *J. Electrochem. Soc.* 2009, *156* (9), A726.

70. W. Luo, and L. Hu. Na metal anode: "Holy Grail" for room-temperature Na-Ion batteries. *ACS Centr. Sci.* 2015, *1* (8), 420.

71. P. Hartmann, C. L. Bender, M. Vračar, A. K. Dürr, A. Garsuch, J. Janek, and P. Adelhelm. A rechargeable room-temperature sodium superoxide ($NaO_2$) battery. *Nat. Mater.* 2013, *12* (3), 228.

72. W. Zhou, Y. Li, S. Xin, and J. B. Goodenough. Rechargeable sodium all-solid-state battery. *ACS Centr. Sci.* 2017, *3* (1), 52.

73. R. Dugas, A. Ponrouch, G. Gachot, R. David, M. R. Palacin, and J. M. Tarascon. Na reactivity toward carbonate-based electrolytes: The effect of FEC as additive. *J. Electrochem. Soc.* 2016, *163* (10), A2333.

74. R. Rodriguez, K. E. Loeffler, S. S. Nathan, J. K. Sheavly, A. Dolocan, A. Heller, and C. B. Mullins. In situ optical imaging of sodium electrodeposition: Effects of fluoroethylene carbonate. *ACS Energy Lett.* 2017, *2* (9), 2051.

75. L. Xiao, Y. Cao, J. Xiao, W. Wang, L. Kovarik, Z. Nie, and J. Liu. High capacity, reversible alloying reactions in SnSb/C nanocomposites for Na-ion battery applications. *Chem. Commun.* 2012, *48* (27), 3321.

76. Z. Wang, S. M. Selbach, and T. Grande. Van der Waals density functional study of the energetics of alkali metal intercalation in graphite. *RSC Adv.* 2014, *4* (8), 3973.

77. K. Nobuhara, H. Nakayama, M. Nose, S. Nakanishi, and H. Iba. First-principles study of alkali metal-graphite intercalation compounds. *J. Power Sources* 2013, *243*, 585.

78. E. Memarzadeh Lotfabad, P. Kalisvaart, A. Kohandehghan, D. Karpuzov, and D. Mitlin. Origin of non-SEI related coulombic efficiency loss in carbons tested against Na and Li. *J. Mater. Chem. A* 2014, *2* (46), 19685.

79. H. Liu, X. Cheng, Z. Jin, R. Zhang, G. Wang, L. Chen, Q. Liu, J. Huang, and Q. Zhang. Recent advances in understanding dendrite growth on alkali metal anodes. *Energy chem.* 2019, *1* (1), 100003.

80. J. Zheng, M. H. Engelhard, D. Mei, S. Jiao, B. J. Polzin, J. Zhang, and W. Xu. Electrolyte additive enabled fast charging and stable cycling lithium metal batteries. *Nat. Energy* 2017, *2* (3), 17012.

81. S. Wei, S. Choudhury, J. Xu, P. Nath, Z. Tu, and L. A. Archer. Highly stable sodium batteries enabled by functional ionic polymer membranes. *Adv. Mater.* 2017, *29* (12), 1605512.

82. N. Xiao, X. Ren, W. D. Mcculloch, G. Gourdin, and Y. Wu. Potassium superoxide: A unique alternative for metal–air batteries. *Accounts Chem. Res.* 2018, *51* (9), 2335.

83. N. Xiao, W. D. Mcculloch, and Y. Wu. Reversible dendrite-free potassium plating and stripping electrochemistry for potassium secondary batteries. *J. Am. Chem. Soc.* 2017, *139* (28), 9475.

84. C. Ling, D. Banerjee, and M. Matsui. Study of the Electrochemical deposition of Mg in the atomic level: Why it prefers the non-dendritic morphology. *Electrochim. Acta* 2012, *76*, 270.

85. M. Matsui. Study on electrochemically deposited Mg metal. *J. Power Sources* 2011, *196* (16), 7048.

86. R. Davidson, A. Verma, D. Santos, F. Hao, C. Fincher, S. Xiang, J. Van Buskirk, K. Xie, M. Pharr, P. P. Mukherjee, and S. Banerjee. Formation of magnesium dendrites during electrodeposition. *ACS Energy Lett.* 2019, *4* (2), 375.

87. R. Davidson, A. Verma, D. Santos, F. Hao, C. D. Fincher, D. Zhao, V. Attari, P. Schofield, J. Van Buskirk, A. Fraticelli-Cartagena, T. E. G. Alivio, R. Arroyave, K. Xie, M. Pharr, P. P. Mukherjee, and S. Banerjee. Mapping mechanisms and growth regimes of magnesium electrodeposition at high current densities. *Mater. Horizons* 2020, *7* (3), 843.

88. M. K. Aslam, Y. Niu, T. Hussain, H. Tabassum, W. Tang, M. Xu, and R. Ahuja. How to avoid dendrite formation in metal batteries: Innovative strategies for dendrite suppression. *Nano Energy* 2021, *86*, 106142.

89. X. Liu, A. Du, Z. Guo, C. Wang, X. Zhou, J. Zhao, F. Sun, S. Dong, and G. Cui. Uneven stripping behavior, an unheeded killer of Mg anodes. *Adv. Mater.* 2022, *34* (31), 2201886.

90. K. Wang, P. Pei, Z. Ma, H. Xu, P. Li, and X. Wang. Morphology control of Zinc regeneration for Zinc-air fuel cell and battery. *J. Power Sources* 2014, *271*, 65.

91. H. Wang, W. Ye, Y. Yang, Y. Zhong, and Y. Hu. Zn-ion hybrid supercapacitors: Achievements, challenges and future perspectives. *Nano Energy* 2021, *85*, 105942.

92. J. Fu, Z. P. Cano, M. G. Park, A. Yu, M. Fowler, and Z. Chen. Electrically rechargeable zinc-air batteries: progress, challenges, and perspectives. *Adv. Mater.* 2017, *29* (7), 1604685.

93. B. Chen, S. R. Shaw, and I. A. Meinertzhagen. Circadian rhythms in light-evoked responses of the fly's compound eye, and the effects of neuromodulators 5-HT and the peptide. *J. Comp. Physiol. A* 1999, *185* (5), 393.

94. E. Peled, D. Golodnitsky, and G. Ardel. Advanced model for solid electrolyte interphase electrodes in liquid and polymer electrolytes. *J. Electrochem. Soc.* 1997, *144* (8), L208.

95. D. Aurbach, Y. Talyosef, B. Markovsky, E. Markevich, E. Zinigrad, L. Asraf, J. S. Gnanaraj, and H. J. Kim. Design of electrolyte solutions for Li and Li-ion batteries: A review. *Electrochim. Acta* 2004, *50* (2–3), 247.

96. E. Endo, M. Ata, K. Tanaka, and K. Sekai. Electron spin resonance study of the electrochemical reduction of electrolyte solutions for lithium secondary batteries. *J. Electrochem. Soc.* 1998, *145* (11), 3757.

97. J. N. Chazalviel. Electrochemical aspects of the generation of ramified metallic electrodeposits. *Phys. Rev. A* 1990, *42* (12), 7355.

98. C. Brissot, M. Rosso, J. N. Chazalviel, P. Baudry, and S. Lascaud. In situ study of dendritic growth in Lithium/PEO-salt/Lithium cells. *Electrochim. Acta* 1998, *43* (10–11), 1569.

99. R. Akolkar. Mathematical model of the dendritic growth during lithium electrodeposition. *J. Power Sources* 2013, *232*, 23.

100. K. Nishikawa, T. Mori, T. Nishida, Y. Fukunaka, and M. Rosso. Li dendrite growth and Li+ ionic mass transfer phenomenon. *J. Electroanal. Chem.* 2011, *661* (1), 84.

101. A. Pei, G. Zheng, F. Shi, Y. Li, and Y. Cui. Nanoscale nucleation and growth of electrodeposited lithium metal. *Nano Lett.* 2017, *17* (2), 1132.

102. H. Che, S. Chen, Y. Xie, H. Wang, K. Amine, X. Liao, and Z. Ma. Electrolyte design strategies and research progress for room-temperature sodium-ion batteries. *Energy Environ. Sci.* 2017, *10* (5), 1075.

103. Z. W. She, J. Sun, Y. Sun, and Y. Cui. A highly reversible room-temperature sodium metal anode. *ACS Centr. Sci.* 2015, *1* (8), 449.

104. Y.-S. Hong, N. Li, H. Chen, P. Wang, W.-L. Song, and D. Fang. In operando observation of chemical and mechanical stability of Li and Na dendrites under quasi-zero electrochemical field. *Energy Storage Mater.* 2018, *11*, 118.

105. A. Rudola, D. Aurbach, and P. Balaya. A new phenomenon in sodium batteries: Voltage step due to solvent interaction. *Electrochem. Commun.* 2014, *46*, 56.

106. L. Xue, H. Gao, W. Zhou, S. Xin, K. Park, Y. Li, and J. B. Goodenough. Liquid K-Na alloy anode enables dendrite-free potassium batteries. *Adv. Mater.* 2016, *28* (43), 9608.

107. J. Xia, B. Fitch, A. Watson, E. Cabaniss, R. Black, and M. Yakovleva. Printed thin lithium foil with flexible thickness and width for industrial battery applications. *ECS Meet. Abstracts* 2020, *MA2020-02* (5), 976.

108. A. J. Louli, M. Genovese, R. Weber, S. G. Hames, E. R. Logan, and J. R. Dahn. Exploring the impact of mechanical pressure on the performance of anode-free lithium metal cells. *J. Electrochem. Soc.* 2019, *166* (8), A1291.

109. X. Fei, Z. Dong, B. Gong, and X. Zhao. Lightweight through-hole copper foil as a current collector for lithium-ion batteries. *ACS Appl. Mater. Interf.* 2021, *13* (35), 42266.

110. L. Ma, J. Cui, S. Yao, X. Liu, Y. Luo, X. Shen, and J.-K. Kim. Dendrite-free lithium metal and sodium metal batteries. *Energy Storage Mater.* 2020, *27*, 522.

111. A. Maraschky, and R. Akolkar. Temperature dependence of dendritic lithium electrodeposition: A mechanistic study of the role of transport limitations within the SEI. *J. Electrochem. Soc.* 2020, *167* (6), 062503.

112. C. T. Love, O. A. Baturina, and K. E. Swider-Lyons. Observation of lithium dendrites at ambient temperature and below. *ECS Electrochem. Lett.* 2014, *4* (2), A24.

113. L. Li, S. Basu, Y. Wang, Z. Chen, P. Hundekar, B. Wang, J. Shi, Y. Shi, S. Narayanan, and N. Koratkar. Self-heating–induced healing of lithium dendrites. *Science* 2018, *359* (6383), 1513.

114. K. Yan, J. Wang, S. Zhao, D. Zhou, B. Sun, Y. Cui, and G. Wang. Temperature-dependent nucleation and growth of dendrite-free lithium metal anodes. *Angew. Chem. Int. Ed.* 2019, *58* (33), 11364.

115. X. Shen, R. Zhang, P. Shi, X. Chen, and Q. Zhang. How does external pressure shape Li dendrites in Li metal batteries. *Adv. Energy Mater.* 2021, *11* (10), 2003416.

116. L. Qin, K. Wang, H. Xu, M. Zhou, G. Yu, C. Liu, Z. Sun, and J. Chen. The role of mechanical pressure on dendritic surface toward stable lithium metal anode. *Nano Energy* 2020, *77*, 105098.

117. Y. Liu, X. Xu, M. Sadd, O. O. Kapitanova, V. A. Krivchenko, J. Ban, J. Wang, X. Jiao, Z. Song, J. Song, S. Xiong, and A. Matic. Insight into the critical role of exchange current density on electrodeposition behavior of lithium metal. *Adv. Sci.* 2021, *8* (5), 2003301.

118. Q. Xu, Y. Yang, and H. Shao. Substrate effects on Li$^+$ electrodeposition in Li secondary batteries with a competitive kinetics model. *Phys. Chem. Chem. Phys.* 2015, *17* (31), 20398.

119. X. Fan, L. Chen, O. Borodin, X. Ji, J. Chen, S. Hou, T. Deng, J. Zheng, C. Yang, S. C. Liou, K. Amine, K. Xu, and C. Wang. Non-flammable electrolyte enables Li-metal batteries with aggressive cathode chemistries. *Nat. Nanotechnol.* 2018, *13* (8), 715.

120. J. Meng, F. Chu, J. Hu, and C. Li. Liquid polydimethylsiloxane grafting to enable dendrite-free Li plating for highly reversible Li-metal batteries. *Adv. Funct. Mater.* 2019, *29* (30), 1902220.

121. Q. Wang, C. Yang, J. Yang, K. Wu, C. Hu, J. Lu, W. Liu, X. Sun, J. Qiu, and H. Zhou. Dendrite-free lithium deposition via a superfilling mechanism for high-performance Li-metal batteries. *Adv. Mater.* 2019, *31* (41), e1903248.

122. Q. Wang, C. Yang, J. Yang, K. Wu, L. Qi, H. Tang, Z. Zhang, W. Liu, and H. Zhou. Stable Li metal anode with protected interface for high-performance Li metal batteries. *Energy Storage Mater.* 2018, *15*, 249.

123. S. Chi, Y. Liu, W. Song, L. Fan, and Q. Zhang. Prestoring lithium into stable 3D nickel foam host as dendrite-free lithium metal anode. *Adv. Funct. Mater.* 2017, *27* (24), 1700348.

124. L. L. Lu, J. Ge, J. N. Yang, S. M. Chen, H. B. Yao, F. Zhou, and S. H. Yu. Free-standing copper nanowire network current collector for improving lithium anode performance. *Nano Lett.* 2016, *16* (7), 4431.

125. C. Niu, H. Pan, W. Xu, J. Xiao, J. G. Zhang, L. Luo, C. Wang, D. Mei, J. Meng, X. Wang, Z. Liu, L. Mai, and J. Liu. Self-smoothing anode for achieving high-energy lithium metal batteries under realistic conditions. *Nat. Nanotechnol.* 2019, *14* (6), 594.

126. R. V. Salvatierra, G. A. Lopez-Silva, A. S. Jalilov, J. Yoon, G. Wu, A. L. Tsai, and J. M. Tour. Suppressing Li metal dendrites through a solid Li-ion backup layer. *Adv. Mater.* 2018, *30* (50), e1803869.

127. K. Yan, Z. Lu, H. W. Lee, F. Xiong, P. C. Hsu, Y. Li, J. Zhao, S. Chu, and Y. Cui. Selective deposition and stable encapsulation of lithium through heterogeneous seeded growth. *Nat. Energy* 2016, *1* (3), 1.

128. K. Tantratian, D. Cao, A. Abdelaziz, X. Sun, J. Sheng, A. Natan, L. Chen, and H. Zhu. Stable Li metal anode enabled by space confinement and uniform curvature through lithiophilic nanotube arrays. *Adv. Energy Mater.* 2019, *10* (5), 1902819.

129. P. Xu, X. Lin, X. Hu, X. Cui, X. Fan, C. Sun, X. Xu, J. K. Chang, J. Fan, R. Yuan, B. Mao, Q. Dong, and M. Zheng. High reversible Li plating and stripping by in-situ construction a multifunctional lithium-pinned array. *Energy Storage Mater.* 2020, *28*, 188.

130. Z. Tu, S. Choudhury, M. J. Zachman, S. Wei, K. Zhang, L. F. Kourkoutis, and L. A. Archer. Fast ion transport at solid–solid interfaces in hybrid battery anodes. *Nat. Energy* 2018, *3* (4), 310.

131. M. Rosso, C. Brissot, A. Teyssot, M. Dollé, L. Sannier, J. M. Tarascon, R. Bouchet, and S. Lascaud. Dendrite short-circuit and fuse effect on Li/Polymer/Li cells. *Electrochim. Acta* 2006, *51* (25), 5334.

132. R. Khurana, J. L. Schaefer, L. A. Archer, and G. W. Coates. Suppression of Lithium dendrite growth using cross-linked polyethylene/poly(ethylene oxide) electrolytes: A New approach for practical lithium-metal polymer batteries. *J. Am. Chem. Soc.* 2014, *136* (20), 7395.

133. Y. Su, X. Rong, H. Li, X. Huang, L. Chen, B. Liu, and Y. S. Hu. High-entropy microdomain interlocking polymer electrolytes for advanced all-solid-state battery chemistries. *Adv. Mater.* 2022 DOI: 10.1002/adma.202209402, e2209402.

134. A. Manthiram, X. Yu, and S. Wang. Lithium battery chemistries enabled by solid-state electrolytes. *Nat. Rev. Mater.* 2017, *2* (4), 1.

135. H. L. Guo, H. Sun, Z. L. Jiang, J. Y. Hu, C. S. Luo, M. Y. Gao, J. Y. Cheng, W. K. Shi, H. J. Zhou, and S. G. Sun. Asymmetric structure design of electrolytes with flexibility and lithium dendrite-suppression ability for solid-state lithium batteries. *ACS Appl. Mater. Interf.* 2019, *11* (50), 46783.

136. X. Xu, D. Zhou, X. Qin, K. Lin, F. Kang, B. Li, D. Shanmukaraj, T. Rojo, M. Armand, and G. Wang. A room-temperature sodium-sulfur battery with high capacity and stable cycling performance. *Nat. Commun.* 2018, *9* (1), 3870.

137. B. Sun, P. Li, J. Zhang, D. Wang, P. Munroe, C. Wang, P. H. L. Notten, and G. Wang. Dendrite-free sodium-metal anodes for high-energy sodium-metal batteries. *Adv. Mater.* 2018, *30* (29), 1801334,

138. L. Ye, M. Liao, T. Zhao, H. Sun, Y. Zhao, X. Sun, B. Wang, and H. Peng. A sodiophilic interphase-mediated, dendrite-free anode with ultrahigh specific capacity for sodium-metal batteries. *Angew. Chem. Int. Ed.* 2019, *58* (47), 17054.

139. P. Hundekar, S. Basu, X. Fan, L. Li, A. Yoshimura, T. Gupta, V. Sarbada, A. Lakhnot, R. Jain, S. Narayanan, Y. Shi, C. Wang, and N. Koratkar. in situ healing of dendrites in a potassium metal battery. *Proc. Natl. Acad. Sci. USA* 2020, *117* (11), 5588.

140. J. Meng, H. Zhu, Z. Xiao, X. Zhang, C. Niu, Y. Liu, G. Jiang, X. Wang, F. Qiao, X. Hong, F. Liu, Q. Pang and L. Mai. Amine-wetting-enabled dendrite-free potassium metal anode. *ACS Nano* 2022, *16* (5), 7291.

141. C. Shen, G. Hu, L. Cheong, S. Huang, J. Zhang, and D. Wang. Direct observation of the growth of lithium dendrites on graphite anodes by operando EC-AFM. *Small Meth.* 2017, *2* (2), 1700298.

142. X. Liu, D. Wang, and L. Wan. Progress of electrode/electrolyte interfacial investigation of Li-ion batteries via in situ scanning probe microscopy. *Sci. Bull.* 2015, *60* (9), 839.

143. M. Kitta, and C. Fukada. Scanning spreading resistance microscopy: A promising tool for probing the reaction interface of Li-ion battery materials. *Langmuir* 2019, *35* (26), 8726.

144. D. E. McCoy, T. Feo, T. A. Harvey, and R. O. Prum. Structural absorption by barbule microstructures of super black bird of paradise feathers. *Nat. Commun.* 2018, *9* (1), 1.

145. F. Sun, L. Zielke, H. Markotter, A. Hilger, D. Zhou, R. Moroni, R. Zengerle, S. Thiele, J. Banhart, and I. Manke. Morphological evolution of electrochemically plated/stripped lithium microstructures investigated by synchrotron x-ray phase contrast tomography. *ACS Nano* 2016, *10* (8), 7990.

146. B. Song, I. Dhiman, J. C. Carothers, G. M. Veith, J. Liu, H. Z. Bilheux, and A. Huq. Dynamic lithium distribution upon dendrite growth and shorting revealed by operando neutron imaging. *ACS Energy Lett.* 2019, *4* (10), 2402.

147. A. J. Ilott, M. Mohammadi, H. J. Chang, C. P. Grey, and A. Jerschow. Real-time 3D imaging of microstructure growth in battery cells using indirect MRI. *Proc. Natl. Acad. Sci. USA* 2016, *113* (39), 10779.

148. S. C. Nagpure, R. G. Downing, B. Bhushan, S. S. Babu, and L. Cao. *Neutron depth profiling technique for studying aging in Li-ion batteries. Electrochim.* Acta 2011, *56* (13), 4735.

149. F. Han, A. S. Westover, J. Yue, X. Fan, F. Wang, M. Chi, D. N. Leonard, N. J. Dudney, H. Wang, and C. Wang. High electronic conductivity as the origin of lithium dendrite formation within solid electrolytes. *Nat. Energy* 2019, *4* (3), 187.

150. S. Frisco, D. X. Liu, A. Kumar, J. F. Whitacre, C. T. Love, K. E. Swider-Lyons, and S. Litster. Internal morphologies of cycled Li-metal electrodes investigated by nano-scale resolution x-ray computed tomography. *ACS Appl. Mater. Interf.* 2017, *9* (22), 18748.

151. P. Pietsch, and V. Wood. X-Ray tomography for lithium ion battery research: A practical guide. *Ann. Rev. Mater. Res.* 2017, *47* (1), 451.

152. D. S. Eastwood, P. M. Bayley, H. J. Chang, O. O. Taiwo, J. Vila-Comamala, D. J. Brett, C. Rau, P. J. Withers, P. R. Shearing, C. P. Grey, and P. D. Lee. Three-dimensional characterization of electrodeposited lithium microstructures using synchrotron x-ray phase contrast imaging. *Chem. Commun.* 2015, *51* (2), 266.

153. K. J. Harry, D. T. Hallinan, D. Y. Parkinson, A. A. MacDowell, and N. P. Balsara. Detection of subsurface structures underneath dendrites formed on cycled lithium metal electrodes. *Nat. Mater.* 2014, *13* (1), 69.

154. A. Same, V. Battaglia, H. Y. Tang, and J. W. Park. In situ neutron radiography analysis of graphite/NCA lithium-ion battery during overcharge. *J. Appl. Electrochem.* 2011, *42* (1), 1.

155. S. Klamor, K. Zick, T. Oerther, F. M. Schappacher, M. Winter, and G. Brunklaus. $^7$Li in situ 1D NMR imaging of a lithium ion battery. *Phys. Chem. Chem. Phys.* 2015, *17* (6), 4458.

156. H. J. Chang, A. J. Ilott, N. M. Trease, M. Mohammadi, A. Jerschow, and C. P. Grey. Correlating microstructural lithium metal growth with electrolyte salt depletion in lithium batteries using $^7$Li MRI. *J. Am. Chem. Soc.* 2015, *137* (48), 15209.

157. T. Waldmann, B. I. Hogg, and M. Wohlfahrt-Mehrens. Li plating as unwanted side reaction in commercial Li-ion cells—A review. *J. Power Sources* 2018, *384*, 107.

158. T. R. Tanim, P. P. Paul, V. Thampy, C. Cao, H. G. Steinrück, J. Nelson Weker, M. F. Toney, E. J. Dufek, M. C. Evans, A. N. Jansen, B. J. Polzin, A. R. Dunlop, and S. E. Trask. Heterogeneous behavior of lithium plating during extreme fast charging. *Cell Rep. Phys. Sci.* 2020, *1* (7), 100114.

159. D. P. Finegan, A. Quinn, D. S. Wragg, A. M. Colclasure, X. Lu, C. Tan, T. M. M. Heenan, R. Jervis, D. J. L. Brett, S. Das, T. Gao, D. A. Cogswell, M. Z. Bazant, M. Di Michiel, S. Checchia, P. R. Shearing, and K. Smith. Spatial dynamics of lithiation and lithium plating during high-rate operation of graphite electrodes. *Energy Environ. Sci.* 2020, *13* (8), 2570.

160. A. Raj, I. A. Shkrob, J. S. Okasinski, M.-T. Fonseca Rodrigues, A. C. Chuang, X. Huang, and D. P. Abraham. Spatially-resolved lithiation dynamics from operando x-ray diffraction and electrochemical modeling of lithium-ion cells. *J. Power Sources* 2021, *484*, 229247.

161. K. P. C. Yao, J. S. Okasinski, K. Kalaga, J. D. Almer, and D. P. Abraham. Operando quantification of (de)lithiation behavior of silicon–graphite blended electrodes for lithium-ion batteries. *Adv. Energy Mater.* 2019, *9* (8), 1803380.

162. A. C. Marschilok, A. M. Bruck, A. Abraham, C. A. Stackhouse, K. J. Takeuchi, E. S. Takeuchi, M. Croft, and J. W. Gallaway. Energy dispersive X-ray diffraction (EDXRD) for operando materials characterization within batteries. *Phys. Chem. Chem. Phys.* 2020, *22* (37), 20972.

163. D. P. Finegan, A. Vamvakeros, L. Cao, C. Tan, T. M. M. Heenan, S. R. Daemi, S. D. M. Jacques, A. M. Beale, M. Di Michiel, K. Smith, D. J. L. Brett, P. R. Shearing, and C. Ban. Spatially resolving lithiation in silicon-graphite composite electrodes via in situ high-energy x-ray diffraction computed tomography. *Nano. Lett.* 2019, *19* (6), 3811.

164. D. P. Finegan, A. Vamvakeros, C. Tan, T. M. M. Heenan, S. R. Daemi, N. Seitzman, M. Di Michiel, S. Jacques, A. M. Beale, D. J. L. Brett, P. R. Shearing, and K. Smith. Spatial quantification of dynamic inter and intra particle crystallographic heterogeneities within lithium ion electrodes. *Nat. Commun.* 2020, *11* (1), 631.

165. V. Zinth, C. von Lüders, M. Hofmann, J. Hattendorff, I. Buchberger, S. Erhard, J. Rebelo-Kornmeier, A. Jossen, and R. Gilles. Lithium plating in lithium-ion batteries at sub-ambient temperatures investigated by in situ neutron diffraction. *J. Power Sources* 2014, *271*, 152.

166. C. Uhlmann, J. Illig, M. Ender, R. Schuster, and E. Ivers-Tiffée. In situ detection of lithium metal plating on graphite in experimental cells. *J. Power Sources* 2015, *279*, 428.

167. H. Zhou, K. An, S. Allu, S. Pannala, J. Li, H. Z. Bilheux, S. K. Martha, and J. Nanda. Probing multiscale transport and inhomogeneity in a lithium-ion pouch cell using in situ neutron methods. *ACS Energy Lett.* 2016, *1* (5), 981.

168. O. Schmidt, A. Hawkes, A. Gambhir, and I. Staffell. The future cost of electrical energy storage based on experience rates. *Nat. Energy* 2017, *2* (8), 1.

169. J. W. Choi, and D. Aurbach. Promise and reality of post-lithium-ion batteries with high energy densities. *Nat. Rev. Mater.* 2016, *1* (4), 1.

170. K. G. Gallagher, S. Goebel, T. Greszler, M. Mathias, W. Oelerich, D. Eroglu, and V. Srinivasan. Quantifying the promise of lithium–air batteries for electric vehicles. *Energy Environ. Sci.* 2014, *7* (5), 1555.

171. E. J. Cheng, A. Sharafi, and J. Sakamoto. Intergranular Li metal propagation through polycrystalline Li6.25Al0.25La3Zr2O12 ceramic electrolyte. *Electrochim. Acta* 2017, *223*, 85.

172. R. Hernandez-Maya, O. Rosas, J. Saunders, and H. Castaneda. Dynamic characterization of dendrite deposition and growth in Li-surface by electrochemical impedance spectroscopy. *J. Electrochem. Soc.* 2015, *162* (4), A687.

173. A. Sharafi, H. M. Meyer, J. Nanda, J. Wolfenstine, and J. Sakamoto. Characterizing the Li–Li7La3Zr2O12 interface stability and kinetics as a function of temperature and current density. *J. Power Sources* 2016, *302*, 135.

174. W. Li, H. Yao, K. Yan, G. Zheng, Z. Liang, Y. M. Chiang, and Y. Cui. The synergetic effect of lithium polysulfide and lithium nitrate to prevent lithium dendrite growth. *Nat. Commun.* 2015, *6*, 7436.

175. J. Qian, W. A. Henderson, W. Xu, P. Bhattacharya, M. Engelhard, O. Borodin, and J. G. Zhang. High rate and stable cycling of lithium metal anode. *Nat. Commun.* 2015, *6* (1), 1.

176. Z. Tu, M. J. Zachman, S. Choudhury, S. Wei, L. Ma, Y. Yang, L. F. Kourkoutis, and L. A. Archer. Nanoporous hybrid electrolytes for high-energy batteries based on reactive metal anodes. *Adv. Energy Mater.* 2017, *7* (8), 1602367.

177. Y. Kato, S. Hori, T. Saito, K. Suzuki, M. Hirayama, A. Mitsui, M. Yonemura, H. Iba, and R. Kanno. High-power all-solid-state batteries using sulfide superionic conductors. *Nat. Energy* 2016, *1* (4), 1.

178. B. D. McCloskey. Attainable gravimetric and volumetric energy density of Li-S and Li ion battery cells with solid separator-protected Li metal anodes. *J. Phys. Chem. Lett.* 2015, *6* (22), 4581.

179. R. Murugan, V. Thangadurai, and W. Weppner. Fast lithium ion conduction in garnet-type Li7La3Zr2O12. *Angew. Chem. Int. Ed.* 2007, *46* (41), 7778.

180. P. Bron, S. Johansson, K. Zick, J. Schmedt auf der Gunne, S. Dehnen, and B. Roling. Li10SnP2S12: An affordable lithium superionic conductor. *J. Am. Chem. Soc.* 2013, *135* (42), 15694.

# Index

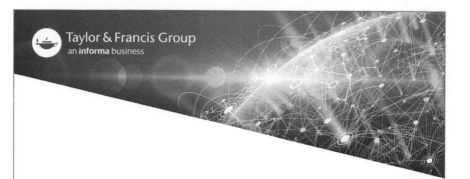